T0189948

Classical Systems in Quantum Mechanics

Pavel Bóna

Classical Systems
in Quantum Mechanics

 Springer

Pavel Bóna
Department of Theoretical Physics
Comenius University
Bratislava, Slovakia

ISBN 978-3-030-45072-4   ISBN 978-3-030-45070-0   (eBook)
https://doi.org/10.1007/978-3-030-45070-0

© Springer Nature Switzerland AG 2020, corrected publication 2020
This work is subject to copyright. All rights are reserved by the Publisher, whether the whole or part
of the material is concerned, specifically the rights of translation, reprinting, reuse of illustrations,
recitation, broadcasting, reproduction on microfilms or in any other physical way, and transmission
or information storage and retrieval, electronic adaptation, computer software, or by similar or dissimilar
methodology now known or hereafter developed.
The use of general descriptive names, registered names, trademarks, service marks, etc. in this
publication does not imply, even in the absence of a specific statement, that such names are exempt from
the relevant protective laws and regulations and therefore free for general use.
The publisher, the authors and the editors are safe to assume that the advice and information in this
book are believed to be true and accurate at the date of publication. Neither the publisher nor the
authors or the editors give a warranty, expressed or implied, with respect to the material contained
herein or for any errors or omissions that may have been made. The publisher remains neutral with regard
to jurisdictional claims in published maps and institutional affiliations.

This Springer imprint is published by the registered company Springer Nature Switzerland AG
The registered company address is: Gewerbestrasse 11, 6330 Cham, Switzerland

Preface

The work contains a description and an analysis of two different approaches determining the connections between quantal and classical theories.

The first approach associates with any quantum-mechanical system with a finite number of degrees of freedom a classical Hamiltonian system 'living' in projective Hilbert space $P(\mathcal{H})$, and it is called here the 'classical projection'.

The second approach deals with 'large' quantal (=quantum mechanical) systems in the limit of an infinite number of degrees of freedom and with their corresponding 'macroscopic limits' described as classical Hamiltonian systems of the system's global (intensive) quantum observables.

The last part of this work contains a series of models describing interactions of the "small" physical (micro) systems with the "macroscopic" ones, in which these interactions lead to a (macroscopic) change of some "classical" parameters of the large systems. These models connect, in a specific way, the two classes of the systems considered earlier in this work by modeling their mutual interactions leading to striking (i.e. theoretically impossible in the framework of finite quantum systems) results.

The projective space $P(\mathcal{H})$ of any complex Hilbert space $\mathcal{H}$ is endowed with a natural symplectic structure, which allows us to rewrite the quantum mechanics of systems with a finite number of degrees of freedom in terms of a classical Hamiltonian dynamics. If a quantum-mechanical system is associated with a continuous unitary representation $U(G)$ of a connected Lie group G on $\mathcal{H}$, the orbits (possibly factorized in a natural way) of the projected action of $U(G)$ in $P(\mathcal{H})$ are naturally mapped onto orbits of the coadjoint representation $Ad^*(G)$ of G. These coadjoint orbits have a canonical symplectic structure which coincides with the one induced from the structure of $P(\mathcal{H})$. For important classes of physical systems, these symplectic spaces are either symplectomorphic to the 'corresponding' classical phase spaces, or they are some extensions of them (describing, e.g. particles with 'classical spin'). Quantal dynamics is projected onto these phase spaces in a natural way, leading to classical Hamiltonian dynamical systems without any limit of Planck constant $\hbar \to 0$.

For a large (infinite) quantal system, an automorphic group action of G on the C^*-algebra $\mathfrak{A}$ of its bounded observables enables us to define a macroscopic subsystem being a classical Hamiltonian system of the same type as we obtained in the case of a finite number of degrees of freedom. There is a difference, however, between the interpretations of 'classical projections' and of these 'macroscopic limits': The classical (mechanical) projection describes classical mechanics of expectation values of quantal observables whereas the macroscopic limit describes a quantal subsystem with classical properties—its observables are elements of a subalgebra $\mathfrak{M}_G$ of the center $\mathfrak{Z}$ of the double dual $\mathfrak{A}^{**}$ of $\mathfrak{A}$. Any state ω on $\mathfrak{A}$ has a unique 'macroscopic limit' $p_M\omega$ which is represented by a probability measure on the corresponding (generalized) classical phase space. This offers us a possibility of deriving a classical (macroscopic) time evolution (which is, in general, in a certain sense stochastic, cf. [29, Sect. III.G]) from the underlying reversible quantal dynamics.

A scheme of 'macroscopic quantization' is outlined, according to which a (nonunique) reconstruction of the infinite quantal system $(\mathfrak{A}; \sigma_G)$ from its macroscopic limit is possible. By determining a classical Hamiltonian function in the macroscopic limit of $(\mathfrak{A}; \sigma_G)$ we can define a 'mean-field' time evolution in the infinite system $(\mathfrak{A}; \sigma_G)$. Our definition of the 'mean-field' evolutions extends the usual ones. The schemes and results developed in the work are applicable to models in the statistical mechanics as well as in gauge-theories (in the 'large N limit'). They might be relevant also in general considerations of 'quantizations' and of foundations of quantum theory.

The last Chapter of this work is devoted to the description of several models of interacting 'microsystems' with 'macrosystems', mimicking a description of the 'process of measurement in QM'. In these models, certain 'quantal properties' of the system, namely a (coherent) superposition of specific vector states (eigenstates of a 'measured' observable), transform by the unitary continuous time evolutions (for $t \to \infty$) into the corresponding 'proper mixtures' of macroscopically different states of the 'macrosystems' occurring in the models.

In this connection we shall shortly discuss the old 'quantum measurement problem', which however, in the light of certain experiments performed in the past decades and suggesting the possibilities of quantum-mechanical interference of several macroscopically different states of a macroscopic system, need not be at all a fundamental theoretical problem; this might mean that the often discussed 'measurement process' can be included into the presently widely accepted model of quantum theory.

Acknowledgements

This work is a revised and completed version of the unpublished text: "Classical Projections and Macroscopic Limits of Quantum Mechanical Systems" written roughly in the years 1985–1986. The author is indebted to Klaus Hepp for his stimulations and the kind help with correcting many formulations of the original text. During the work on the old text, encouragements by Elliott Lieb and by the late Walter Thirring were also stimulating, and the discussions with the colleagues, the late Ivan Korec, Milan Noga, Peter Prešnajder and in particular with Jiří Tolar, were useful and the author expresses his gratitude to all of them. Thankfulness for several times repeated encouragements to publish the old text should be expressed to Nicolaas P. Landsman. Author's thanks belong also, last but not least, to his colleague Vladimír Balek for his kind help with the final formulation of important parts of this book.

Technical Notes

(a)   This book contains several technical concepts which are not introduced here in detail. The readers needing to get a brief acquaintance with some additional elementary concepts and facts of topology, differential geometry (also in infinite dimensions), group theory, or theory of Hilbert space operators and theory of operator algebras, could consult, e.g. the appendices of the freely accessible publication [37], and the literature cited in our Bibliography. Due to connections of many places in the text of this book with the content of the work [37] it is recommended to keep the cited [37], and perhaps also [37, Textbook], as a handbook. The frequent citations from [37] contain usually references to specific places of the cited work.

(b)   Two kinds of quotation marks are used: Either the ones which stress some "*standard expressions*", or those which indicate '*intuitive denotations*'. The difference between these two is not, however, very sharp.

(c)   Many symbols appearing in mathematical formulas are introduced in various places of the text and repeatedly used in the rest of the book. For easier revealing of their meanings, they are included in Index and their first appearance in the text is stressed, sometimes in a not quite usual manner, by **boldface form** of a part of the surrounding text.

Bratislava, Slovakia   Pavel Bóna

The original version of the book was revised: For detailed information please see Correction. The correction to the book is available at https://doi.org/10.1007/978-3-030-45070-0_8

Contents

Chapter 1
Introduction

1.1 Motivation and Summary

1.1.1 Successful communication and manipulation with 'objects' requires construction of some adequate theoretical models ($\approx$ theories) of some classes of 'objects', resp. 'phenomena'. Different phenomena might be described by different theoretical schemes. These schemes should be, however, mutually consistent in the sense giving the same results for phenomena lying in the common domain of applicability of different theories. If one of the theories is considered to be 'more general' then a second one, then the whole domain of applicability of the second theory has to be contained in the domain of the first one. This is the case of *quantum mechanics (QM)*, which is believed to be a 'covering theory' of the more special *classical mechanics (CM)*—to the extent of measurement precision of apparatuses determining of 'classical systems'. Hence we can ask **how to describe phenomena belonging to the domain of applicability of CM in the framework of QM.**

1.1.2 Any single phenomenon, which is unambiguously and reproducibly determined by a specification of an empirical situation is, however, expressible in terms of parameters (resp. variables) occurring in *CM*: coordinates of positions and velocities of points distinguished and measured by 'macroscopic bodies' and various correlations between these variables. Hence also any experimentally realizable situation described in QM (which need not be a consequence of laws of CM, e.g. observation of spectra of atoms) is expressible in terms of CM (e.g. preparation of sources of radiation and measurement of positions of spectral lines displayed on screens). Quantal (:= quantum mechanical) phenomena are not only observed on a 'background' and 'from the point of view' of quantities describing states of macroscopic bodies (resp. of such parameters, the behaviour of which is adequately described by laws of CM), but also specific theoretical models for description of such phenomena in the framework of QM are constructed under strong influence of existing models in CM (e.g. the quantal models of atoms compared to classical planetary motions, or, more generally, some systems of canonically conjugated observables in the sense of the Hamiltonian CM correspond isomorphically to a subset of quantal observables).

© Springer Nature Switzerland AG 2020, corrected publication 2020
P. Bóna, *Classical Systems in Quantum Mechanics*,
https://doi.org/10.1007/978-3-030-45070-0_1

Quantal models of many systems, on the other hand, might be constructed from classical models of the same systems (which are adequate in a certain range of conditions, e.g. classical gases in some intervals of temperature and density) by a more or less standard procedure of 'quantization', compare, e.g. [19, 93, 117], [225, 228, 314], and works quoted therein. The (vaguely stated) question arising from these considerations is: **What is a 'physically justified way' of correct determination of quantal models from their classical approximates ?**

1.1.3 One of the remarkable features of QM is the occurence of the universal (Planck) constant $\hbar$, which might be used to measure mutual 'deviation' of quantal and classical descriptions of a given physical system (we shall not discuss here the nontrivial methodological question: how to determine a 'physical system' and what is its dependence on theoretical concepts used in the process of the determination). Consequently, an approximate description of processes in the framework of QM that are characterized by some quantities S large compared to the Planck constant (S being of the same physical dimension as $\hbar$) is often reached in the limit of large values of $S\hbar^{-1}$ ('short wave asymptotics'). If, however, the system described by QM has some features ('variables' etc.) which are adequately described by CM too, then **the description of this 'classical subsystem' has to be contained in QM** with the fixed value of Planck constant (i.e. the classical description should be exact consequence of QM without any approximation procedure, which is often formally performed by the limit $\hbar \to 0$ [1]). We shall introduce a standard procedure of obtaining classical systems from quantal ones. Such a classical system is called here a 'classical projection' of the quantal system (contrary to the 'classical limit' obtained in some way by $\hbar \to 0$).

1.1.4 This work is considered as a conceptually and intuitively (however, not always technically) simple way to give some insight into the indicated questions. Much more complete and extensive overview of these and related technical topics is given in the recent book [192] by Landsman. Many relevant questions are discussed in the author's work [37], containing also a detailed discussion of possible extensions of the QM formalism to its nonlinear versions; these nonlinear quantum motions are closely connected with the theory presented in our Chap. 6, corresponding to the motions of a single "microsystem" moving in the "mean-field" acting on it by interaction with infinite number of similar microsystems; the dynamics of the whole infinite collection of "microsystems" is, however, linear. Such a nonlinear quantum dynamics is also discussed by S. Weinberg in [328], whose work is also discussed and reformulated in [37, Sect. 3.6].

The mentioned work of S. Weinberg is not intrinsically consistent in the case of nonlinear motions of nontrivial density matrices, resp. "mixtures". To obtain successful picture of nonlinear quantum dynamics of "mixed states" together with their physically satisfactory quantal interpretation, one has to introduce **two kinds of "mixed states"**: The usual one used in (linear) QM are described in the standard way

---

[1]Consider here macroscopic quantal effects (e.g. superconductivity, superfluidity) vanishing for $\hbar \to 0$.

by density matrices (called there "elementary mixtures"), and others are called "genuine mixtures" (or. also "proper mixtures")—these correspond to the states which arose by a real 'mixing' of different quantal states, as it appears in classical statistical mechanics in ensembles of systems occurring in different states—different points of the phase space of the described system; they are introduced in [37, Sect. 2.1-e] and difference of these two kinds of mixtures is illustrated e.g. in [37, Sect. 3.3-e].

In the remaining sections of this introductory chapter it is specified briefly what we mean here by QM, CM and by the 'quantum theory of large systems'. The second chapter is devoted to a detailed study of geometry of the *projective Hilbert space* $P(\mathcal{H})$, where $\mathcal{H}$ is the Hilbert space used in description of a quantal system. We emphasize there the natural **symplectic structure** on $P(\mathcal{H})$, cf. e.g. [7, 37, 214, 231]. This structure is used in Sect. 2.3 to description of QM in terms of infinite dimensional CM, i.e. of classical Hamiltonian field theory with, however, the standard quantum statistical interpretation.

1.1.5 The Chap. 3 "Classical Mechanical Projections" is devoted to a general construction of Hamiltonian CM from a given quantal system (provided that an interpretation of its 'basic quantities' is specified by a unitary representation $U(G)$ of a *Lie group* G; for Lie groups see e.g. [13, 50, 209, 247]). The scheme of this construction is very simple: Take the *orbit* $O_\varrho := G.\varrho$ through a point $\varrho \in P(\mathcal{H})$ of the action of $U(G)$ on $P(\mathcal{H})$ corresponding to the action of $U(G)$ on $\mathcal{H}$ and restrict the natural symplectic form on $P(\mathcal{H})$ onto O_ϱ. For properly chosen ϱ the orbit O_ϱ is an immersed (and regularly embedded, cf. [37, Proposition 2.1.5(iv)] completed by [47]) submanifold of $P(\mathcal{H}) \subset \mathfrak{T}(\mathcal{H})$ (cf. 1.2.3), hence the restriction is well defined. The obtained two-form on the manifold O_ϱ might be degenerate, but after a natural factorization of the orbit we obtain a symplectic manifold which is symplectomorphic to an orbit of the coadjoint representation $Ad^*(G)$. Symplectic manifolds obtained in this way are interpreted as classical phase spaces. In some cases, if the generator of time evolution (the Hamiltonian operator) belongs to the generators of $U(G)$, we can obtain from the symplectic structure of $P(\mathcal{H})$ a *contact structure* on O_ϱ which reproduces an '*extended phase space*' (odd dimensional) of classical mechanics. If the Hamiltonian is not a generator of $U(G)$ (i.e. if G is only a 'kinematical group' without representing any time evolution), the quantal dynamics might be in some cases naturally projected onto the obtained classical phase space as a *globally Hamiltonian complete vector field*; this situation is analyzed in Sect. 3.3.

Although such a construction of CM from QM is equally applicable to any quantal system (specified by some $U(G)$), the interpretation of the obtained classical system depends on the specific physical system, and also on the physical quantal state ϱ from which the orbit O_ϱ is constructed. In any case, it is obtained a formal procedure for construction of 'classical projections' from arbitrary (finite) quantal systems.

Chapter 4 provides some simple examples of this formal procedure. In the Subsect. 4.1.6, we obtain from a simple nonrelativistic quantal system with the potential energy V the corresponding classical system (in the conventional sense) with a modified potential energy, where the modification depends on the choice of the 'initial state'

$\rho \in P(\mathcal{H})$ (for the orbit O_ρ) and can be made arbitrarily small (in the sense of weak convergence of distributions to the distribution V).

For a general time evolution, the orbits O_ρ are not invariant with respect to the quantal time evolution, and also on various orbits of the same quantal system the projected classical evolutions are mutually different. This brings in mind an idea of some stochastic time evolution on a classical phase space reflecting the underlying quantal evolution.

Such an idea is not, however, realizable for systems with finite number degrees of freedom (briefly: finite systems) because their density matrices have not unique decomposition into convex combinations of pure states $\varrho \in P(\mathcal{H})$. This is just a crude intuition which was not clearly formulated and realized in the following text.[2]

1.1.6 Quantal systems with infinite number of degrees of freedom (briefly: infinite systems) are considered in the Chap. 5. A physical motivation for such a consideration connected with our investigation of the relations between QM and CM consists in the fact, that 'macroscopicality' and 'classicality' are almost synonyma: most of physical systems containing an operationally well defined classical subsystem are compound of a large number (say: of the order 10^{20} and more) of microscopic constituents (like atoms) and vice versa.[3] Described approximately as infinite quantal systems, these systems have some characteristic properties distinguishing them from finite ones: the existence of nontrivial sets of 'classical observables' in given representations of observable algebra (this fact is a consequence of the existence of various inequivalent unitary representations), the existence of quite a rich simplexes (in the sense of Choquet) in the state space of the system allowing (in the presence of some additional assumptions) unique decomposition of their elements into extremal elements etc. This enables us to describe their *'classical subsystems'* directly in terms of the quantal description—hence the name *'macroscopic limit'*. This means that, contrary to the case of finite systems,[4] in the case of infinite systems quantal and classical interpretations of the 'macroscopic observables' coincide (at least on a G-invariant subset of states): classical, resp. macroscopic quantities are represented by operators belonging to the *center* of the weak closure of the algebra of observables in some representations.[5]

1.1.7 The Chap. 5 is divided into two sections. In the first one we consider the system consisting of denumerably infinite number of quantal subsystems, each of which is described by a G-covariant representation of its algebra of bounded observables. To be more specific, we consider a sequence of copies of the same finite system in the

---

[2]Some more specific hints on this possible classical stochastic evolutions from quantal time development could be found perhaps in [29].

[3]The macroscopic quantal effects like superfluidity and superconductivity are additional effects observed in these *'classical subsystems'* of the large quantal systems.

[4]where the quantal interpretation of classical quantities (i.e. expectation values of generators of $U(G)$ in corresponding states) was different from the classical interpretation (i.e. sharp values of corresponding classical generators).

[5]The **center** $\mathcal{Z}(\mathfrak{A})$ of a C^*-algebra $\mathfrak{A}$ is the commutative C^*-subalgebra of $\mathfrak{A}$ consisting of all elements of $\mathfrak{A}$, each commuting with all elements of $\mathfrak{A}$: $\mathcal{Z}(\mathfrak{A}) := \{z \in \mathfrak{A} : z \cdot x - x \cdot z = 0, \forall x \in \mathfrak{A}\}$.

(infinite) complete tensor product representation on a nonseparable Hilbert space $\mathcal{H}_\Pi$. The representation $U(G)$ describing an elementary subsystem determines a unitary (discontinuous) representation $U_\Pi(G)$ on $\mathcal{H}_\Pi$ which, in turn, determines an automorphism group σ_G of the algebra $\mathfrak{A}^\Pi$ of quasilocal observables of the infinite system. A natural definition of a classical subsystem of the large quantal system $(\mathfrak{A}^\Pi; \sigma_G)$ appearing in this case can be extended to the case of arbitrary systems $(\mathfrak{A}; \sigma_G)$, as it is shown in Sect. 5.2. The arising classical (macroscopic) subsystem $(\mathfrak{M}; \sigma_G)$ is naturally mapped into the classical Poisson system $(G^*; Ad^*(G))$, or to its generalizations.

1.1.8 Chapter 6 is devoted to an application of Sects. 5.1 and 5.2:

It is shown that 'mean-field' type time evolutions can be determined on a large quantal system $(\mathfrak{A}; \sigma_G)$ by specification of a Hamiltonian dynamics of a classical (macroscopic) Poisson system—the macroscopic limit of $(\mathfrak{A}; \sigma_G)$.

This is a perhaps simplest example of (infinite-)long-range interactions in many body systems. The correspondence between classical and quantum descriptions of systems appear there '**selfconsistently**': The quantum theory of the entering 'elementary subsystems' is built 'on the background' of the classical 'environment' what is compound of the infinite collection of those 'elementary subsystems'.[6] The dynamics of a general class of such systems is described in Sect. 6.3, and the statistical thermodynamics of equilibrium states is introduced in Sect. 6.4.

A slightly alternative approach to these quantum mean-field theories is described in the papers [40, 41].

1.1.9 Finally, the Chap. 7 contains four exactly solved models of interaction of a microscopic quantal system with a 'macroscopic' one. Due to this interaction the macroscopic quantal system changes its classical state to a different one. Such a change of a macroscopic (classical) state can be interpreted as a change of a 'pointer position', hence these models could be considered as models of 'quantum measurements' in the sense of Klaus Hepp [153]. The change of the macroscopic state is reached in the limit $t \to \infty$ of infinite time, and the convergence in the first three models is very slow.

In the last of the described models (in Sect. 7.6) the 'macroscopic' quantal system is described as a **finite collection** of 'small' quantal systems. This leads to problems with an unambiguous definition of 'macroscopic states', since it is possible (formally, in this abstract theory) to observe interference effects between such different 'macroscopic states'. To make clear the correspondence of quantum theory with observations, it would be necessary to introduce also quantum models of observation apparatuses used for detection of states of such a large but finite 'macroscopic system'. Some discussion on this problem (including reports of observations of 'macroscopic interference phenomena') appeared in literature in last decades, cf. e.g. [55, 56, 190–192, 195, 196]. In the model of Sect. 7.6, the (large but finite)

---

[6]Ideas of this kind could, perhaps, reconcile the basic idea of Niels Bohr [26, 27] on fundamental role of a "classical background" in formulations of QM with the postulate that QM is *the* basic theory.

'apparatus' radiates a Fermi particle escaping to infinity and, contrary to the other above mentioned models of Chap. 7, *it converges very quickly* to the final 'almost macroscopic' state.

1.1.10 Bibliographical notes.

The canonical symplectic structure (in the case of finite dimensions) on complex projective Hilbert spaces is described in [7]; in context of QM it appeared, e.g. in [17, 37, 69, 268]. Orbits of $U(G)$ in the Hilbert space were introduced in the special case of Heisenberg group G in [125], and in general case in [176, 239] under the name '(generalized) coherent states'. John Klauder obtained CM on such orbits (or even on more general submanifolds of Hilbert space) from the quantal Hamilton principle restricted to corresponding orbits (resp. to 'overcomplete sets of unit vectors'), see [176]. The orbits $G.\varrho$ in $P(\mathcal{H})$, and the functions $\nu \mapsto f_A(\nu) := Tr(\nu A)$ $(\nu \in G.\varrho)$, named (in the case of one-dimensional ϱ) 'covariant symbols' by Berezin [18] or 'lower symbols' by Simon [291], were used for determination of bounds for quantum partition functions (see [199, 291]), in time dependent Hartree-Fock theory [268], and also for description of specific types of unitary representations of Galilean and Poincaré groups [2]. Some essential properties of generalized coherent states are described in [84]. The natural symplectic orbits of coadjoint representations was introduced in [174].

A further development of these (mathematical, as well as physical) ideas is also contained in the work [37, 47], which contains also a nonlinear extension of the formulation of QM. This nonlinear extension is also compared in [37, Sect. 3.6] with the Weinberg attempt [328] to formulate a nonlinear version of QM.

Some of the main ideas on connections between QM and CM leading to the present work are implicitly contained already in the classical work [330]. The idea and techniques used for transition to infinite systems was gained mainly from works by Haag, Hepp, Lieb, Neumann, Ruelle and others (see e.g. [139, 155, 227, 271], and for a review compare [53, 54, 106]). A transition to macroscopic limit ('statistical quasiclassics') is described in [17] for a specific choice of the group G and a mean-field type interaction. A review of works on macroscopic limits ('large N limits') is given in [342]. An attempt of the description of classical quantities of large quantal systems analogous to the here presented one is described in works by Rieckers with collaborators [101, 265], and by Morchio with Strocchi [221, 222]; see also the works [317–320] by Thomas Unnerstall. A preliminary outline of a part of this work is contained in [32], and also in [40, 41]. The necessary mathematics can be found in the cited monographs, cf. also Appendices in [37].

An alternative way of description of thermodynamics and dynamics of quantum mean-field systems was later proposed in the work of the group around R.F. Werner, see e.g. [100].

A new approach to the theoretical description of classical (macroscopic) systems in the framework of quantum theory in a unique mathematical formalism is presented in a series of papers by Jean-Bernard Bru and collaborators [60].

1.2 Quantum Mechanics

1.2.1 In formal schemes of all theories considered in this work, the basic concepts are 'states', 'observables' and their transformations ascribed to a considered physical system. We shall not discuss here details of the empirical meaning of these concepts. Roughly, **states** are prepared by some standard empirical procedures and represent the situation, what has to be measured, **observables** describe (equivalence classes of) measuring apparatuses (i.e. the role of their function in the theory) giving certain empirically obtained responses if applied to states, and transformations include time evolution of the system in given conditions as well as various changes of equivalent descriptions of the system (**symmetries**).

In this section, we shall outline a simple standard scheme of the formalism of non-relativistic (resp. Galilean-relativistic) quantum mechanics of finite systems (QM), i.e. the nonrelativistic view on physical systems containing only finite number of their further indecomposable elementary constituents (particles, spins,...).[7]

1.2.2 Observables: A separable complex Hilbert space $\mathcal{H}$ corresponds to any physical system in QM. Let $\mathcal{L}(\mathcal{H})$ denote the set of all bounded linear operators from $\mathcal{H}$ to $\mathcal{H}$, where the boundedness (equiv. continuity) is defined with respect to the norm of $\mathcal{H}$ coming from the scalar product (x, y), $(x, y \in \mathcal{H})$, which is linear in the second factor y. **Observables in QM** (i.e. physical quantities empirically identifiable by some realizable(?) measuring devices) are represented by selfadjoint operators on $\mathcal{H}$ (in general unbounded). It is useful to consider along with any self-adjoint operator A (corresponding to an equally denoted observable A) its spectral measure E_A defined on Borel subsets of the real line $\mathbb{R}$ with values in projectors $E_A(\bullet)$ in $\mathcal{L}(\mathcal{H})$, $E_A(\mathbb{R}) = I := id_{\mathcal{H}}$ (:= the identity of the algebra $\mathcal{L}(\mathcal{H})$), cf. [37, Appendices B & C].

It is important to stress here, that in the conventional QM of finite systems (atoms, molecules, and finite collections of them) the set of observables contains **the whole set $\mathcal{L}(\mathcal{H})$ of operators** representing these observables. Hence, the algebra $\mathcal{L}(\mathcal{H})$ acts on $\mathcal{H}$ by the **irreducible** manner (i.e. no nontrivial subspace of $\mathcal{H}$ is by the actions of the whole $\mathcal{L}(\mathcal{H})$ left invariant). This also implies the impossibility, resp. inadequacy, of interpretation of the "mixed states" as representing some statistical mixture of systems occurring in the states decomposing the corresponding "mixture" (cf. 1.2.3) in this QM of finite systems.

1.2.3 States in QM are conventionally represented by *density matrices*, i.e. **positive** *trace class operators* ϱ on $\mathcal{H}$ **with unit trace** (=the trace norm): $Tr(\varrho) = 1$. Density matrices form a convex subset in the *linear space* $\mathfrak{T}(\mathcal{H})$ of all trace class operators which is closed in the **trace norm** $\|A\|_1 := Tr\sqrt{A^*A}$. Denote this *set of states $\mathcal{S}_*$*. The **extreme points of $\mathcal{S}_*$** are represented by the one-dimensional orthogonal projectors $P_x \in \mathcal{L}(\mathcal{H})$ (projecting $\mathcal{H}$ onto one-dimensional subspaces $\mathbf{x}$ containing x, $0 \neq x \in \mathcal{H}$). Any $\varrho \in \mathcal{S}_*$ can be expressed as a weak limit of finite convex

---

[7] The concepts of "system", and "physical system" are taken here to be as intuitively clear.

combinations of elements $P_j \in P(\mathcal{H}) := \{P_x : x \in \mathcal{H}, x \neq 0\}$ of the **projective Hilbert space** $P(\mathcal{H})$. We can write

$$\varrho = \sum_j \lambda_j P_j, \quad \sum_j \lambda_j = 1, \ \lambda_j \geq 0. \tag{1.2.1}$$

The states from $P(\mathcal{H})$ are called **pure states**. The decomposition (1.2.1) of an arbitrary state ϱ into pure states is highly nonunique if ϱ does not belong to $P(\mathcal{H})$, hence the state-space $\mathcal{S}_*$ is not a **simplex**, cf. [73], what is an important difference with respect to classical mechanics. This have important consequences for interpretation of the 'mixed states' described by density matrices $\varrho \notin P(\mathcal{H})$: The nonunique decompositions (1.2.1) show that **these quantum states cannot be interpreted as representations of statistical ensembles** each element of which (i.e. a copy of the considered physical system) occurs in a definite pure state, because pure states appearing in certain mutually different decompositions of the same density matrix are in general incompatible, i.e. they are eigenstates of mutually noncommuting (hence simultaneously nonmeasurable) observables, cf. [37, 1.5-b].[8] We have $\varrho \in P(\mathcal{H})$ iff $\varrho^2 = \varrho$ and $\varrho \in \mathcal{S}_*$.

1.2.4 Quantum theories are 'intrinsically (or irreducibly) statistical', i.e. experimentally verifiable assertions can be expressed in general in terms of probabilities only in the frame of these theories. Results of repeated measurements of a given quantity (observable) applied to the same state (which should be, however, repeatedly prepared for each single measurement because of its unavoidable disturbance by the interaction with the measuring apparatuses) have a nonzero dispersion for a general quantity. The **expectation value** of measured values of a given bounded observable (represented by the operator) $A = A^* \in \mathcal{L}(\mathcal{H})$ in the state (represented by the density matrix) $\varrho \in \mathcal{S}_*$ is in QM expressed by

$$\omega_\varrho(A) := Tr(\varrho A). \tag{1.2.2}$$

ω_ϱ can be considered here as a positive linear functional on $\mathcal{L}(\mathcal{H})$, which is **normalized** (i.e. $\omega_\varrho(I_\mathcal{H}) = 1$) and **normal** (i.e. ultraweakly continuous), compare, e.g. [53, 54, 274]; the set of all such functionals ω might be identified with $\mathcal{S}_*$: to each ω corresponds a unique density matrix $\varrho =: \varrho_\omega$, for which $\omega = \omega_\varrho$ according to (1.2.2). For an arbitrary selfadjoint (not necessarily bounded) operator A, the probability of obtaining of its value in a Borel set $B \subset \mathbb{R}$, if measured in the state $\omega \in \mathcal{S}_*$, is

$$\omega(E_A(B)) .$$

---

[8]This point was important also in the discussion about (im-)possibility of deducing the linearity of QM-time evolutions from mere quantal kinematics together with the so called "No-Signaling Condition", cf. [46].

Here $E_A : B \mapsto E_A(B)$, is the unique *projector valued measure* of A, or its *spectral measure*, characterizing any selfadjoint operator A, [37, B & C]. We shall define also

$$\omega(A) := \int_{\mathbb{R}} \lambda \, \omega(E_A(\mathrm{d}\lambda)) \tag{1.2.3}$$

if the integral converges absolutely. This is a generalization, resp. an alternative form of (1.2.2). If $\omega_x \in S_*$ corresponds to $P_x \in P(\mathcal{H})$ and for a given $A = A^*$ the quantity $\omega_x(A^2)$ is defined (i.e. is finite), then $x \in D(A)$ (:= the domain of A), and vice versa.[9]

1.2.5 Any observable A determines a strongly continuous *one-parameter group* $t \mapsto \exp(-itA)$ of unitary transformations of $\mathcal{H}$ of which A is its **generator**. This induces a *weakly*-continuous* ($\equiv$ w*-*continuous*) *group* τ^A of **-automorphisms* of the von Neumann algebra $\mathcal{L}(\mathcal{H})$ (cf. [37, B.2.1(v)]), $B \mapsto \tau_t^A(B) := e^{itA} B e^{-itA}$, $B \in \mathcal{L}(\mathcal{H})$, $t \in \mathbb{R}$, i.e. the functions

$$t \mapsto \omega(\tau_t^A B) := \omega(e^{itA} B e^{-itA}) \tag{1.2.4}$$

are continuous for all $B \in \mathcal{L}(\mathcal{H})$ and all $\omega \in S_*$. The observable A represents in this way a one-parameter group of **symmetries** of the physical system. Conversely, any w*-continuous one-parameter group of *-automorphisms of $\mathcal{L}(\mathcal{H})$ is given by an observable (determined up to an arbitrary additive real constant) in the above described manner (see e.g. [53, Example 3.2.35]). If A is bounded, $t \mapsto \exp(-itA)$ is norm-continuous.

1.2.6 To obtain an empirical meaning of the formal scheme outlined above, it is necessary to specify how to measure quantities corresponding to specific operators. As far as the present author knows, this type of *interpretation* for arbitrary selfadjoint operators was not realized for any physical system (except, perhaps, of some systems consisting of spins only). It might be, however, sufficient to ascribe a certain empirical meaning to 'sufficiently many' operators. We can use, for such an identification of operators and empirical manipulations, the above mentioned connection between one-parameter groups of automorphisms τ^A and operators A. We shall take into account, moreover, that also 'microscopic systems' described adequately in the framework of quantum mechanics are only empirically specified by manipulations with 'macroscopic bodies', which are well described by CM. Let a physical system preserve its identity if the surrounding macroscopic bodies undergo some group of motions. Then we obtain a group of symmetry transformations of that system.[10] To any one-parameter subgroup of such 'macroscopically determined' transformations corresponds in our formalism a selfadjoint operator, which in turn corresponds in

---

[9]Let us remember here that no unbounded symmetric linear operator A acting on a Hilbert space $\mathcal{H}$ can be defined on the whole space $\mathcal{H}$: $D(A) \subsetneqq \mathcal{H}$.

[10]This is so called "passive symmetry transformation", contrasted to the "active" one, when the 'physical system' is moved in the fixed environment; these two ways of understanding of transformations applied to a system are mathematically equivalent.

some way (we shall not specify it here) to a measurable quantity connected with the macroscopic motions. We shall assume (and this is really fulfilled for many finite systems) that the group G obtained in this way is large enough to determine all the 'basic observables'; all the other observables are supposed to be functions of these basic ones (see the following subsections).

1.2.7 We shall assume that a w*-continuous representation σ of a connected Lie group G in the group of *-automorphisms of $\mathcal{L}(\mathcal{H})$ is given and that the group $\{\sigma_g \in$ *- Aut $\mathcal{L}(\mathcal{H}) : g \in G\}$ acts on $\mathcal{L}(\mathcal{H})$ irreducibly: there is no nontrivial von Neumann subalgebra of $\mathcal{L}(\mathcal{H})$ which is left invariant by the all σ_g ($g \in G$). One-parameter subgroups of G are in bijective correspondence with elements ξ of the Lie algebra $\mathfrak{g}$ of G to which, in turn, correspond selfadjoint generators X_ξ of unitary groups determined by $\sigma_{exp(t\xi)}$, cf. [37, A.4.8].

Since the unitary operators $U(g)$ determined by automorphisms σ_g ($g \in G$) via the relation

$$U(g)^*BU(g) = \sigma_g(B), \quad \forall B \in \mathcal{L}(\mathcal{H}) \tag{1.2.5}$$

are only defined up to a phase factor, in general case, the representation σ leads only to a **projective representation** $g \mapsto U(g)$ of G in the unitary group of $\mathcal{H}$, i.e.

$$U(g_1 g_2) = m(g_1, g_2)U(g_1)U(g_2), \tag{1.2.6}$$

where $m : G \times G \mapsto S^1$ (:= the complex numbers of unit modulus) is a **multiplier** of the projective representation, cf. [37, 3.3.6]. Such a representation can be always extended to a unitary representation of a group G_m, which is the **central extension** of G [174, 15.2, Thm. 1] by the multiplicative group S^1 corresponding to the multiplier m, [37, 1.5-c]. The group multiplication in G_m (which can be identified, as a set, with $G \times S^1$) is

$$(g_1; \lambda_1)(g_2; \lambda_2) = (g_1 g_2; m(g_1, g_2)\lambda_1\lambda_2), \quad \lambda_j \in S^1. \tag{1.2.7}$$

In the unitary extension of the projective representation $U(G)$ the elements of the center of G_m are represented by the numbers from S^1 ('phase factors') acting by multiplication of the vectors $x \in \mathcal{H}$. All the extensions G_m of G (corresponding to various multipliers m) are classified by the second cohomology group $H^2(G, S^1)$ of the group G with values in S^1, for details see [174, 321]. We shall assume that the unitary representation $U(G_m)$ corresponding to the representation σ of G according to (1.2.5) can be (and really is) chosen strongly continuous. In the following we shall usually write G instead of G_m.

A natural consequence of irreducibility of σ is the irreducibility of corresponding unitary representation U. Hence, the weak-operator closure of the linear hull of the subset $\{U(g) : g \in G\}$ of $\mathcal{L}(\mathcal{H})$ in the von Neumann algebra $\mathcal{L}(\mathcal{H})$ is $\mathcal{L}(\mathcal{H})$ itself.

1.2.8 The interpretation of G as a group of (empirically defined) physical symmetries of the system leads to a natural interpretation of generators X_ξ ($\xi \in G$) of the unitary representation U. Since any bounded operator is weakly approximated by linear combinations $\sum \lambda_j U(g_j)$ we can hope to obtain some insight into possible interpretations of other operators. The complete answer to this problem of interpretation needs, probably, an analysis of possible interactions of the system under consideration with all other systems, or, at least with systems which could be used in the role of measuring instruments. The choice of G together with (eventually) some other assumptions on the physical properties of the system (e.g. the value of spin) might also determine the dimension of $\mathcal{H}$.

The proper choice of the representation of G depends on comparison of consequences of the chosen 'interpretation U' with empirical data; this step contains, e.g. the choice of the correct value of the Planck constant, if G is the Heisenberg group (i.e. a *central extension* of the classical phase space $\mathbb{R}^{2n}$ considered as the commutative group of translations).

1.2.9 It will be further assumed that the time evolution of the system is either a one-parameter subgroup of G, or it is separately defined as a one-parameter w*-continuous subgroup τ of the group of *-automorphisms of $\mathcal{L}(\mathcal{H})$ $t \mapsto \tau_t \in$ *-aut$(\mathcal{L}(\mathcal{H}))$, $\tau_{t+u} = \tau_t \circ \tau_u$ $(t, u \in \mathbb{R})$, $\tau_o :=$ identity. Note that for each automorphism $\alpha \in$ *-aut$(\mathcal{L}(\mathcal{H}))$ there is some unitary $U_\alpha \in \mathcal{L}(\mathcal{H})$ such that for all $A \in \mathcal{L}(\mathcal{H})$: $\alpha(A) \equiv U_\alpha A U_\alpha^*$, i.e. the automorphisms of $\mathcal{L}(\mathcal{H})$ are *inner automorphisms*, cf. e.g. [274, Corollary 2.9.32].

1.3 Classical Hamiltonian Mechanics

1.3.1 In this section, we shall outline the formal scheme of classical Hamiltonian mechanics (CM) parallel to the exposition of QM in the preceding section. We shall restrict our considerations to the case of systems with finite number of degrees of freedom. We shall use the language of *differential geometry* (for pedagogically well written course of differential geometry we refer to [111]). A technically more complicated quantum theory of systems with infinite number of degrees of freedom will be described later. For classical theory of infinite systems, i.e. *classical field theory*, see the corresponding monographs, or also e.g. [1, II.5.5], [7, Append.2], [37].

1.3.2 To any physical system there corresponds in CM a **symplectic manifold** $(M; \Omega)$ (cf. [1, 7, 178]). M is here an (even dimensional) infinitely differentiable Hausdorff second countable connected *manifold modeled by* $\mathbb{R}^{2n}$ and Ω is a nondegenerate closed *two-form on M*, the **symplectic form**, cf. also [37, A.3]. Observables in CM are represented by real-valued functions f on M; for technical convenience, we shall assume usually f to be infinitely differentiable, $f \in C^\infty(M, \mathbb{R})$. These observables constitute a real associative algebra $\mathfrak{F}(M)$ with respect to the ordinary multiplication of functions: $f.g(x) := f(x)g(x)$ $(f, g \in \mathfrak{F}(M),$ $x \in M)$. This algebra has the natural complexification $\mathfrak{F}_\mathbb{C}(M)$.

Remark: In the larger algebra $\mathfrak{B}(M)$ of all bounded Borel functions on M we can associate to any $f \in \mathfrak{B}(M)$ the projector-valued measure E_f defined on Borel subsets of $\mathbb{R}$ (for a real-valued f):

$E_f : B \mapsto \chi_{f^{-1}(B)}$ for any Borel $B \subset \mathbb{R}$, where χ_N is the characteristic function of the Borel subset $N \subset M$.

It is clear that $E_f(B) := \chi_{f^{-1}(B)}$ are projectors in $\mathfrak{B}(M)$ and the association $B \mapsto E_f(B)$ is σ-additive, with $E_f(\mathbb{R}) = \chi_M = $ the unit element of $\mathfrak{B}(M)$. The real-valued Borel functions f can be also considered as selfadjoint operators on a Hilbert space $\mathcal{H} := L^2(\mathbb{R}, \mu)$ acting as the multiplication operators, and E_f's are their canonical spectral measures.

1.3.3 States in CM are probability Borel measures μ on M, which form a convex set $\mathcal{S}_{cl}$ with extremal points consisting of all measures concentrated at one-point sets in M, i.e. of all *Dirac measures* on M. Hence, *pure states* are identified with points $x \in M$. Any measure $\mu \in \mathcal{S}_{cl}$ has a *unique decomposition* into (an integral of) Dirac measures, i.e. it is a **simplex**, contrary to the state space of QM. This has serious consequences for different possibilities of statistical interpretations of states in CM and QM, cf. 1.2.3, also footnote 8.

1.3.4 According to CM, the disturbance of the state connected with the measurement of arbitrary observables can be made negligibly small. Because of uniqueness of the decomposition of an arbitrary state to its extremal components we can interpret any $\mu \in \mathcal{S}_{cl}$ as a representative of a statistical ensemble of a large number of copies of the considered system, each being in an (its own) pure state. Repeated measurements on the state μ have to be understood now as a repeated random choice (with probability corresponding to the probability measure μ on M) from the ensemble of a system appearing in a pure state $x \in M$ and measuring precise values $f(x)$ of observables $f \in \mathfrak{F}(M)$ afterwards. For such a measurement procedure the probability of finding the value of an observable f in a Borel set $B \subset \mathbb{R}$ is $\mu(E_f(B))$ (compare Remark in 1.3.2), where $\mu(f)$ for $f \in \mathfrak{B}(M)$ means the integral of f with the measure μ on M. The value $\mu(f)$ for $f \in \mathfrak{F}(M)$ is then the expectation value of f in the state $\mu \in \mathcal{S}_{cl}$. The mapping $\mu : f \mapsto \mu(f)$ is a positive normalized linear functional on $\mathfrak{F}(M)$ (and also on $\mathfrak{B}(M)$), which is continuous with respect to the usual sup-norm on $\mathfrak{B}(M)$. Better continuity properties have, e.g. functionals μ which are absolutely continuous (as measures) with respect to the natural measure Ω^n on the *symplectic manifold* $(M; \Omega)$.

1.3.5 A **symmetry** of a system in CM is defined as a **symplectomorphism** F of $(M; \Omega)$, i.e. F is such a diffeomorphism of M onto itself which leaves the symplectic form Ω unchanged: $F^*\Omega = \Omega$, where F^* is the pull-back on M, see e.g. [1], or also [37, A]. For $f \in \mathfrak{F}(M)$ let $F^*f := f \circ F$; such an action of F onto the algebra $\mathfrak{F}(M)$ is an automorphism. It conserves, moreover, another structure on $\mathfrak{F}(M)$—the *Poisson algebra structure* defined below.

Let $t \mapsto F_t$ be a one-parameter group of symmetries, which is differentiable with respect to $t \in \mathbb{R} : F_{t+s} = F_t \circ F_s$ $(t, s, \in \mathbb{R})$ and the derivative

$$\left. \frac{\mathrm{d}}{\mathrm{d}t} \right|_{t=0} f(F_t x) =: \mathrm{d}_x f(\sigma_F) \tag{1.3.1}$$

exists for all $f \in \mathfrak{F}(M), \forall x \in M$, and the functions

$$\mathrm{d}f(\sigma_F) : x \mapsto \mathrm{d}_x f(\sigma_F) \; (\in \mathbb{R}) \tag{1.3.2}$$

are infinitely differentiable, $\mathrm{d}f(\sigma_F) \in \mathfrak{F}(M)$. Here σ_F is the vector field on M corresponding to the flow $x \mapsto F_t x$, $(x \in M)$. Let $\mathfrak{X}(M)$ be the set of all infinitely differentiable vector fields on M. Let $i(\sigma)\Omega$ be the one-form on M defined by: $i(\sigma)\Omega(\varphi) := \Omega(\sigma, \varphi)$ for any $\sigma, \varphi \in \mathfrak{X}(M)$, i.e. $i(\sigma)\Omega$ is the **inner product** [37, A.3.10] of the vector field σ with the two-form Ω. For the vector field σ_F we have:

$$\mathrm{d}i(\sigma_F)\Omega = 0. \tag{1.3.3}$$

Vector fields σ_F and the corresponding flows of symplectomorphisms F_t are called **locally Hamiltonian**. If there is $f_F \in \mathfrak{F}(M)$ such that

$$i(\sigma_F)\Omega = -\mathrm{d}f_F \text{ on } M, \tag{1.3.4}$$

then σ_F is (globally) Hamiltonian and f_F is its **Hamiltonian function**. To any $f \in \mathfrak{F}(M)$, we can unambiguously define a **Hamiltonian vector field** σ_f with the Hamiltonian function f by the formula

$$i(\sigma_f)\Omega = -\mathrm{d}f. \tag{1.3.5}$$

Uniqueness of σ_f is a consequence of nondegeneracy of Ω. Two functions $f, g \in \mathfrak{F}(M)$ give the same vector field $\sigma_f = \sigma_g$ iff $f - g = const$. We can introduce now a *Lie algebra structure* into $\mathfrak{F}(M)$, the structure of **Poisson bracket** multiplication: $(f; g) \mapsto \{f, g\} \in \mathfrak{F}(M)$ for all $f, g \in \mathfrak{F}(M)$. We define

$$\{f, g\} := \Omega(\sigma_f, \sigma_g), \tag{1.3.6}$$

where σ_f (resp.σ_g) is given in (1.3.5). If we denote by $\pounds_\sigma$, $\sigma \in \mathfrak{X}(M)$, the **Lie derivative** [37, A.3.7,A.3.8] in the direction of σ of tensor fields on M ($\pounds_\sigma$ acting on the differential forms has the expression $\pounds_\sigma = i(\sigma)\mathrm{d} + \mathrm{d}i(\sigma)$), then, according to (1.3.5) :

$$\{f, g\} = \pounds_{\sigma_f} g = -\pounds_{\sigma_g} f. \tag{1.3.7}$$

The properties of Ω are reflected in the following properties of the Poisson bracket:

(i) $\{f, g + \lambda h\} = \{f, g\} + \lambda\{f, h\}$,

(ii) $\{f, g\} = -\{g, f\}$, (*bilinearity and antisymmetry of* Ω),

(iii) $\{f, \{g, h\}\} + \{g, \{h, f\}\} + \{h, \{f, g\}\} = 0$, (*closedness* $\mathrm{d}\Omega = 0$),

(iv) $\{f, g \cdot h\} = \{f, g\} \cdot h + g \cdot \{f, h\}$, (*derivation property* (1.3.7)),

(v)   *If* $\{f, g\} = 0 \ \ \forall g \in \mathfrak{F}(M) \Rightarrow f \equiv const.$ (*nondegeneracy of* Ω).

It is not difficult to prove for the commutator of Hamiltonian vector fields:

$$[\sigma_f, \sigma_g] = \sigma_{\{f,g\}}. \tag{1.3.8}$$

1.3.6 In CM, all the observables are functions of points $x \in M$, hence locally can be expressed as functions of a finite number $2n$ coordinate functions. In accordance with the 'philosophy' of 1.2.6, we shall look for an interpretation of a finite number of observables which contain systems of coordinate functions for a neighbourhood of any point of M. This can be naturally done, if M is a homogeneous space of a connected *Lie group* G (cf. [37, A.4]) corresponding to a group of empirical manipulations with objects relevant to the determination of the considered system. Since the symplectic structure Ω on M reflects important physical properties of many physical systems, it is desirable for the group action on M to conserve this structure. In this case, one parameter subgroups of symmetries correspond to Hamiltonian flows which can be physically interpreted.

1.3.7 From now on, we shall assume that $(M; \Omega)$ is a *homogeneous space* of a connected *Lie group* G, on which the group G acts as an infinitely differentiable group of symplectomorphisms F_g ($g \in G$): $F_g^* \Omega = \Omega$, $F_{gh} = F_g \circ F_h$ ($g, h, \in G$), and functions $g \mapsto f(F_g x)$ are in $C^\infty(G, \mathbb{R})$ for all $f \in \mathfrak{F}(M)$ and all $x \in M$. If $e \in G$ is the unit element of G, then $F_e := id_M$. To any $\xi \in \mathfrak{g}$ (:= the **Lie algebra of** G) there is a one-parameter group of symplectomorphisms $t \mapsto F_{\exp(t\xi)}$ of M generated by the vector field σ_ξ (compare with (1.3.1)). If $[\xi, \eta]$ denotes the **commutator in** $\mathfrak{g}$, and $[\sigma_\xi, \sigma_\eta] \in \mathfrak{X}(M)$ the *commutator of vector fields* on M, then (see [1, Proposition 4.1.26])

$$[\sigma_\xi, \sigma_\eta] = -\sigma_{[\xi,\eta]}. \tag{1.3.9}$$

Every homogeneous symplectic manifold has universal covering symplectic homogeneous manifold with respect to the universal covering group of G. On any simply connected homogeneous symplectic manifold of a connected Lie group G, the functions f_ξ ($\xi \in \mathfrak{g}$) determined up to additive constants by the formula

$$i(\sigma_\xi)\Omega = -\mathrm{d}f_\xi \tag{1.3.10}$$

are defined globally on the manifold M, $f_\xi \in \mathfrak{F}(M)$, i.e. the vector fields σ_ξ ($\xi \in \mathfrak{g}$) are globally Hamiltonian. We shall assume that this is the case for our $(M; \Omega)$. Then arbitrary additive constants in the definitions of f_ξ's can be chosen such that the mapping $\xi \mapsto f_\xi$ from $\mathfrak{g}$ to $\mathfrak{F}(M)$ will be linear. Then

$$\{f_\xi, f_\eta\} = -f_{[\xi,\eta]} + C(\xi, \eta), \tag{1.3.11}$$

where C is a bilinear antisymmetric mapping from $\mathfrak{g} \times \mathfrak{g}$ to real constants on M called a **two-cocycle on** $\mathfrak{g}$ with values in $\mathbb{R}$. Any change of constants in f_ξ's (conserving the linearity of $\xi \mapsto f_\xi$) leads to an *equivalent cocycle* $C'(\xi, \eta) = C(\xi, \eta) + a([\xi, \eta])$, where $a \in \mathfrak{g}^*$ (:= the dual of $\mathfrak{g}$). Equivalence classes of two-cocycles form the commutative (additive)

$$\text{2-cohomology group } H^2(\mathfrak{g}, \mathbb{R})$$

of $\mathfrak{g}$ with values in $\mathbb{R}$. This group is isomorphic to $H^2(G, S^1)$ if G is simply connected (compare [321, Chap. 10.4.]). This isomorphism determines a canonical bijection between classes of irreducible projective representations and symplectic transitive actions of a simply connected Lie group. This bijection associates the class of all representations corresponding to the given (similarity class of a) multiplier with the class of symplectic actions with the corresponding (equivalence class of a) cocycle, compare also [28, 139]. If the *multiplier m corresponds to the cocycle C* from (1.3.11), then the *central extension* G_m of G (cf. [174, 15.2, Thm. 1]) acts on M in such a way, that

$$\{f_\xi, f_\eta\} = -f_{[\xi,\eta]} \quad \text{for all } \xi, \eta \in \mathfrak{g}_m, \tag{1.3.12}$$

if the added vector fields act on M trivially and constants in f_ξ's are properly chosen. If the action of G on M satisfies (1.3.11) with $C \equiv 0$, then it is called a **Poisson action** [7], and the symplectic manifold M is called *exactly homogeneous* [174].

1.3.8 Any observable $f \in \mathfrak{F}(M)$ on the homogeneous symplectic manifold M with globally defined Hamiltonian functions f_ξ ($\xi \in \mathfrak{g}$) can be expressed as a function of the 'basic observables f_ξ'. Hence measurement of any $f \in \mathfrak{F}(M)$ can be reduced to the measurements of f_ξ's. This does not make easier, however, of an ascribing a direct physical (i.e. empirical) interpretation to an arbitrary $f \in \mathfrak{F}(M)$ and the situation is similar to that one of QM, see 1.2.8.

1.3.9 A time evolution on $(M; \Omega)$ is defined in CM as a differentiable one-parameter group of symplectomorphisms with a globally defined Hamiltonian function $h \in \mathfrak{F}(M)$. This one-parameter group might be either a subgroup of G, or it is separately defined. In each case the group G might contain an invariance subgroup of h—the symmetry group of the dynamics (determining integrals of motion—conservation laws).

1.4   Quantum Theory of Large Systems

1.4.1 Models of systems with infinite number of degrees of freedom enter to quantum theory when we want to describe either processes accompanied with changes of numbers of particles (resp. quasiparticles) present in the physical system (what also occurs each time if we try to describe quantal analogues of classical continuous media, resp. fields), or systems with actual infinity of particles (the 'thermodynamic limit' necessary e.g. for clear conceptual description and abstract investigation of phase transitions). In standard models of infinite systems in quantum theory the algebras of bounded observables (e.g, CCR or CAR algebras for infinite number of degrees of freedom or algebras of spin systems on infinite lattices) have many mutually unitarily inequivalent physically relevant representations as algebras of bounded operators in some Hilbert spaces. These inequivalent representations might correspond e.g. to various states on the algebra of observables representing situations with various values of some macroscopic–global parameters of the large syatem. It often happens, moreover, that for description of some processes (time evolution, symmetry transformations), we are not able to work in the framework of only one (even faithful) representation. It is, consequently, useful to formulate theoretical scheme for the quantum theory of large systems (QTLS) in a representation independent, algebraic language. As basic sources of most of the here necessary mathematics and its application to description of large quantal systems could be taken, e.g. [53, 54, 84, 223, 235, 274, 286, 289]; a very brief summary can be found also in [37, Sect. 3.4].

1.4.2 A C^*-algebra $\mathfrak{A}$ [37, B.2] (details on C^*-algebras can be found in [53, 54, 90, 91, 106, 235, 274, 275, 305, 306]) corresponds to any physical system in QTLS. $\mathfrak{A}$ is a Banach algebra over complex numbers with involution $x \mapsto x^*$, $x \in \mathfrak{A}$, and with special (C^*) property. This means that it is a norm-closed linear space endowed with associative and distributive multiplication, and for any $x, y \in \mathfrak{A}$, $\lambda \in \mathbb{C}$, and with $\|x\| \geq 0$—the norm of $x \in \mathfrak{A}$, it is: $\|xy\| \leq \|x\| \cdot \|y\|$, the **involution** $x \mapsto x^*$ is antilinear: $(x + \lambda y)^* = x^* + \bar{\lambda} y^*$, where $\bar{\lambda}$ is the complex conjugate of λ, with $(xy)^* = y^* x^*$, $\|x^*\| = \|x\|$ ($= 0$ iff $x = 0$), and the C^*-property means: $\|x^* x\| = \|x\|^2$, $\forall x \in \mathfrak{A}$. $\mathfrak{A}$ is called **unital** C^*-algebra if it contains unit element $e \in \mathfrak{A} : ex = xe = x$, $\forall x \in \mathfrak{A}$. Selfadjoint elements $x = x^* \in \mathfrak{A}$ represent bounded **observables** of the system. The algebra $\mathfrak{A}$ is the **algebra of observables** of the system. For many interesting systems, $\mathfrak{A}$ is constructed as a C^*-**inductive limit** of a *net of local algebras* of finite (sub)systems (see [53, 106, 274], and specifically [274, 1.23]); in this case, these finite systems are interpreted e.g. as systems located in bounded space (-time) regions. Quasilocal algebras used in QTLS have such a structure (see [53, Definition 2.6.3]). It will be useful in our considerations to connect the *quasilocal structure of* $\mathfrak{A}$ with an action of a (usually abelian) group Π (Π **is an infinite set**—it might be a locally compact noncompact group) on $\mathfrak{A}$: For any $p \in \Pi$ let $\pi(p) \in$ *- Aut $\mathfrak{A}$, $\pi(p_1 p_2) = \pi(p_1)\pi(p_2)$ ($p_1, p_2 \in \Pi$). Let Π act transitively on a noncompact locally compact space V and let to any bounded open subset $v \subset V$ (denote the set of all such subsets by $\mathcal{B}(V)$) corresponds a C^*-subalgebra $\mathfrak{A}_v$ of $\mathfrak{A}$, the **local**

subalgebra of $\mathfrak{A}$ corresponding to $v \subset V$. If $v_1 \subset v_2 \subset V$, then $\mathfrak{A}_{v_1} \subset \mathfrak{A}_{v_2}$. All the $\mathfrak{A}_v$ ($v \in \mathcal{B}(V)$) have common unit $\equiv$ the unit $e := \mathrm{id}_{\mathfrak{A}}$ of $\mathfrak{A}$, and

$$\overline{\bigcup_{v \in \mathcal{B}(V)} \mathfrak{A}_v} = \mathfrak{A}, \tag{1.4.1}$$

where the over-bar denotes the uniform closure. We assume further that $\pi(p)(\mathfrak{A}_v) = \mathfrak{A}_{p \cdot v}$, where $p \cdot v := \{\lambda' \in V : \lambda' = p \cdot \lambda, \lambda \in v\}$, and $p \cdot \lambda$ denotes the action of $p \in \Pi$ on the point $\lambda \in V$. This action is supposed continuous and bounded: $p \cdot \mathcal{B}(V) \subset \mathcal{B}(V)$ for all $p \in \Pi$. We can assume (for simplicity) that for mutually disjoint $v, u \in \mathcal{B}(V)$, $v \cap u = \emptyset$, we have

$$[x, y] = 0, \text{ for all } x \in \mathfrak{A}_v, \ y \in \mathfrak{A}_u. \tag{1.4.2}$$

(The anticommutativity of Fermi systems can also be included, cf. [53, Sect. 2.6]).[11] We shall characterize this situation by saying that the algebra $\mathfrak{A}$ is **quasilocal with respect to the action** of the group Π. We shall use another technical assumption, that all the local subalgebras $\mathfrak{A}_v$ are W^*-algebras: A W^*-algebra $\mathfrak{A}$ is such a C^*-algebra which is (isomorphic to a) Banach space topological dual of another B-space $\mathfrak{A}_*$ called the **predual of** $\mathfrak{A}$; such an $\mathfrak{A}$ is always unital and generated by its projectors. W^*-algebras were introduced originally as weakly closed symmetric subalgebras of bounded operators in a Hilbert space containing identity and named **von Neumann algebras** after their originator.

1.4.3 Mathematically defined **states** on a C^*-algebra $\mathfrak{A}$ are any positive normalized linear functionals ω on $\mathfrak{A}$, i.e. such $\omega \in \mathfrak{A}^*$ (:= the dual of $\mathfrak{A}$), that

$$\omega(x^*x) \geq 0, \quad \|\omega\| = 1 \ (= \omega(\mathrm{id}_{\mathfrak{A}})). \tag{1.4.3}$$

Not all mathematical states, however, can be used as adequate descriptions of physical situations. As physical states on a quasilocal algebra $\mathfrak{A}$ are usually used **locally normal states**, i.e. such states ω on $\mathfrak{A}$, the restrictions of which to all the local W^*-subalgebras $\mathfrak{A}_v$ ($v \in \mathcal{B}(V)$) are $\sigma(\mathfrak{A}_v, (\mathfrak{A}_v)_*)$-*continuous* (here $(\mathfrak{A}_v)_*$ is the predual Banach space of $\mathfrak{A}_v$); the local normality of ω means that the restriction of ω to any $\mathfrak{A}_v$ is expressible by a density matrix in a faithful W^*-representation of $\mathfrak{A}_v$. We shall denote by $\mathcal{S}(\mathfrak{A})$ the set of all mathematical states on $\mathfrak{A}$ and by $\mathcal{S}_{ph} := \mathcal{S}_{ph}(\mathfrak{A})$ the set of (properly defined) physical states of the system. The subset $\mathcal{S}_{ph}(\mathfrak{A}) \subset \mathcal{S}(\mathfrak{A})$ has to satisfy some natural requirements, e.g. invariance with respect to transformations of physical symmetries (cf. below), convexity, local normality and (eventually) to form a stable face (see [53, Sect. 4.1]).

The set $\mathcal{S}(\mathfrak{A})$ is convex and compact in the w^*-topology of $\mathfrak{A}^*$ (i.e. in $\sigma(\mathfrak{A}^*, \mathfrak{A})$-*topology*). The set $\mathcal{E}\mathcal{S}(\mathfrak{A})$ of extreme points of $\mathcal{S}(\mathfrak{A})$ consists of **pure states** on

---

[11]Our formalism is built for the nonrelativistic situations. If the space V was the Minkowski space and our considerations were Einstein-Lorentz–relativistic, the condition for the commutativity in (1.4.2) would be the space–like separation instead of the disjointness of the domains $u, v \subset V$.

$\mathfrak{A}$: $\omega \in \mathcal{ES}(\mathfrak{A}) \Leftrightarrow \{\omega = \frac{1}{2}\omega_1 + \frac{1}{2}\omega_2 \ (\omega_{1,2} \in \mathcal{S}(\mathfrak{A})) \Rightarrow \omega_1 = \omega_2 = \omega\}$. Although the decomposition of a general $\omega \in \mathcal{S}(\mathfrak{A})$ into its extremal components ($\in \mathcal{ES}(\mathfrak{A})$) is not unique if $\mathfrak{A}$ is noncommutative, there are other physically relevant convex compact subsets of $\mathcal{S}(\mathfrak{A})$ (*Choquet simplexes*) allowing unique extremal decompositions of their elements into extremal components of these simplexes, cf. [73, 218] for basic mathematics, or also [53, Ch.4], [235, Ch.4], [274, Ch.3] for broader contexts.

1.4.4 The expectation value of a bounded observable $x = x^* \in \mathfrak{A}$ in the state $\omega \in \mathcal{S}(\mathfrak{A})$ (in accordance with comments in 1.2.4) is expressed by the value $\omega(x)$ of the functional ω on the element x. For calculations of probability distributions of values of $x = x^* \in \mathfrak{A}$ in the states $\omega \in \mathcal{S}(\mathfrak{A})$ it is used, however, the spectral decomposition of x. If $\mathfrak{A}$ is a general C^*-algebra, its selfadjoint elements need not have their spectral resolutions in $\mathfrak{A}$. The spectral resolutions in $\mathfrak{A}$ exist, however, if $\mathfrak{A}$ is a W^*-algebra, [274]: $x = x^* \in \mathfrak{A} \Rightarrow x = \int_{\mathbb{R}} \lambda \, E_x(d\lambda)$, $E_x(B)^* = E_x(B) = E_x(B)^2 \in \mathfrak{A}$, $B \subset \mathbb{R}$ Borel, ..., hence E_x is the projector valued spectral measure in the W^*-algebra $\mathfrak{A}$. Any C^*-algebra is naturally embedded into a W^*-algebra - the bidual $\mathfrak{A}^{**}$ of $\mathfrak{A}$, and any state $\omega \in \mathcal{S}(\mathfrak{A})$ can be uniquely extended to a state (equally denoted) $\omega \in \mathcal{S}_*(\mathfrak{A}^{**})$. For any state $\omega \in \mathcal{S}(\mathfrak{A})$, we can construct by the **GNS-algorithm** corresponding *cyclic representation* π_ω of $\mathfrak{A}$ in a Hilbert space $\mathcal{H}_\omega$ with a *cyclic vector*, Ω_ω (i.e. the norm-closure $\overline{\pi_\omega(\mathfrak{A})\Omega_\omega} = \mathcal{H}_\omega$), cf. [53, 223, 274], or also [37, Textbook], characterized (up to the unitary equivalence) by

$$\omega(x) = (\Omega_\omega, \pi_\omega(x)\Omega_\omega), \quad \forall x \in \mathfrak{A}. \tag{1.4.4}$$

The representation π_ω is irreducible iff $\omega \in \mathcal{ES}(\mathfrak{A})$. If we generalize the concept of observables to all operators from the bicommutant $\pi_\omega(\mathfrak{A})''$ in $\mathcal{L}(\mathcal{H}_\omega)$ (what is a W^*-subalgebra in $\mathcal{L}(\mathcal{H}_\omega)$), we can obtain spectral resolutions of selfadjoint elements of $\mathfrak{A}$ in such (extended) representations and the corresponding expressions for probability distributions, compare 1.2.4. In specific representations, we can define also unbounded observables as such selfadjoint operators on $\mathcal{H}_\omega$ the spectral projectors of which belong to $\pi_\omega(\mathfrak{A})''$, cf. [274].

We shall need later in this work to distinguish between states which are mutually *macroscopically distinguishable*. Mathematically are such states mutually **disjoint** together with the mutual disjointness of their GNS representations. It might be useful, for a characterization of this difference, to reproduce a theorem from [235, Thm. 3.8.11]:

Theorem: Let $\{\pi_1; \mathcal{H}_1\}$, $\{\pi_2; \mathcal{H}_2\}$ be two nondegenerate representations of a C^*-algebra $\mathfrak{A}$ with their central supports (equiv. central covers) s_1, s_2, cf. [235, 3.8.1]. The following conditions are equivalent:

(i) $s_1 \perp s_2$.
(ii) $((\pi_1 \oplus \pi_2)(\mathfrak{A}))'' = \pi_1(\mathfrak{A})'' \oplus \pi_2(\mathfrak{A})''$.
(iii) $((\pi_1 \oplus \pi_2)(\mathfrak{A}))' = \pi_1(\mathfrak{A})' \oplus \pi_2(\mathfrak{A})'$.
(iv) There are no unitarily equivalent subrepresentations of $\{\pi_1; \mathcal{H}_1\}$ and $\{\pi_2; \mathcal{H}_2\}$.

Here $\mathfrak{C}'$ for a subset $\mathfrak{C} \subset \mathcal{L}(\mathcal{H})$ denotes the commutant of $\mathfrak{C}$ in $\mathcal{L}(\mathcal{H})$: $\mathfrak{C}' :=$ $\{B \in \mathcal{L}(\mathcal{H}) : [B, A] \equiv BA - AB = 0, \forall A \in \mathfrak{C}\}$, and $\mathfrak{C}'' := (\mathfrak{C}')'$. The representations π_1, π_2 satisfying the conditions of the Theorem are called mutually **disjoint representations**. If the GNS representations determined by the two states ω_1, ω_2: $\{\pi_{\omega_1}; \mathcal{H}_{\omega_1}\}$, $\{\pi_{\omega_2}; \mathcal{H}_{\omega_2}\}$, are mutually disjoint, then we call these two states also mutually **disjoint** : $\omega_1 \perp \omega_2$.

1.4.5 An abstractly defined **symmetry** of the system in QTLS is any *-automorphism of the algebra $\mathfrak{A}$ of bounded observables. Let τ be a representation of the group $\mathbb{R}$ as a group of symmetries, i.e. a homomorphism $t(\in \mathbb{R}) \mapsto \tau_t \in$ *- Aut $\mathfrak{A}$, which is 'conveniently continuous', e.g. functions $t \mapsto \omega(\tau_t x)$ are continuous for all $x \in \mathfrak{A}$ and all $\omega \in \mathcal{S}_{ph}(\mathfrak{A})$. It is often assumed, that the group τ corresponding to a one-parameter group of empirically defined transformations is $\sigma(\mathfrak{A}, \mathfrak{A}^*)$-continuous (i.e. $\mathcal{S}_{ph}$ replaced by $\mathcal{S}(\mathfrak{A})$ in the last mentioned case), but this assumption might be too stringent. Let $\mathcal{S}$ be a 'sufficiently large' subset of states containing $\mathcal{S}_{ph}$ and denote by $\sigma(\mathfrak{A}, \mathcal{S})$ **the topology on** $\mathfrak{A}$ determined by functions $x(\in \mathfrak{A}) \mapsto \omega(x)$ for all $\omega \in \mathcal{S}$. We shall assume, that τ is $\sigma(\mathfrak{A}, \mathcal{S})$-continuous in the sense:

(i)   *functions* $t \mapsto \omega(\tau_t x)$ *are continuous for all* $\omega \in \mathcal{S}$, $x \in \mathfrak{A}$,

(ii)   *functions* $x \mapsto \tau_t x$ *are* $\sigma(\mathfrak{A}, \mathcal{S}) - \sigma(\mathfrak{A}, \mathcal{S}) - continuous$
       for all $t \in \mathbb{R}$,

(iii) $\omega \in \mathcal{S} \Rightarrow \omega \circ \tau_t \in \mathcal{S}$ *for all* $t \in \mathbb{R}$.

The last condition (iii) allows us to define a $\sigma(\mathcal{S}, \mathfrak{A})$-continuous group of transformations of $\mathcal{S}$ by

$$\tau_t^* \omega := \omega \circ \tau_t \quad (\text{for all } t \in \mathbb{R}), \ \omega \in \mathcal{S}. \tag{1.4.5}$$

Any selfadjoint element $a \in \mathfrak{A}$ generates a $\sigma(\mathfrak{A}, \mathcal{S})$ (i.e. $\sigma(\mathfrak{A}, \mathfrak{A}^*)$)-continuous group of inner *-automorphisms of $\mathfrak{A}$, τ^a, by

$$\tau_t^a x := \exp(ita) x \exp(-ita), \quad \text{for all } x \in \mathfrak{A}. \tag{1.4.6}$$

A one-parameter group of **inner automorphisms** of $\mathfrak{A}$ cannot represent some of physically important global transformations of *quasilocal algebras*, e.g. *Euclidean or Poincaré transformations*, cf. e.g. [106, Ch.4,Thm.3]. For a general (sufficiently continuous) one-parameter group τ of automorphisms of $\mathfrak{A}$ we can define a generator δ_τ - a densely defined **derivation on** $\mathfrak{A}$, [53, 274].[12] The connection of such generators with physically measurable quantities is in general in QTLS less transparent then it is in QM or in CM. If the state $\omega \in \mathcal{S}$ is τ−invariant, i.e. $\tau_t^* \omega \equiv \omega$, then there is unique weakly continuous unitary group U^ω acting on $\mathcal{H}_\omega$ (cf. 1.4.4) such that [271]:

$$\pi_\omega(\tau_t x) = U_{-t}^\omega \pi_\omega(x) U_t^\omega, \ U_t^\omega \Omega_\omega = \Omega_\omega \ \text{for all } t \in \mathbb{R}. \tag{1.4.7}$$

---

[12]A densely defined linear mapping $\delta : D(\delta) \subset \mathfrak{A} \to \mathfrak{A}$ is a **derivation on** $\mathfrak{A}$ if it satisfies the *Leibniz rule* : $\delta(xy) = \delta(x)y + x\delta(y) \ \forall \, x, y \in D(\delta) \subset \mathfrak{A}$.

Representations of $\mathfrak{A}$ in which τ is unitarily implemented in the sense of the first relation of (1.4.7) are called $\tau-covariant\ representations$. In such representations, the action of τ is given by an 'observable'—the selfadjoint generator of the corresponding unitary group acting on the Hilbert space of the representation. An interpretation of such a generator might be dependent, however, on the choice of the covariant representation.

Chapter 2
Geometry of the State Space of Quantum Mechanics

2.1 Manifold Structure of $P(\mathcal{H})$

2.1.1 Let $\mathcal{H}$ be a complex separable Hilbert space with the *scalar product* $(x, y) \in \mathbb{C}$ $(x, y, \in \mathcal{H})$, which is linear in the second factor y. Let $P(\mathcal{H}) := \mathcal{H}/\mathbb{C}^*$ be the *factor-space of* $\mathcal{H}$ by the multiplicative group $\mathbb{C}^*$ of nonzero complex numbers acting on $\mathcal{H}$ by multiplications by scalars. Any element $\mathbf{x} \in P(\mathcal{H})$ has the form

$$\mathbf{x} := \{y \in \mathcal{H} : y = \lambda x, \lambda \in \mathbb{C}^*\}, \ 0 \neq x \in \mathcal{H}. \tag{2.1.1}$$

The natural topology on $P(\mathcal{H})$ is the *factor-topology* coming from the norm-topology in $\mathcal{H}$. This topological space $P(\mathcal{H})$ is the **projective Hilbert space** of $\mathcal{H}$. The space $P(\mathcal{H})$ can be considered as the set of all one-dimensional complex subspaces of $\mathcal{H}$, or the set of all one-dimensional projectors $P_x \in \mathcal{L}(H)$ $(0 \neq x \in \mathcal{H})$, $P_x^* = P_x = P_x^2$, $P_x x = x$, with the natural bijective correspondence $P_x \leftrightarrow \mathbf{x}$. It is known that there is a natural **Kähler structure** on complex projective spaces. We shall describe it in some details in the case of $P(\mathcal{H})$.[1]

2.1.2 Let us define two natural (mutually equivalent) metrices (i.e. *distance functions*) $\mathrm{d}_1, \mathrm{d}_2$ on $P(\mathcal{H})$ (as usual: $\|x\|^2 := (x, x), \ x \in \mathcal{H}$):

$$\mathrm{d}_1(\mathbf{x}, \mathbf{y}) := \sqrt{2} \inf \left\{ \left\| \frac{x}{\|x\|} - e^{i\lambda} \frac{y}{\|y\|} \right\| : \lambda \in \mathbb{R} \right\}, \tag{2.1.2}$$

$$\mathrm{d}_2(\mathbf{x}, \mathbf{y}) := \sqrt{2} \, \|P_x - P_y\|. \tag{2.1.3}$$

---

[1] Another, more intuitive and more detailed approach to the structure of quantum state space can be found in [16]. For geometry and dynamics (also nonlinear) of general—not only pure—states see also [37, Sect. 2.1].

© Springer Nature Switzerland AG 2020, corrected publication 2020
P. Bóna, *Classical Systems in Quantum Mechanics*,
https://doi.org/10.1007/978-3-030-45070-0_2

It is not difficult to see that

$$d_1(\mathbf{x}, \mathbf{y}) = 2 \left(1 - (Tr(P_x P_y))^{1/2}\right)^{1/2}. \tag{2.1.4}$$

In (2.1.3), $\|A\|$ denotes the usual C^*-norm of the operator $A \in \mathcal{L}(H)$; if $|A| := \sqrt{A^*A} \in \mathcal{L}(H)$ is its *absolute value*, then one can prove

$$\sqrt{2}\, d_2(\mathbf{x}, \mathbf{y}) = Tr|P_x - P_y| = 2 \left(1 - Tr(P_x P_y)\right)^{1/2},$$
$$= \left(1 + (Tr(P_x P_y))^{1/2}\right)^{1/2} d_1(\mathbf{x}, \mathbf{y}), \tag{2.1.5}$$

what proves the equivalence of d_1 and d_2.

We shall examine now relations between various natural topologies on $P(\mathcal{H})$. We shall prove first.

2.1.3 Lemma. *The factor-topology on $P(\mathcal{H})$ coming from the Hilbert-space norm-topology of $\mathcal{H}$ is equivalent to the metric topology defined on $P(\mathcal{H})$ by the distance function d_1 (equiv.: by d_2).*

Proof. Let $\mathrm{Pr}: x \mapsto \mathbf{x}$ be the natural projection of $\mathcal{H}$ onto $P(\mathcal{H})$. The factor-topology on $P(\mathcal{H})$ is generated by projections of open balls $B(x; \varepsilon) := \{y \in \mathcal{H} : \|x - y\| < \varepsilon\}$ for $\varepsilon > 0$, $x \neq 0$. But $\mathrm{Pr}\, B(x; \varepsilon) = \{\mathbf{y} \in P(\mathcal{H}) : \inf\{\|\lambda y - x\| : \lambda \in \mathbb{C}\} < \varepsilon\}$, and $\inf\{\|\lambda y - x\| : \lambda \in \mathbb{C}\} = \|z_\mathbf{y} - x\|$ with $z_\mathbf{y} := \frac{(x,y)}{\|y\|^2} y$ if $y \neq 0$. Hence $\mathrm{Pr}\, B(x; \varepsilon) = \{\mathbf{y} : \|z_\mathbf{y} - x\| < \varepsilon\} = \{\mathbf{y} : 1 - Tr(P_x P_y) < \frac{\varepsilon^2}{\|x\|^2}\} = \{\mathbf{y} \in P(\mathcal{H}) : d_2(\mathbf{x}, \mathbf{y}) < \sqrt{2}\, \frac{\varepsilon}{\|x\|}\}$, which is an open ball in the metric topology and the desired equivalence of topologies follows. $\square$

2.1.4 Proposition. *All the following natural topologies on $P(\mathcal{H})$ are mutually equivalent:*

(i) *the factor-topology coming from the Hilbert space norm-topology on $\mathcal{H}$;*
(ii) *the metric topology defined by the distance functions on $P(\mathcal{H})$ from 2.1.2;*
(iii) *the Hilbert-Schmidt topology of $\mathfrak{H} \subset \mathcal{L}(H)$ of Hilbert–Schmidt operators;*
(iv) *the trace-norm topology of $\mathfrak{T}(\mathcal{H})$;*
(v) *$\sigma(P(\mathcal{H}), \mathcal{L}(H))$-topology;*
(vi) *$\sigma(P(\mathcal{H}), \mathfrak{C}(\mathcal{H}))$-topology.*

[In (v), resp. (vi), the topologies are determined by the functions $\mathbf{x} \mapsto Tr(P_x A)$ for all $A \in \mathcal{L}(H)$, resp. for all $A \in \mathfrak{C}(\mathcal{H}) :=$ the set of all compact operators on $\mathcal{H}$.]

Proof. The equivalence of the first four topologies follows from the Lemma 2.1.3 and from the formulas (2.1.3), (2.1.5), since the *Hilbert-Schmidt operator topology* is given by the norm

$$\|P_x - P_y\|_{HS}^2 := Tr(P_x - P_y)^2 = 2(1 - Tr(P_x P_y)) = [d_2(\mathbf{x}, \mathbf{y})]^2. \tag{2.1.6}$$

The equivalence of the trace-norm topology and the $\sigma(P(\mathcal{H}), \mathfrak{C}(\mathcal{H}))$-topology follows from [53, Proposition 2.16.15], and the 'stronger' $\sigma(P(\mathcal{H}), \mathcal{L}(H))$-topology coincides with the w^*-topology from $\mathcal{L}(H)^*$, which is 'weaker' than the norm-topology of $\mathcal{L}(H)^*$. The last mentioned topology coincides on $P(\mathcal{H})$ with the trace-norm topology given by the metric $d_2(\mathbf{x}, \mathbf{y}) \propto Tr|P_x - P_y|$, what finishes the proof. $\qquad\Box$

2.1.5 We shall introduce now a manifold structure on $P(\mathcal{H})$ consistent with the topology of $P(\mathcal{H})$. Let for $0 \neq x \in \mathcal{H}$

$$N_{\mathbf{x}} := \{\mathbf{y} \in P(\mathcal{H}) : Tr(P_x P_y) \neq 0\} \qquad (2.1.7)$$

be an open *neighbourhood of* $\mathbf{x} \in P(\mathcal{H})$, and let $[x]^{\perp}$ be the complex orthogonal complement of $\mathbf{x}$ in $\mathcal{H}$. We shall define the mapping $\Psi_x : N_{\mathbf{x}} \mapsto [x]^{\perp}$ by the formula

$$\Psi_x(\mathbf{y}) := \frac{\|x\|^2}{(x, y)}(I - P_x)y, \qquad (2.1.8)$$

where $y \in \mathbf{y}$.

2.1.6 Proposition. *The mapping* Ψ_x *is a homeomorphism of* $N_{\mathbf{x}}$ *onto* $[x]^{\perp}$ *(with the norm-topology of* $\mathcal{H}$*). The set*

$$\{(N_{\mathbf{x}}; \Psi_x; [x]^{\perp}) : 0 \neq x \in \mathcal{H}\} \qquad (2.1.9)$$

is an **atlas on** $P(\mathcal{H})$ *defining a complex-analytic* manifold structure *consistent with the topology of* $P(\mathcal{H})$ *(defined in 2.1.1).*

Proof. Let $0 \neq x \in \mathcal{H}$. For any $\mathbf{y}_j \in N_{\mathbf{x}}$ and any $y_j \in \mathbf{y}_j$ $(j = 1, 2)$ it is $\mathbf{y}_1 \neq \mathbf{y}_2$ iff $(x, y_2)y_1 \neq (x, y_1)y_2$, hence Ψ_x is injective. For any $z \in [x]^{\perp}$ and $y := z + x$ we have $\mathbf{y} \in N_{\mathbf{x}}$ (since $x \neq 0$) and $\Psi_x(\mathbf{y}) = z$, hence Ψ_x is bijective. For $\|x\| = 1$ and $z_j \in [x]^{\perp}$, $y_j := z_j + x$ $(j = 1, 2,)$ the identity

$$1 - Tr(P_{y_1} P_{y_2}) = \frac{1}{(\|z_1\|^2 + 1)(\|z_2\|^2 + 1)}\left(\|z_1 - z_2\|^2 + \|z_2\|^2 \|(1 - P_{z_2})(z_1 - z_2)\|^2\right) \qquad (2.1.10)$$

implies the bicontinuity of Ψ_x. For $z \in \Psi_{x_1}(N_{\mathbf{x}_1} \cap N_{\mathbf{x}_2})$ it is

$$\Psi_{x_2} \circ \Psi_{x_1}^{-1}(z) = \|x_2\|^2 \frac{x_1 + z}{(x_2, x_1 + z)} - x_2$$

and we see that the mapping

$$\Psi_{x_2} \circ \Psi_{x_1}^{-1} : \Psi_{x_1}(N_{\mathbf{x}_1} \cap N_{\mathbf{x}_2}) \to \Psi_{x_2}(N_{\mathbf{x}_1} \cap N_{\mathbf{x}_2}) \qquad (2.1.11)$$

is a complex analytic function, compare e.g. [51, 71]. $\qquad\Box$

2.1.7 Let $T_{\mathbf{x}}P(\mathcal{H})$ be the **tangent space of** $P(\mathcal{H})$ **at x**, elements of which can be represented in the usual way (see e.g. [1, 74]) by (classes of mutually tangent) differentiable curves at **x**. If c is such a curve (i.e. $c : J \rightarrow P(\mathcal{H})$ for an open interval J in $\mathbb{R}$ containing $0 \in \mathbb{R}$, $c(0) = \mathbf{x}$ and $t \mapsto \Psi_{\mathbf{y}}(c(t))$ is differentiable for $\mathbf{x} \in N_{\mathbf{y}}$) denote by $\dot{c}_{\mathbf{x}} := \dot{c}(0)$ (or simply $\dot{c}$ if the point **x** is fixed) the corresponding equivalence class, $\dot{c}_{\mathbf{x}} \in T_{\mathbf{x}}P(\mathcal{H})$. With any $x \in \mathbf{x}$, we associate an identification of $T_{\mathbf{x}}P(\mathcal{H})$ with $[x]^{\perp}$ by the mapping

$$T_{\mathbf{x}}\Psi_x : T_{\mathbf{x}}P(\mathcal{H}) \rightarrow [x]^{\perp}, \quad \dot{c} \mapsto T_{\mathbf{x}}\Psi_x(\dot{c}) := \frac{d}{dt}\bigg|_{t=0} \Psi_x(c(t)). \qquad (2.1.12)$$

In (2.1.12), we identify, in the usual way, the tangent space $T_v[x]^{\perp}$ of the linear space $[x]^{\perp}$ at any of its points $v \in [x]^{\perp}$ with the base space $[x]^{\perp}$ itself. The mapping $T_{\mathbf{x}}\Psi_x$ is a linear isomorphism for any $x \in \mathbf{x}$, and also $T_{\mathbf{x}}\Psi_{\lambda x} = \lambda T_{\mathbf{x}}\Psi_x$ ($\lambda \in \mathbb{C}$). The derivative in (2.1.12) is taken with respect to the Hilbert space norm in $[x]^{\perp}$.

2.1.8 Let us mention two simple examples of the representation of elements $\dot{c} \in T_{\mathbf{x}}P(\mathcal{H})$ by curves c and of the corresponding identification of $T_{\mathbf{x}}P(\mathcal{H})$ with $[x]^{\perp}$. Each vector $\dot{c} \in T_{\mathbf{x}}P(\mathcal{H})$ can be represented by a curve of any of the following forms (the expressions written by bold typeface represent the projections to $P(\mathcal{H})$ of the corresponding elements of $\mathcal{H}$, i.e. $z(\in \mathcal{H}) \mapsto \mathbf{z} \equiv P_z(\in P(\mathcal{H}))$):

$$c_1(t) := \boldsymbol{\lambda x + ty} \equiv P_{\lambda x + ty} \quad (\lambda \in \mathbb{C}, \ y \in \mathcal{H}, \ x \in \mathbf{x}), \quad t \in \mathbb{R}, \qquad (2.1.13)$$

$$c_2(t) := \exp(\boldsymbol{it B})\boldsymbol{x} \equiv P_{\exp(itB)x} \quad (B = B^* \in \mathcal{L}(H), \ x \in \mathbf{x}), \ t \in \mathbb{R}. \qquad (2.1.14)$$

If we denote corresponding tangent vectors by $\dot{c}_1$ and $\dot{c}_2$, then

$$T_{\mathbf{x}}\Psi_x(\dot{c}_1) = \lambda^{-1}(1 - P_x)y, \qquad (2.1.15)$$

$$T_{\mathbf{x}}\Psi_x(\dot{c}_2) = i(1 - P_x)Bx. \qquad (2.1.16)$$

Clearly $\dot{c}_1 = \dot{c}_2$ iff the right hand sides of (2.1.15) and (2.1.16) coincide as vectors in $[x]^{\perp}$. This is the case if e.g. $y = i\lambda Bx$ in (2.1.15). The representants (c_1, or c_2, or ...) of a given $\dot{c}$ can be chosen in many various ways. We shall use notation:

$$v_x := T_{\mathbf{x}}\Psi_x(v) \in [x]^{\perp}$$

for $v \in T_{\mathbf{x}}P(\mathcal{H})$; $v_{\lambda x} = \lambda v_x$.

2.1.9 We shall consider $P(\mathcal{H})$ as a **real manifold** of the dimension $\dim P(\mathcal{H}) = 2\dim_{\mathbb{C}} \mathcal{H} - 2$, (if $\mathcal{H}$ is finite dimensional) where $\dim_{\mathbb{C}}$ means the complex dimension. On this manifold, we introduce a **metric** Q, i.e. a real-analytic symmetric

2-covariant tensor field $\mathbf{x} \mapsto Q_{\mathbf{x}}$ defining an isomorphism between $T_{\mathbf{x}} P(\mathcal{H})$ and its dual $T_{\mathbf{x}}^* P(\mathcal{H})$ at any point $\mathbf{x} \in P(\mathcal{H})$:

$$v \, (\in T_{\mathbf{x}} P(\mathcal{H})) \mapsto Q_{\mathbf{x}}(v, \cdot) \in T_{\mathbf{x}}^* P(\mathcal{H}), \tag{2.1.17}$$

where the linear functional $Q_{\mathbf{x}}(v, \cdot) : w \, (\in T_{\mathbf{x}} P(\mathcal{H})) \mapsto Q_{\mathbf{x}}(v, w) \in \mathbb{R}$ depends linearly on v, and for any $F \in T_{\mathbf{x}}^* P(\mathcal{H})$ there is a unique $v_F \in T_{\mathbf{x}} P(\mathcal{H})$ such, that $F = Q_{\mathbf{x}}(v_F, \cdot)$. Let the metric be given by

$$Q_{\mathbf{x}}(v, v) := \frac{2}{\|x\|^2} (v_x, v_x) = \frac{2}{\|x\|^2} \|v_x\|^2, \quad v_x := T_{\mathbf{x}} \Psi_x(v). \tag{2.1.18}$$

Since $v_{\lambda x} = \lambda v_x$, the definition does not depend on the choice of $0 \neq x \in \mathbf{x}$ in the mapping Ψ_x. The nondegeneracy is a consequence of the Riesz theorem applied to the Hilbert space $[x]^{\perp}$ and analyticity is also straightforward. From the bilinearity and symmetry we have

$$Q_{\mathbf{x}}(v, w) = \frac{2}{\|x\|^2} \operatorname{Re}(v_x, w_x), \quad \forall v, w \in T_{\mathbf{x}} P(\mathcal{H}). \tag{2.1.19}$$

It is possible to prove by straightforward calculations of lengths of differentiable curves in $P(\mathcal{H})$ (compare also [1, 262]):

2.1.10 Proposition. *The metric Q from* (2.1.19) *endows $P(\mathcal{H})$ with a distance function* d *(calculated as the minimal length of differentiable curves joining two points) different from* d_j, $j = 1, 2;$ (2.1.2), (2.1.3). *Both the distance functions* d_1, d_2 *give (by differentiation) the metric Q from* (2.1.19) *on $P(\mathcal{H})$.*

2.2 Symplectic Structure

2.2.1 Let us define a *complex structure J on $P(\mathcal{H})$* induced by that of $\mathcal{H}$. For each $\mathbf{x} \in P(\mathcal{H})$ and $v \in T_{\mathbf{x}} P(\mathcal{H})$, we define

$$J v := (T_{\mathbf{x}} \Psi_x)^{-1} \circ i \circ (T_{\mathbf{x}} \Psi_x)(v), \tag{2.2.1}$$

where i is the multiplication by the imaginary unit $i \in \mathbb{C}$ in the complex subspace $[x]^{\perp} \subset \mathcal{H}$. The definition (2.2.1) of J does not depend on the choice of $x \in \mathbf{x}$. Clearly: $(J v)_x = i \, v_x$. We define now a *two-form Ω on $P(\mathcal{H})$* :

$$\Omega_{\mathbf{x}}(v, w) := Q_{\mathbf{x}}(v, J w), \quad \forall \mathbf{x} \in P(\mathcal{H}), \ v, w \in T_{\mathbf{x}} P(\mathcal{H}). \tag{2.2.2}$$

We shall use charts Ψ_x with $\|x\| = 1$ in the following. In such a chart, the form Ω is written

$$\Omega_{\mathbf{x}}(v, w) = -2 \operatorname{Im}(v_x, w_x). \tag{2.2.3}$$

The just introduced structures lead to the standard *symplectic*, and also metric (known as the "*Fubini-Study metric*") structures on the space of pure quantum states $P(\mathcal{H})$. If this both structures are connected as in (2.2.2) by a complex structure J (coming, in this case, from that in the underlying Hilbert space $\mathcal{H}$), we obtain a structure on the manifold $P(\mathcal{H})$ which is called the **Kähler structure**.

2.2.2 Lemma. *The form Ω is nondegenerate.*

Proof. If $\Omega_{\mathbf{x}}(w, v) = 0$ for all $w \in T_{\mathbf{x}} P(\mathcal{H})$, then also $\Omega_{\mathbf{x}}(Jv, v) = 2 \|v_x\|^2 = 0$, hence $v = 0$.   □

2.2.3 Lemma. *For any unitary transformation U of $\mathcal{H}$ onto itself, the form Ω is invariant with respect to the projected mapping $U : P(\mathcal{H}) \to P(\mathcal{H})$, $\mathbf{x} \mapsto U(\mathbf{x}) := U\mathbf{x}$, i.e.*

$$(U^*\Omega)_{\mathbf{x}}(v, w) := \Omega_{U\mathbf{x}}(U_* v, U_* w) = \Omega_{\mathbf{x}}(v, w). \tag{2.2.4}$$

Here $U^*\Omega$ is the pull-back of Ω by U, and $U_* : T_{\mathbf{x}} P(\mathcal{H}) \to T_{U\mathbf{x}} P(\mathcal{H})$ maps the equivalence class $\dot{c}$ containing the curve $c : t \mapsto c(t)$ at $\mathbf{x}$ (i.e. $\mathbf{x} = c(0)$) into the class Uc containing the curve $Uc : t \mapsto Uc(t)$ at $U(\mathbf{x})$.

Proof. According to 2.1.8, the vector v_x corresponds to the class containing the curve $c : t \mapsto \mathbf{x} + t\mathbf{v}_x$, hence the vector $(U_* v)_{Ux}$ corresponds to the class $Uc \in T_{U\mathbf{x}} P(\mathcal{H})$ containing the curve $Uc : t \mapsto U\mathbf{x} + tU\mathbf{v}_x$, and since U conserves orthogonality in $\mathcal{H}$ we have

$$(U_* v)_{Ux} = Uv_x. \tag{2.2.5}$$

Substitution into the expression (2.2.3) from (2.2.5) gives the result.   □

2.2.4 Proposition. *The two-form Ω on $P(\mathcal{H})$ is closed: $d\Omega = 0$; it is a **symplectic form on $P(\mathcal{H})$**, hence strongly nondegenerate (cf. [37, A.3.14]).*

Proof. The skew symmetry and bilinearity is trivial and (strong) nondegeneracy is proved in Lemma 2.2.2. The proof of closedness used in an appendix of the Arnold's book [7, Appendix 3 B] in the finite-dimensional case is literally applicable for any complex Hilbert space and its projective space, because of the validity of Lemma 2.2.3.[2] Hence Ω is symplectic.   □

---

[2]For an alternative proof valid also for unitary orbits of density matrices see [37, Theorem 2.1.19].

2.2.5 According to a *theorem by Wigner*, any bijective transformation **F** of $P(\mathcal{H})$ which conserves the '*transition probabilities*', i.e.:

$$Tr(P_x P_y) = Tr(\mathbf{F}(P_x)\mathbf{F}(P_y)), \quad \forall x, y \in \mathcal{H}, \ x \neq 0 \neq y, \tag{2.2.6}$$

can be extended to a transformation F of $\mathcal{H}$ onto itself, which is either unitary or antiunitary, compare [53, 3.2.1 and 3.2.14]:

$$Tr(P_{Fx} P_{Fy}) = Tr(P_x P_y). \tag{2.2.7}$$

Such transformations conserve also distances and the metric Q, see 2.1.2 and 2.1.9. Bijections of $P(\mathcal{H})$ onto itself conserving the metric Q will be called the **Wigner maps**.

On the other hand, *antiunitary transformations* F of $\mathcal{H}$ do not conserve the symplectic form $\Omega : \mathbf{F}_*\Omega = -\Omega$. Transformations **F** of $P(\mathcal{H})$ conserving Ω are called **symplectic transformations**.

2.2.6 Lemma. *Let **F** be any symplectic transformation of $P(\mathcal{H})$ the restriction of which to $T_x P(\mathcal{H})$ for any $\mathbf{x} \in P(\mathcal{H})$ (i.e. the mappings $\mathbf{F}_* : T_x P(\mathcal{H}) \to T_{\mathbf{Fx}} P(\mathcal{H})$) are complex linear with respect to the complex structure J, cf. 2.2.1. Then **F** can be extended to a unitary transformation $F \in \mathcal{L}(H)$.*

Proof. Symplecticity and complex linearity of **F** give

$$Q_{\mathbf{x}}(v, w) = -\Omega_{\mathbf{x}}(v, Jw) = -\Omega_{\mathbf{Fx}}(\mathbf{F}_*v, J\mathbf{F}_*w) = Q_{\mathbf{Fx}}(\mathbf{F}_*v, \mathbf{F}_*w), \tag{2.2.8}$$

i.e. $Q = \mathbf{F}_*Q$, what implies the invariance of distances:

$$d(\mathbf{Fx}, \mathbf{Fy}) = d(\mathbf{x}, \mathbf{y}),$$

which in turn implies the invariance of $Tr(P_x P_y)$. Hence **F** can be extended either to a unitary or to an antiunitary transformation. Since antiunitary transformations have nonsymplectic projections in $P(\mathcal{H})$, extension F of **F** must be unitary. □

2.2.7 Proposition. *Any symplectic isometry $\mathbf{F} : P(\mathcal{H}) \to P(\mathcal{H})$ is an analytic diffeomorphism of $P(\mathcal{H})$.*

Proof. **F** is a symplectic Wigner map, hence extendable to a unitary $F \in \mathcal{L}(H)$. With the help of the charts Ψ_x, analyticity follows for the projection **U** of any unitary $U \in \mathcal{L}(H)$. The same considerations apply to the inverse map $\mathbf{F}^{-1}$, and the assertion follows. □

2.3   Quantum Mechanics as a Classical Hamiltonian Field Theory

2.3.1 After introducing the symplectic structure Ω on the set $P(\mathcal{H})$ of all pure states of conventional QM (compare Sect. 1.2), we shall try to reformulate also other concepts of QM into the form analogous to that of CM as it was outlined in Sect. 1.3. It will be shown that this is possible to a large extent. There are, however, certain important differences. The main technical difference consists in infinite dimensionality of the 'phase space' $P(\mathcal{H})$ what implies e.g. *nonexistence of a (Liouville) measure on* $P(\mathcal{H})$, invariant with respect to all symplectic Wigner maps. The main physical difference consists, however, in the interpretation of basic quantities in QM. This *difference between QM and CM* does not vanish even for finite dimensional Hilbert space $\mathcal{H}$.

2.3.2 Let A be a *selfadjoint operator*[3] on the Hilbert space $\mathcal{H}$ with *domain* $D(A) \subset \mathcal{H}$. Let $PD(A) \subset P(\mathcal{H})$ be the projection of $D(A)$ into $P(\mathcal{H})$:

$$PD(A) := \{\mathbf{x} \in P(\mathcal{H}) : x \in D(A), \ x \in \mathbf{x}\}. \tag{2.3.1}$$

Define a real-valued function f_A on $PD(A)$:

$$f_A(\mathbf{x}) := Tr(P_x A) \equiv \frac{(x, Ax)}{\|x\|^2}, \ 0 \neq x \in \mathbf{x} \in PD(A). \tag{2.3.2}$$

The function f_A determines the operator A in an unambiguous way by the *polarization identity*:

$$(x, Ay) = \frac{1}{4} \sum_{\lambda=\pm1,\pm i} \lambda \|\lambda x + y\|^2 f_A(\lambda x + y). \tag{2.3.3}$$

For bounded $A \in \mathcal{L}(H)$, the function $f_A : P(\mathcal{H}) \to \mathbb{R}$ is real analytic. Since for arbitrary selfadjoint $A, B \in \mathcal{L}(H)$ there need not be any selfadjoint operator C on $\mathcal{H}$ such, that $f_C := f_A \cdot f_B$ (:= pointwise multiplication of functions), the set of 'classical observables' f_A ($A^* = A \in \mathcal{L}(H)$) does not form an associative algebra.

Remark: Corresponding to the spectral decomposition of A,[3] we have the decomposition of f_A:

$$f_A(\cdot) = \int_{\mathbb{R}} \lambda E_A^f(\mathrm{d}\lambda)(\cdot), \quad \text{where} \ E_A^f(B)(\mathbf{x}) := Tr(P_x E_A(B)) \tag{2.3.4}$$

for any Borel set $B \subset \mathbb{R}$, with E_A the *spectral measure of* A. Contrary to the case of classical mechanics 1.3.2, the functions $\mathbf{x} \mapsto E_A^f(B)(\mathbf{x})$ are not characteristic

---

[3] A brief review of the theory of unbounded operators is present in [37, C], or in [37, Textbook] in detail.

(indicator) functions on $P(\mathcal{H})$. The decomposition into characteristic functions similar to that in 1.3.2 does not correspond to any decomposition into quantal observables.

2.3.3 The function $f_\varrho(\mathbf{x}) := Tr(P_x \varrho)$ might remind us of a probability distribution on the "phase space" $P(\mathcal{H})$ representing a Gibbs ensemble in the sense of classical statistical physics, cf. e.g. [271, 272], or any textbook on statistical physics.

Although any density matrix ϱ is uniquely reconstructed from the corresponding function $f_\varrho(\mathbf{x})$ on the phase space $P(\mathcal{H})$ with the help of (2.3.3), the function f_ϱ cannot be interpreted as a *probability distribution* of systems occurring in the pure states $P_x \equiv \mathbf{x}$ in a statistical ensemble described by ϱ. The function f_ϱ is interpreted to give the probability $f_\varrho(\mathbf{x})$ of positive result (i.e. of the number $= 1$) by measuring of the observable P_x (with just two possible outcomes $\in \{0, 1\}$ of any of its measurements) in the state ϱ. Because of the existing nonuniqueness of decompositions of ϱ into pure states, mentioned in 1.2.3, a *classical interpretation* of any probability measure on $P(\mathcal{H})$ representing ϱ would be inadequate in general. In the following, we shall restrict our attention (mainly) to pure states.[4]

For a quantal observable A, the numbers $f_A(\mathbf{x})$ are interpreted as *expectation values* for (real valued) results of measurements of the observable A in the state $\mathbf{x} \in P(\mathcal{H})$. Also the functions f_A will be called here 'the (quantal) observables'.

2.3.4 In the setting of this section, it is natural to define a symmetry of the system as a *symplectic isometry of* $P(\mathcal{H})$. According to the Sect. 2.2, any such symmetry can be extended to a unitary transformation of $\mathcal{H}$. Let $t \mapsto \mathbf{F}_t$ be a one-parameter group of symplectic isometries of $P(\mathcal{H})$ which is weakly continuous, i.e. the functions

$$t \mapsto \mathbf{F}_t \mathbf{x}, \quad \forall \mathbf{x} \in P(\mathcal{H}) \tag{2.3.5}$$

are continuous. Such a group can be extended to a *weakly continuous unitary group* on $\mathcal{H}$ (compare [53, 3.2.35]), which corresponds to uniquely defined selfadjoint operator A on $\mathcal{H}$ (by Stone's theorem, [37, C3 & Textbook]). In this way, for the group $\mathbf{F}_t$, we obtain the expression:

$$\mathbf{F}_t \mathbf{x} = \exp(-itA)\mathbf{x}, \quad i.e. \ \mathbf{F}_t \mathbf{x} = \exp(-itA) P_x \exp(itA) \in P(\mathcal{H}). \tag{2.3.6}$$

The operator A in (2.3.6) is defined by $\mathbf{F}_t$ up to an additive real constant multiple of identity I of $\mathcal{L}(H)$, i.e. any other A' satisfying (2.3.6) has the form $A' = A + \lambda I$, $(\lambda \in \mathbb{R})$. Conversely, any selfadjoint operator A on $\mathcal{H}$ determines, according to (2.3.6), a weakly continuous one-parameter group of *symplectic isometries* of $P(\mathcal{H})$. The flow $\mathbf{F}_t$ and its unitary extension $F_t := \exp(-itA)$ are related by

$$\mathbf{F}_t(P_x) = P_{F_t x} = F_t P_x F_{-t}, \quad P_x \in P(\mathcal{H}). \tag{2.3.7}$$

---

[4] A certain, more detailed, account of the geometry and interpretation questions of the set of density matrices is given in [37, 2.1-e].

The functions (2.3.5) for specific $\mathbf{x}$'s are differentiable if the corresponding generator A has domain $D(A)$ containing $x \in \mathbf{x} : x \in D(A)$. If $A \in \mathcal{L}(H)$, then functions (2.3.5) are analytic in $t \in \mathbb{C}$, $\forall \mathbf{x} \in P(\mathcal{H})$. It is clear from the group property of $t \mapsto \mathbf{F}_t$, that differentiability of (2.3.5) in any point $\mathbf{x}$ for $t = 0$ implies differentiability on the whole curve (2.3.5), i.e. for all $t \in \mathbb{R}$.

2.3.5 We have obtained a set of differentiable curves lying densely in $P(\mathcal{H})$ for any one-parameter weakly continuous group $\mathbf{F}_t$ of symmetries of $(P(\mathcal{H}), \Omega)$ (since $PD(A)$ is dense in $P(\mathcal{H})$ for any selfadjoint A). For $\mathbf{x} \in PD(A)$ (A is a generator of F_t), the curve (2.3.5) determines a vector $\sigma_A(\mathbf{x}) \in T_{\mathbf{x}}P(\mathcal{H})$. The set of vectors $\sigma_A(\mathbf{x})$ is defined for $\mathbf{x} \in PD(A)$ only, and for unbounded A it is not a differentiable vector field on $P(\mathcal{H})$ (it is differentiable only in directions of some curves lying densely in $PD(A)$, and in $P(\mathcal{H})$). We shall call it, nevertheless, '*the vector field σ_A*'. Its value in $\mathbf{x}$ is expressed in $[x]^{\perp}$ according to (2.1.16) :

$$T_{\mathbf{x}}\Psi_x(\sigma_A(\mathbf{x})) = -i(I - P_x)Ax \quad \text{for} \quad x \in D(A). \tag{2.3.8}$$

For $A \in \mathcal{L}(H)$, σ_A is an analytic vector field on $P(\mathcal{H})$. But also for an unbounded A, the vector field σ_A determines its flow $\mathbf{F}_t =: \mathbf{F}_t^A$ uniquely: it can be integrated along a densely in $P(\mathcal{H})$ lying set of differentiable curves (this is just the solution of Schrödinger equation with the Hamiltonian A), and afterwards the obtained (densely defined) flow extended to the whole $P(\mathcal{H})$ by continuity.

2.3.6 Let $x \in D(A) \cap D(B)$ for two selfadjoint operators A and B on $\mathcal{H}$ and $\|x\| = 1$. Then the value of the symplectic form Ω on vectors $\sigma_A(\mathbf{x})$ and $\sigma_B(\mathbf{x})$ is, according to (2.2.3) and (2.3.8),

$$\Omega_{\mathbf{x}}(\sigma_A, \sigma_B) = -2\operatorname{Im}(Ax, (I - P_x)Bx). \tag{2.3.9}$$

If, moreover, $Bx \in D(A)$ and $Ax \in D(B)$ (e.g. if A and B have a common invariant set $D \subset D(A) \cap D(B)$ and $x \in D$), then we can write

$$\Omega_{\mathbf{x}}(\sigma_A, \sigma_B) = i\, Tr(P_x[A, B]) \tag{2.3.10}$$

where $[A, B] := AB - BA$.

Let f be a real-valued function defined on a dense subset $\mathbf{D}$ of $P(\mathcal{H})$. Let c be a differentiable curve in $P(\mathcal{H})$ at $\mathbf{x} \in \mathbf{D}$ such, that $c(t) \in \mathbf{D}$ for some open interval of reals t containing $t = 0$, $c(0) = \mathbf{x}$. Let $\dot{c} \in T_{\mathbf{x}}P(\mathcal{H})$ be the corresponding tangent vector. Denote

$$d_{\mathbf{x}}f(\dot{c}) := \left.\frac{d}{dt}\right|_{t=0} f(c(t)) \tag{2.3.11}$$

if the derivative on the right hand side exists. Assume, that (2.3.11) is well defined for a dense set of vectors $\dot{c} \in T_{\mathbf{x}} P(\mathcal{H})$. The function

$$\mathrm{d}_{\mathbf{x}} f : \dot{c} \mapsto \mathrm{d}_{\mathbf{x}} f(\dot{c}) \tag{2.3.12}$$

is linear. If it is bounded, it can be extended by continuity to the whole $T_{\mathbf{x}} P(\mathcal{H})$, hence it defines an element $\mathrm{d}_{\mathbf{x}} f \in T_{\mathbf{x}}^* P(\mathcal{H})$ which will be called the **exterior differential** of f in $\mathbf{x}$.

2.3.7 Proposition. *Let A be a selfadjoint operator on $\mathcal{H}$, f_A is given by (2.3.2), and the corresponding vector field σ_A is defined in 2.3.5. Then, for any $\mathbf{x} \in PD(A)$, the exterior differential $\mathrm{d}_{\mathbf{x}} f_A \in T_{\mathbf{x}}^* P(\mathcal{H})$ exists, and for all $v \in T_{\mathbf{x}} P(\mathcal{H})$ we have*

$$\Omega_{\mathbf{x}}(\sigma_A(x), v) = -\mathrm{d}_{\mathbf{x}} f_A(v), \qquad \forall \mathbf{x} \in PD(A). \tag{2.3.13}$$

Proof. Let $\{v_x\}^{\perp}$ be defined according to (2.1.18) for $v \in T_{\mathbf{x}} P(\mathcal{H})$. Define the selfadjoint $B(v) \in \mathcal{L}(H)$:

$$B(v)y := i(v_x, y)x - i(x, y)v_x, \qquad \forall y \in \mathcal{H}. \tag{2.3.14}$$

Assume $\|x\| = 1$. Then, according to (2.1.14) and (2.1.16), the curve

$$t \mapsto c_v(t) := \exp(it\mathbf{B}(\mathbf{v}))\mathbf{x} \tag{2.3.15}$$

corresponds to $v = \dot{c}_v$. Let v be such that $v_x \in D(A)$. Then it is seen that $\mathrm{d}_{\mathbf{x}} f_A(v)$ defined in (2.3.11) exists and has the form

$$\mathrm{d}_{\mathbf{x}} f_A(v) = -i\, Tr(P_x[B(v), A]) = -\Omega_{\mathbf{x}}(\sigma_A(\mathbf{x}), v), \tag{2.3.16}$$

where, in the second equality, we used (2.3.10) and $\sigma_{B(v)}(\mathbf{x}) = -v$. Because $(I - P_x)D(A) \subset D(A)$ is dense in $\{x\}^{\perp}$, we have proved (2.3.13) for a dense linear subset $D \subset T_{\mathbf{x}} P(\mathcal{H})$, $v \in D$. The boundednes is clear either from our construction, or from the boundednes of the left hand side of (2.3.13) for a well defined $\sigma_A(\mathbf{x})$. $\qquad\square$

2.3.8 We can see from the Proposition 2.3.7, how to reconstruct the vector field σ_A from f_A with the help of the symplectic form Ω. Hence, σ_A is globally Hamiltonian vector field on (the dense subset of) $P(\mathcal{H})$ corresponding to the Hamiltonian function f_A (compare with 1.3.5—up to domain differences).

Let two selfadjoint A, B have a common dense domain $D \subset D(A) \cap D(B)$. Then the function (the Poisson bracket)

$$\mathbf{x} \mapsto \{f_A, f_B\}(\mathbf{x}) := \Omega_{\mathbf{x}}(\sigma_A, \sigma_B), \quad \mathbf{x} \in PD, \tag{2.3.17}$$

is densely defined. If, moreover, the operator $i[A, B]$ is selfadjoint and D is its core[5] then, according to (2.3.10), we have

$$\{f_A, f_B\} = f_{i[A,B]}. \tag{2.3.18}$$

Remember that this is a quantummechanical formula corresponding to (1.3.8).

2.3.9 Assume that a weakly continuous unitary representation U of a connected Lie group G in the Hilbert space $\mathcal{H}$ is given:

$$U(g_1 g_2) = U(g_1)U(g_2), \ g_1, g_2 \in G. \tag{2.3.19}$$

Then U is projected onto a weakly continuous realization of G by a group of symplectic isometries $U(g)$ $(g \in G)$ of $(P(\mathcal{H}), \Omega)$. To any element ξ of the Lie algebra $\mathfrak{g}$ of G corresponds the selfadjoint generator X_ξ of the one-parameter subgroup $U(\exp(t\xi))$:

$$X_\xi x := i \left. \frac{\mathrm{d}}{\mathrm{d}t} \right|_{t=0} U(\exp(t\xi))x, \ x \in D(X_\xi), \tag{2.3.20}$$

and $U(\exp(t\xi)) = \exp(-it X_\xi)$. By a use of the **adjoint representation** $Ad : G \to \mathcal{L}(\mathfrak{g})$,

$$Ad(g)\xi := \left. \frac{\mathrm{d}}{\mathrm{d}t} \right|_{t=0} [g \exp(t\xi)g^{-1}]$$

we obtain:

$$[X_\xi, X_\eta] := X_\xi X_\eta - X_\eta X_\xi = i X_{[\xi,\eta]}. \tag{2.3.21}$$

The mapping $\xi \mapsto X_\xi$ is linear. It is known (compare [13]), that the *Gårding domain* $\mathcal{D}_G$, as well as the *analytic domain* $\mathcal{A}_G$ of the representation $U(G)$ are common dense invariant sets of all the generators X_ξ $(\xi \in \mathfrak{g})$ and they are also common cores of all these selfadjoint operators (cf. also 3.1.1). Let us **define the vector fields** σ_ξ $(\xi \in \mathfrak{g})$ **on** $P\mathcal{D}_G \subset P(\mathcal{H})$ corresponding to the flows $U(\exp(t\xi))$ on $P(\mathcal{H})$ according to the definition of σ_A in 2.3.5. Let $f_\xi(\mathbf{x}) := Tr(P_x X_\xi)$ for $x \in \mathcal{D}_G$. Then 2.3.8 is applicable to these quantities. All the formulas of 1.3.7 are valid on $P\mathcal{D}_G$. Difference w.r.t. the classical case is that neither $P(\mathcal{H})$ nor $P\mathcal{D}_G$ are homogeneous spaces even for irreducible $U(G)$.

2.3.10 Up to now, we used *charts* $(N_\mathbf{x}; \Psi_\mathbf{x}; [x]^\perp)$ for identification of $T_\mathbf{x} P(\mathcal{H})$ with $[x]^\perp$, and for each point $\mathbf{x} \in P(\mathcal{H})$ it was used its own chart. Let us rewrite now the

---

[5]A **core** $D \subset \mathcal{H}$ **of a closable operator** C is such a subset $D \subset D(C) \subset \mathcal{H}$, that the closure of the restriction $\overline{C \restriction D} = \overline{C}$, cf. also [37, C1].

evolution equation corresponding to the one-parameter flow $\mathbf{F}_t^A$ on $P(\mathcal{H})$ generated by the Hamiltonian A, i.e. the *Schrödinger equation*

$$i\frac{d}{dt}x(t) = Ax(t), \ x(0) := x \in \mathcal{H}, \tag{2.3.22}$$

projected onto $P(\mathcal{H})$, with the help of the chart $(N_\mathbf{y}; \Psi_\mathbf{y}; [y]^\perp)$, $\mathbf{x} \in N_\mathbf{y}$. Let us denote by $c : t \mapsto c(t)$ a differentiable curve in $P(\mathcal{H})$, and by $\dot{c}(t)$ its tangent vector: $\dot{c}(t) \in T_{c(t)}P(\mathcal{H})$. The curve c will be a solution of our problem, if for some $\mathbf{x} \in P(\mathcal{H}) : c(t) = \mathbf{F}_t^A\mathbf{x}$ for all $t \in \mathbb{R}$. For $\mathbf{x} \in PD(A)$, we then obtain by differentiation

$$\dot{c}(t) = \sigma_A(c(t)), \tag{2.3.23}$$

which is an abstract form of Hamilton equations on $P(\mathcal{H})$ corresponding to the Hamiltonian function f_A, cf. (2.3.13). The correspondence with (2.3.22) consists in that, that $c(t) = \mathbf{x}(t)$ if $c(0) = \mathbf{x}$, where $x(t) \ (\in \mathbf{x}(t))$ is the solution of (2.3.22) with the initial value $x \in \mathbf{x}$. Let us fix $y \in \mathcal{H}$, $\|y\| = 1$, and choose the chart $(N_\mathbf{y}; \Psi_y; [y]^\perp)$ defined in (2.1.8). Denote

$$\Psi(t) := \Psi_y(c(t)) \text{ for a curve c in } N_\mathbf{y}. \tag{2.3.24}$$

The curve Ψ in $[y]^\perp$ will correspond to a solution c of (2.3.23) iff it satisfies the equation

$$i\frac{d}{dt}\Psi(t) = [A - (y, A(y + \Psi(t)))](y + \Psi(t)), \ \Psi(0) \in [y]^\perp. \tag{2.3.25}$$

The equation (2.3.25) describes the wanted projection of (2.3.22) onto $P(\mathcal{H})$ in the chart Ψ_y. It is a *nonlinear (field-) equation* in the Hilbert space $[y]^\perp$, in which different vectors correspond to different physical states.

If we denote by v_y the representative of a vector $v \in T_\mathbf{x}P(\mathcal{H})$ for $\mathbf{x} \in N_\mathbf{y}$ in the chart Ψ_y $(\|y\| = 1)$, then the symplectic form Ω in this chart has the expression:

$$\Omega_\mathbf{x}(v, w) = -2\,Tr(P_x P_y)\,\text{Im}(v_y, (I - P_x)w_y). \tag{2.3.26}$$

Remember, that $v_y, w_y \in [y]^\perp := (I - P_y)\mathcal{H}$.

Let us write $f^t := f \circ \mathbf{F}_t^A$ for any differentiable function f on $P(\mathcal{H})$. Then, for $\mathbf{x} \in PD(A)$, we obtain the wanted form of the Schrödinger equation:

$$\frac{d}{dt}f^t(\mathbf{x}) = \{f_A, f^t\}(\mathbf{x}) := \Omega_\mathbf{x}(\sigma_A, \sigma_f), \tag{2.3.27}$$

where σ_f is a vector field defined on the whole $P(\mathcal{H})$ by

$$\Omega_{\mathbf{x}}(\sigma_f(\mathbf{x}), v) := -d_{\mathbf{x}} f(v), \ \forall v \in T_{\mathbf{x}} P(\mathcal{H}). \tag{2.3.28}$$

The equation (2.3.27) has the form of evolution equation of classical mechanics in terms of Poisson brackets.

2.3.11 Let us add a note concerning possible generalizations of the here presented dynamics. Since $P(\mathcal{H})$ is a symplectic manifold, more general Hamiltonian evolutions can be defined on it than the evolutions corresponding to linear Schrödinger equations (2.3.22). We can choose instead of the function $f_A : P(\mathcal{H}) \to \mathbb{R}$ as a ('classical') Hamiltonian an arbitrary 'sufficiently differentiable' function $h : P(\mathcal{H}) \to \mathbb{R}$. Then we obtain from the corresponding Hamiltonian dynamics on the infinite dimensional symplectic manifold $P(\mathcal{H})$ evolution of QM-vector states in $\mathcal{H}$, which cannot be described (in general) by a linear Schrödinger equation. This situation is described in many details in [37].

Chapter 3
Classical Mechanical Projections of QM

3.1 Orbits of Lie Group Actions on $P(\mathcal{H})$

3.1.1 Let U be a weakly continuous unitary representation of a connected Lie group G in the Hilbert space $\mathcal{H}$, and X_ξ be the selfadjoint generator of the one-parameter subgroup of $U(G)$ corresponding to an arbitrary element ξ of the Lie algebra $\mathfrak{g}$ of G, as it was defined in (2.3.20). Let $\mathcal{D}_G \subset \mathcal{H}$ be the *Gårding domain of* $U(G)$ [13, 11.1.8.], i.e. a dense $U(G)$-invariant set of vectors $x \in \mathcal{H}$, for which the functions $g \mapsto U(g)x$ ($g \in G$) are infinitely differentiable. **We shall denote by** $\mathcal{A}_G$ ($\subset \mathcal{H}$) the dense set of *analytic vectors of* $U(G)$ invariant with respect to the action of $U(G)$. For $x \in \mathcal{A}_G$, not only the functions $g \mapsto U(g)x$ are real analytic (resp. the functions $t \mapsto U(\exp(t\xi))x$ are complex analytic in a neighbourhood[1] of real axis for any $\xi \in \mathfrak{g}$) in the norm of $\mathcal{H}$, but also $\mathcal{A}_G$ is invariant and analytic with respect to the **Lie algebra** $U(\mathfrak{g})$ **of generators** X_ξ ($\xi \in \mathfrak{g}$); for $x \in \mathcal{A}_G$, also $X_\xi x \in \mathcal{A}_G$ ($\forall \xi \in U(\mathfrak{g})$) and for any basis $\{X_j \in U(\mathfrak{g}) : j = 1, 2, \dots d := \dim G\} \subset U(\mathfrak{g})$ and $x \in \mathcal{A}_G$ there is some $t \neq 0$ such that

$$\sum_{n=0}^{\infty} \frac{|t|^n}{n!} \sum_{j_1,\dots,j_n=1}^{d} \|X_{j_1} \dots X_{j_n} x\| < \infty, \tag{3.1.1}$$

compare [13, Chap. 11, §3].

Let $U(G)$ be the projection of $U(G)$ onto $P(\mathcal{H})$, i.e. $U(G)$ is a realization of G in a continuous group of symplectic isometries of $(P(\mathcal{H}), \Omega)$.[2] For any $\mathbf{x} \in P(\mathcal{H})$, **define the orbit** $O_{\mathbf{x}} := G \cdot \mathbf{x}$ (we shall use also the notation $g \cdot \mathbf{x} := U(g)\mathbf{x}$):

$$O_{\mathbf{x}} = O_{g \cdot \mathbf{x}} := \{\mathbf{z} \in P(\mathcal{H}) : \mathbf{z} = g \cdot \mathbf{x}, \ g \in G\}. \tag{3.1.2}$$

---

[1] In the following, if not explicitly mentioned different, the word 'neighbourhood' in a topological space will mean 'an open neighbourhood'.

[2] Remember that if $\mathbf{x} \equiv P_x \in P(\mathcal{H})$, then $U(g)\mathbf{x} \equiv P_{U(g)x} \equiv U(g)P_x U(g^{-1})$.

© Springer Nature Switzerland AG 2020, corrected publication 2020
P. Bóna, *Classical Systems in Quantum Mechanics*,
https://doi.org/10.1007/978-3-030-45070-0_3

Let $K_z^\circ := K^\circ := \{h \in G : U(h)\mathbf{z} = \mathbf{z}\}$ **be the stability (or 'isotropy') group** of the point $\mathbf{z} \in O_\mathbf{x} = O_\mathbf{z}$. Because $P(\mathcal{H})$ is a Hausdorff space and U is continuous, the group K° is closed, hence it is a Lie subgroup of G. The space G/K° of left cosets $gK^\circ \subset G$ is an analytic manifold (with the analytic structure coming from G via the natural projection, [152, Ch.II, Theorem 4.2]) and it is bijectively and continuously mapped onto $O_\mathbf{z}$ by the mapping $\boldsymbol{u} : m := gK^\circ \mapsto U(g)\mathbf{z}$. The orbit is not, in general, closed in $P(\mathcal{H})$ and it need not be a submanifold of $P(\mathcal{H})$, cf. also [37, Proposition 2.1.5] & [47]. The mapping $\boldsymbol{u}$ induces, however, a manifold structure on $O_\mathbf{z}$ from the analytic manifold G/K°. This manifold structure is not in general consistent with the relative topology of $O_\mathbf{z}$ in $P(\mathcal{H})$. If the map is differentiable, then we have:

3.1.2 Proposition. *Let $\boldsymbol{u}$ (defined as above) be continuously differentiable in a neighbourhood of a point $m \in G/K^\circ$, where $K^\circ := \{h \in G : U(h)\mathbf{z} = \mathbf{z}\}$. Then there is a neighbourhood N_m of m such, that the restriction of $\boldsymbol{u}$ on N_m is a diffeomorphism of N_m onto the submanifold $\boldsymbol{u}(N_m)$ of $P(\mathcal{H})$. If $\mathbf{z} \in P\mathcal{D}_G$ (resp. $\mathbf{z} \in P\mathcal{A}_G$), then each point $m \in G/K^\circ$ has a neighbourhood N_m, which is C^∞–diffeomorphic (resp. analytically diffeomorphic) to $\boldsymbol{u}(N_m)$, with the submanifold structure from $P(\mathcal{H})$; in this case, the orbit $O_\mathbf{z}$ is an immersed submanifold of $P(\mathcal{H})$.*

Proof. Bijectivity of $\boldsymbol{u} : G/K^\circ \to O_\mathbf{z}$ and differentiability in a neighbourhood of m imply, that the tangent mapping $T_m\boldsymbol{u} : T_m(G/K^\circ) \to T_{\boldsymbol{u}(m)}P(\mathcal{H})$ is an isomorphism onto a finite dimensional subspace of the tangent space of $P(\mathcal{H})$ at $\boldsymbol{u}(m)$. Since each finite dimensional real subspace of a Banach space is *complementable*, the restriction of $\boldsymbol{u}$ to a neighbourhood is an immersion. Hence, there is a neighbourhood N_m of m satisfying the first statement, compare [74, p. 549]. The rest is a consequence of the invariance of $P\mathcal{D}_G$ and $P\mathcal{A}_G$ as well as of the inverse mapping theorem, see also [51]. $\qquad\square$

3.1.3 We shall assume in the following that $\mathbf{z} \in P\mathcal{A}_G$, for the orbit $O_\mathbf{z}$ which we shall consider. Many of the following considerations are valid, however, also for orbits passing through $\mathbf{z} \in P\mathcal{D}_G$. Let σ_ξ ($\xi \in \mathfrak{g}$) be the (densely defined) vector field on $P(\mathcal{H})$ corresponding to the generator X_ξ according to 2.3.9 and 2.3.5. According to the definition of $O_\mathbf{z}$, for any $\mathbf{x} \in O_\mathbf{z}$, the vectors $\sigma_\xi(\mathbf{x})$ ($\xi \in \mathfrak{g}$) are well defined, they span $T_\mathbf{x} O_\mathbf{z}$ and depend analytically on $\mathbf{x} \in O_\mathbf{z}$. (Note: Here and in the following, we use without comments the topology on $O_\mathbf{z}$ inherited from G/K° via the mapping $\boldsymbol{u}$ introduced in 3.1.1) Let $K_\mathbf{x}^\circ$ be the stability subgroup of G at the point $\mathbf{x} \in O_\mathbf{z}$, and let **its Lie algebra** be $\mathfrak{k}_\mathbf{x}^\circ$. Then the Lie algebra $\mathfrak{g}$ of G is the direct sum

$$\mathfrak{g} = \mathfrak{m}_\mathbf{x}^\circ \oplus \mathfrak{k}_\mathbf{x}^\circ \tag{3.1.3}$$

of two vector spaces (the choice of $\mathfrak{m}_\mathbf{x}^\circ \subset \mathfrak{g}$ is nonunique). If $\{\xi_j \in \mathfrak{g} : j = 1, 2, \ldots n :=$ $\dim O_\mathbf{z}\}$ is a basis of $\mathfrak{m}_\mathbf{x}^\circ$, then σ_{ξ_j} span tangent spaces to $O_\mathbf{z}$ in any point $\mathbf{y}$ lying in some neighbourhood of $\mathbf{x}$ in $O_\mathbf{z}$. Then integral curves of σ_{ξ_j} ($j = 1, 2, \ldots n$) can be used to introduce a natural coordinate system on $O_\mathbf{z}$ in a neighbourhood of $\mathbf{x}$ (see [152, Ch.II. Lemma 4.1]). In these coordinates, the point

$$\mathbf{y}(t) := U(\exp(t_1\xi_1 + t_2\xi_2 + \cdots + t_n\xi_n))\mathbf{x} \in O_\mathbf{z} \qquad (3.1.4)$$

corresponds to the point $t \in \mathbb{R}^n$. We would like to interpret physically the coordinates as possible values of 'quantities' ξ_j (where the choice of lengths of vectors ξ_j corresponds to a choice of units). If we, however, take such a point of view that only the expectation values

$$F_\mathbf{x}(\xi) := Tr(P_x X_\xi), \ \xi \in \mathfrak{g}, \qquad (3.1.5)$$

of quantal observables X_ξ in states $\mathbf{x} \in P(\mathcal{H})$ are measurable, then, for a general orbit $O_\mathbf{z}$ and a group G, not all values $t \in \mathbb{R}^n$ (neither all t in any open neighbourhood of $0 \in \mathbb{R}^n$) are physically distinguishable. From this point of view the most natural coordinates of $\mathbf{x} \in P\mathcal{A}_G$ are just the values $F_\mathbf{x}(\xi)$ for a conveniently chosen subset of $\xi \in \mathfrak{g}$. These values need not distinguish points of a neighbourhood of $\mathbf{x} \in O_\mathbf{z}$:

$$\left.\frac{\mathrm{d}}{\mathrm{d}t}\right|_{t=0} F_{\exp(t\eta)\cdot\mathbf{x}}(\xi) = F_\mathbf{x}([\xi, \eta]), \quad \xi, \eta \in \mathfrak{g}, \qquad (3.1.6)$$

(compare (2.3.21)), and the derivative might be zero for some nonvanishing $\eta \in \mathfrak{m}_\mathbf{x}^\circ$ and for all $\xi \in \mathfrak{g}$. If it is the case, then the derivative in (3.1.6) vanish on the whole curve $t \mapsto \exp(t\eta) \cdot \mathbf{x}$ $(t \in \mathbb{R})$. This is easily seen with a help of the next Lemma, cf. Proposition 3.1.6:

3.1.4 Lemma. *For all $g \in G$ and $\xi \in \mathfrak{g}$, we have:*

$$U(g)X_\xi U(g^{-1}) = X_{Ad(g)\xi}, \qquad (3.1.7)$$

where the adjoint representation Ad of G is defined in 2.3.9.

Proof. According to the definition of Ad, the curve $t \mapsto g \exp(t\xi)g^{-1}$ at the identity e of G determines the tangent vector $Ad(g)\xi \in T_eG$, and this one, in turn, according to the definition of the Lie algebra $\mathfrak{g}$, determines a unique curve $t \mapsto \exp(t\,Ad(g)\xi)$ in G at e. Hence,

$$g \exp(t\xi)\, g^{-1} = \exp(t\,Ad(g)\xi) \ \ \forall t \in \mathbb{R}, g \in G, \xi \in \mathfrak{g}. \qquad (3.1.8)$$

From the definition (2.3.20) of the generators X_ξ of the representation $U(G)$, we then obtain

$$U(g)\exp(-itX_\xi)U(g^{-1}) = U(g \exp(t\xi)g^{-1}) = U(\exp(t\,Ad(g)\xi)), \qquad (3.1.9)$$

and after differentiation at $t = 0$ we obtain (3.1.7).   □

3.1.5 Suppose now that $F_\mathbf{x}([\xi, \eta]) = 0$ for all $\xi \in \mathfrak{g}$ at some $\mathbf{x} \in O_\mathbf{z}$. Substitution of $\exp(t\eta) \cdot \mathbf{x}$ to the place of $\mathbf{x}$ gives according to the preceding lemma:

$$F_{\exp(t\eta)\cdot\mathbf{x}}([\xi,\eta]) = Tr(U(\exp(t\eta))P_x U(\exp(-t\eta))X_{[\xi,\eta]}) \qquad (3.1.10)$$
$$= -i\,Tr\left(P_x U(\exp(-t\eta))[X_\xi, X_\eta]U(\exp(t\eta))\right) \quad (3.1.11)$$
$$= -i\,Tr\left(P_x[U(\exp(-t\eta))X_\xi U(\exp(t\eta)), X_\eta]\right) \quad (3.1.12)$$
$$= -i\,Tr\left(P_x[X_{Ad(\exp(-t\eta))\xi}, X_\eta]\right) \qquad (3.1.13)$$
$$= F_{\mathbf{x}}([Ad(\exp(-t\eta))\xi,\eta]). \qquad (3.1.14)$$

We used (2.3.7) in (3.1.10), it was used the formula (2.3.21) in (3.1.11), we considered commutativity of $U(\exp(t\eta))$ with X_η in (3.1.12), and in the last step the Lemma 3.1.4 was used. According to the assumption, the expression (3.1.14) vanishes for all $\xi \in \mathfrak{g}$, since $Ad(g) : \mathfrak{g} \to \mathfrak{g}$. Hence we have obtained:

3.1.6 Proposition. *For all* $\mathbf{x} \in P\mathcal{A}_G$, $\xi, \eta \in \mathfrak{g}$, *and all* $t \in \mathbb{R}$, *it is*

$$\frac{d}{dt} F_{\exp(t\eta)\cdot\mathbf{x}}(\xi) = F_{\mathbf{x}}([Ad(\exp(-t\eta))\xi,\eta]). \qquad (3.1.15)$$

If, in particular, the derivative vanishes for all $\xi \in \mathfrak{g}$ *at one value of* $t \in \mathbb{R}$, *then it vanishes for all* $\xi \in \mathfrak{g}$ *at all* $t \in \mathbb{R}$, *for the given* η.

3.1.7 From the preceding considerations, we see that the numbers $F_{\mathbf{x}}(\xi)$ cannot distinguish points $\mathbf{x}$ on the integral curves of the vector fields σ_η passing through $\mathbf{x}$ iff $F_{\mathbf{x}}([\xi,\eta]) = 0$ for all $\xi \in \mathfrak{g}$. Physical states lying on such curves should be identified mutually, if we could measure only expectations of observables X_ξ, $(\xi \in \mathfrak{g})$. Such an identification of points of orbits O_z ($\mathbf{z} \in P\mathcal{A}_G$) will be performed in the next section. After the identification, we obtain from each orbit an even-dimensional manifold endowed with canonical symplectic structure obtained from the symplectic structure Ω on $P(\mathcal{H})$.

3.1.8 Note that, for an irreducible representation $U(G)$, there can occur in $P\mathcal{A}_G$ mutually nonhomeomorphic orbits. But any such an orbit O_z, if it is considered in the Hilbert space $\mathcal{H}$ as the union of equivalence classes $\mathbf{x} = \{z \in \mathcal{H} : z = \lambda x, \lambda \in \mathbb{C}\} \subset \mathcal{H}$ for all $\mathbf{x} \in O_z$, contains total sets of vectors in $\mathcal{H}$. Such '*overcomplete families of vectors*' in $\mathcal{H}$ were discussed e.g. in [18, 84, 176, 239] and they are interesting from the point of view of representation theory, as it is explained e.g. in [91], and used in [2].

3.2 Classical Phase Spaces from the Quantal State Space

3.2.1 We have constructed orbits O_z of the action of G, $U(G)$, on $P(\mathcal{H})$ from pure states of conventional QM. We shall construct now *symplectic homogeneous spaces* of G from these orbits, of which the symplectic structure is a canonical restriction of the form Ω defined on $P(\mathcal{H})$ in Sect. 2.2. The obtained symplectic manifolds are all symplectomorphic to the orbits of G in the *coadjoint representation* $Ad^*(G)$ on the space $\mathfrak{g}^*$ dual to the Lie algebra $\mathfrak{g}$ endowed with the natural *Kirillov-Kostant symplectic form*.

3.2.2 Let Ω° denotes the restriction of the form Ω onto the immersed submanifold O_z ($z \in P\mathcal{A}_G$) of $P(\mathcal{H})$. Since the vector fields σ_ξ ($\xi \in \mathfrak{g}$) span $T_x O_z$ at each point $x \in O_z$, the form Ω° is uniquely defined by its values on vectors $\sigma_\xi(x)$ ($\xi \in \mathfrak{g}, x \in O_z$):

$$\Omega^\circ_x(\sigma_\xi, \sigma_\eta) := \Omega_x(\sigma_\xi, \sigma_\eta) = i\, Tr(P_x[X_\xi, X_\eta]), \qquad (3.2.1)$$

where we used formula (2.3.10) and the restrictions of the fields σ_ξ onto O_z are equally denoted as the unrestricted fields. According to the definition (3.1.5) and with the use (2.3.21), we can write

$$\Omega^\circ_x(\sigma_\xi, \sigma_\eta) = -F_x([\xi, \eta]). \qquad (3.2.2)$$

If we denote by

$$u_\circ : O_z = u(G/K^\circ) \to P(\mathcal{H})$$

the inclusion of the orbit into $P(\mathcal{H})$, then the form Ω° is simply the pull-back of Ω by $u_\circ$:

$$\Omega^\circ = u_\circ^* \Omega. \qquad (3.2.3)$$

Since exterior derivative commutes with any pull-back, e.g. [74, p.204], we see that the *two-form Ω° on O_z is closed*. It is clear from (3.2.2), that Ω° is degenerate iff for some $\eta \neq 0$ and for all $\xi \in \mathfrak{g}$ the term $F_x([\xi, \eta]) = 0$ for some x in the orbit. This is, however, the situation discussed in 3.1.7.

3.2.3 The mapping $F_x : \mathfrak{g} \to \mathbb{R}, \xi \mapsto F_x(\xi)$ is linear because of linearity of $\xi \mapsto X_\xi$, hence $F_x \in \mathfrak{g}^*$ for any $x \in O_z$. Define the action of G on the functionals F_x ($x \in O_z$) by

$$g \cdot F_x := F_{g \cdot x}, \quad \text{for all } g \in G. \qquad (3.2.4)$$

Then analogous computations to those in 3.1.5 lead to:

$$F_{g \cdot x}(\xi) = F_x(Ad(g^{-1})\xi), \text{ what means: } g \cdot F_x = Ad^*(g)F_x. \qquad (3.2.5)$$

Let now K_x be, as above, the stability subgroup of G of the coadjoint action at the point $F_x \in \mathfrak{g}^*$. Since Ad^* is continuous, K_x is closed. Let $\mathfrak{k}_x$ be the Lie algebra of K_x. Then it is clear, that:

3.2.4 Lemma. *Let* $x \in P\mathcal{A}_G$, $y := g \cdot x$. *Then* $K_y = gK_xg^{-1}$, $\mathfrak{k}_y = Ad(g)\mathfrak{k}_x$, *and* $K_x^\circ \subset K_x$ *for all* x *and all* $g \in G$. *It is* $\xi \in \mathfrak{k}_x$ *iff*

$$F_{\mathbf{x}}([\xi, \eta]) = 0, \quad \forall \eta \in \mathfrak{g}. \tag{3.2.6}$$

A trivial consequence of this is, according to (3.2.2), the

3.2.5 Proposition. $\Omega_{\mathbf{x}}^{\circ}(\sigma_{\xi}, \sigma_{\eta}) = 0$ *for all* $\eta \in \mathfrak{g}$ iff $\xi \in \mathfrak{k}_{\mathbf{x}}$.

3.2.6 We can decompose $O_{\mathbf{z}}$ into **equivalence classes**

$$[\mathbf{x}] := \{g \cdot \mathbf{x} : g \in K_{\mathbf{x}}\}, \; \mathbf{x} \in O_{\mathbf{z}} \; (\mathbf{z} \in P\mathcal{A}_G). \tag{3.2.7}$$

The action of G on $O_{\mathbf{z}}$ is analytic, and $[\mathbf{x}]$ **are analytic submanifolds of** $O_{\mathbf{z}}$ (if $O_{\mathbf{z}}$ is endowed with the topology of $G/K_{\mathbf{x}}^{\circ}$) which are mutually diffeomorphic for all $\mathbf{x} \in O_{\mathbf{z}}$. Hence $O_{\mathbf{z}}$ can-be considered as a fibred manifold with a typical fibre diffeomorphic to $K_{\mathbf{z}} \cdot \mathbf{z} = [\mathbf{z}]$, which is in turn diffeomorphic to $K_{\mathbf{x}}/K_{\mathbf{x}}^{\circ}$ ($\mathbf{x} \in O_{\mathbf{z}}$). Let us denote the **base space** by $M = M_{\mathbf{z}}$:

$$M := M_{\mathbf{z}} := \{[\mathbf{x}] : \mathbf{x} \in O_{\mathbf{z}}\}, \tag{3.2.8}$$

which is endowed with the natural factor topology given by the continuity and openness condition on the projection

$$p_M : O_{\mathbf{z}} \to M_{\mathbf{z}}, \; \mathbf{x} \mapsto p_M(\mathbf{x}) := [\mathbf{x}]. \tag{3.2.9}$$

From the definitions (3.1.5) of $F_{\mathbf{x}}$ and of the action of G on $F_{\mathbf{x}}$ in (3.2.4), we see that $[\mathbf{x}]$ are exactly those subsets of $O_{\mathbf{z}}$, on which expectations of all the observables X_{ξ} ($\xi \in \mathfrak{g}$) remain constant.

3.2.7 Lemma. $\Omega_{h \cdot \mathbf{x}}^{\circ}(\sigma_{\xi}, \sigma_{\eta}) = \Omega_{\mathbf{x}}^{\circ}(\sigma_{\xi}, \sigma_{\eta})$ *for all* $h \in K_{\mathbf{x}}$ *and all* $\eta, \xi \in \mathfrak{g}$.

Proof. Immediate from (3.2.2) and the definition of $K_{\mathbf{x}}$.   □

3.2.8 Let $p_{M*} := Tp_M : TO_{\mathbf{z}} \to TM_{\mathbf{z}}$ be the tangent mapping corresponding to the natural projection (3.2.9). For a general *vector field* σ on $O_{\mathbf{z}}$, the vectors $Tp_M \sigma(\mathbf{x})$ are mutually different for various choices of $\mathbf{x} \in [\mathbf{z}]$. Let, however, $t \mapsto g(t)$ be any differentiable curve in G. Then curves $t \mapsto g(t) \cdot \mathbf{x}$ and $c_h : t \mapsto g(t)h \cdot \mathbf{x}$ for any $h \in K_{\mathbf{x}}$ are projected by p_M onto the same curve $t \mapsto [g(t) \cdot \mathbf{x}]$ in $M_{\mathbf{z}}$. This is true due to the validity of

$$[g \cdot \mathbf{x}] = K_{g \cdot \mathbf{x}} g \cdot \mathbf{x} = g K_{\mathbf{x}} g^{-1} g \cdot \mathbf{x} = g K_{\mathbf{x}} \cdot \mathbf{x}, \tag{3.2.10}$$

for all $g \in G$,

$$[gh \cdot \mathbf{x}] = gh K_{\mathbf{x}} \cdot \mathbf{x} = g K_{\mathbf{x}} \cdot \mathbf{x} = [g \cdot \mathbf{x}], \; \forall h \in K_{\mathbf{x}}, \; g \in G. \tag{3.2.11}$$

Hence tangent vectors $\dot{c}_h \in T_{h \cdot \mathbf{x}} O_{\mathbf{z}}$ corresponding to the curves c_h with $g(t = 0) := e$ have identical projections $Tp_M(\dot{c}_h) = Tp_M(\dot{c}_e) \in T_{[\mathbf{x}]} M_{\mathbf{z}}$ for all $h \in K_{\mathbf{x}}$. If we set $g(t) := \exp(t\xi)$, i.e. $\dot{c}_h = \sigma_{\xi}(h \cdot \mathbf{x})$, then we have obtained:

3.2.9 Lemma. *All the vector fields* σ_ξ *(*$\xi \in \mathfrak{g}$*) on* O_z *are projected onto unambiguously defined (analytic, if* $\mathbf{z} \in P\mathcal{A}_G$*) vector fields* σ_ξ^M *on* M_z:

$$\sigma_\xi^M([\mathbf{x}]) := T p_M \sigma_\xi (h \cdot \mathbf{x}) \qquad (3.2.12)$$

for all $h \in K_\mathbf{x}$.

3.2.10 Proposition. *There is a unique symplectic form* Ω^M *on* M_z *satisfying*

$$\Omega_{[\mathbf{x}]}^M (\sigma_\xi^M, \sigma_\eta^M) = \Omega_\mathbf{x}^\circ (\sigma_\xi, \sigma_\eta) = (p_M^* \Omega^M)_\mathbf{x} (\sigma_\xi, \sigma_\eta) \qquad (3.2.13)$$

for all $\xi, \eta \in \mathfrak{g}$ *and all* $\mathbf{x} \in O_z$. p_M^* *in (3.2.13) is the* pull-back *corresponding to the projector* p_M *(compare [1], resp. also [37, A.3.11] for the definition).*

Proof. The first equality can be considered as a definition of a two-form Ω^M, which is correct due to two preceding lemmas and the fact, that vectors $\sigma_\xi^M(p_M\mathbf{x})$ ($\xi \in \mathfrak{g}$) contain a basis of $T_{[\mathbf{x}]}M_z$: $\sigma_\eta^M(p_M\mathbf{x}) = 0$ implies $\eta \in \mathfrak{k}_\mathbf{x}$ and M_z is diffeomorphic to $G/K_\mathbf{x}$. This ensures also the uniqueness of Ω^M. The second equality is a consequence of the definition (3.2.12) of σ_ξ^M and it shows, how Ω° can be reconstructed from Ω^M.

The bilinearity of Ω^M follows from linearity of the mapping $T_\mathbf{x} p_M$ and the bilinearity of Ω°, antisymmetry is trivial and closedness holds due to commutativity of the exterior derivative with the pull-bacs: $\mathrm{d} p_M^* = p_M^* \mathrm{d}$, and due to closedness of Ω°. Nondegeneracy follows from (3.2.1) and 3.2.4, which completes the proof.   □

3.2.11 As it was pointed out, the manifold $M := M_z$ ($\mathbf{z} \in P\mathcal{A}_G$) is diffeomorphic to G/K_z, where K_z is the stability group of the point $F_z \in \mathfrak{g}^*$ with respect to the coadjoint representation of G. On the other hand, the form Ω^M on M has the expression

$$\Omega_{[z]}^M (\sigma_\xi^M, \sigma_\eta^M) = -F_\mathbf{z}([\xi, \eta]), \qquad (3.2.14)$$

which follows from (3.2.2). This is, up to the sign, the canonical symplectic form on the orbit of $Ad^*(G)$ passing through F_z and diffeomorphic to G/K_z. Hence the symplectic manifold $(M; \Omega^M)$ is symplectomorphic to a *Kirillov-Kostant symplectic orbit*, compare [174]. This manifold is here interpreted as a *classical phase space* obtained by the above described canonical procedure from a given quantal system, in which *interpretation of observables* is (at least partly) determined by a Lie group action $U(G)$. This action is projected on the coadjoint action $Ad^*(G)$ on M, see (3.2.5). Almost obvious is the following

3.2.12 Proposition. *The vector fields* σ_ξ^M *(*$\xi \in \mathfrak{g}$*) are globally Hamiltonian vector fields on the symplectic manifold* $(M; \Omega^M)$ *corresponding to Hamiltonian functions*

$$f_\xi : [\mathbf{x}] \mapsto f_\xi([\mathbf{x}]) := F_\mathbf{x}(\xi). \qquad (3.2.15)$$

They generate Hamiltonian flows F_t^ξ *on* M:

$$F_t^\xi : [\mathbf{x}] = p_M\mathbf{x} \mapsto F_t^\xi(p_M\mathbf{x}) := p_M\left(U(\exp(t\xi))\mathbf{x}\right). \tag{3.2.16}$$

Proof. From the definition of σ_ξ^M in 3.2.8 and 3.2.9 the relation (3.2.16) follows. Differentiation of f_ξ according to (3.1.15) and (3.2.2) gives

$$\mathrm{d}f_\xi = -i(\sigma_\xi^M)\Omega^M, \tag{3.2.17}$$

compare (1.3.4). This proves the first statement. $\square$

With the usual definition of Poisson brackets on $(M; \Omega^M)$, we obtain the obvious (compare also (1.3.11) + (1.3.12))

3.2.13 Lemma. $\{f_\xi, f_\eta\} = -f_{[\xi,\eta]}$ *for all* $\xi, \eta \in \mathfrak{g}$.

3.2.14 This shows, that the action of $Ad^*(G)$ is *strictly Hamiltonian*. Since for the generators of $U(G)$ in $\mathcal{H}$ we have $X_{[\xi,\eta]} = -i[X_\xi, X_\eta]$, (2.3.21), the Lemma 3.2.13 establishes the usual *correspondence between classical and quantal observables* associated with generators of the group action.

3.3   Classical Mechanical Projections of Quantal Dynamics

3.3.1 Let the time evolution of a given system in QM be described by a one parameter subgroup of $U(G)$ corresponding to an element $\chi \in \mathfrak{g}$. Then, for a given $\mathbf{z} \in P(\mathcal{H})$, the flow $U(\exp(t\chi))$ leaves the orbit $O_\mathbf{z}$ invariant. If $\mathbf{z} \in P\mathcal{A}_G$, then this flow is projected onto the Hamiltonian flow on $M_\mathbf{z}$ generated by the Hamiltonian function f_χ with the corresponding Hamiltonian vector field σ_χ^M, as it was described above. Models one frequently encounters are, however, in which the time evolution is given by a one parameter group of unitaries $U_A(\mathbb{R})$:

$$U_A : t \mapsto U_A(t) := \exp(-itA), \ A = A^*, \tag{3.3.1}$$

where the generator A has not the form X_χ for any $\chi \in \mathfrak{g}$. The orbits $O_\mathbf{z}$ are then in general not invariant with respect to the action of $U_A(\mathbb{R})$. We shall be interested here in the question whether and how such an action $U_A(\mathbb{R})$ can be projected onto a Hamiltonian flow on $M_\mathbf{z}$.

3.3.2 Let A be any selfadjoint operator on $\mathcal{H}$ and E_A the corresponding projector-valued measure on $\mathbb{R}$. Assume that $\mathbf{z} \in P\mathcal{A}_G$ (defined in 3.1.1) and that $O_\mathbf{z} := U(G)\mathbf{z}$ is contained in the form domain of A, i.e. the integral in

$$f_A(\mathbf{x}) := Tr(P_x A) := \int_\mathbb{R} \lambda\, Tr(P_x E_A(\mathrm{d}\lambda)) \tag{3.3.2}$$

converges absolutely for all $\mathbf{x} \in O_z$. In an analogy with the constructions of the preceding sections, the function f_A will be considered as a candidate for a *classical observable corresponding to the quantal observable A*. We shall require that

$$f_A \in C^\infty(O_z). \tag{3.3.3}$$

This requirement is fulfilled in the following situation:

3.3.3 Lemma. *Let $\mathcal{E}(\mathfrak{g})$ be the linear space of all polynomials in selfadjoint generators X_ξ ($\xi \in \mathfrak{g}$) of $U(G)$ with complex coefficients. Assume that for a fixed $\mathbf{z} \in P\mathcal{A}_G$ and for any $\mathbf{x} \in O_z$ and any $E \in \mathcal{E}(\mathfrak{g})$ there is an open neighbourhood $N(\mathbf{x}, E)$ of the identity $e \in G$ such, that the function*

$$g \mapsto \|AEU(g)x\| \quad (x \in \mathbf{x}) \tag{3.3.4}$$

is uniformly bounded on $N(\mathbf{x}, E)$. Here A is a given symmetric operator on $\mathcal{H}$ containing $\mathcal{E}(\mathfrak{g})U(G)z := \{EU(g)z : E \in \mathcal{E}(\mathfrak{g}), g \in G\}$ in its domain $D(A)$, $z \in \mathbf{z}$. Set $f_A(\mathbf{x}) := (x, Ax)$ for $\|x\| = 1$, $x \in \mathbf{x} \in O_z$. Then f_A is infinitely differentiable on O_z.

Proof. It suffices to prove infinite differentiability of the function $g \mapsto f_A(g \cdot \mathbf{x})$ defined on G. For any $E_1, E_2 \in \mathcal{E}(\mathfrak{g})$ the functions $g \mapsto E_j U(g)x$ ($j = 1, 2$) are norm-analytic according to (3.1.1), see also [13]. Consequently, the function

$$(g_1; g_2) \mapsto (E_1 U(g_1)x, AE_2 U(g_2)x) \text{ from } G \times G \text{ to } \mathbb{C} \tag{3.3.5}$$

is infinitely differentiable in each variable g_1, g_2 separately and any partial derivative (in the direction of some one parameter subgroup of G) has the form (3.3.5) (with some other E_j's). To prove differentiability of

$$g \mapsto (E_1 U(g)x, AE_2 U(g)x), \tag{3.3.6}$$

it suffices to prove simultaneous continuity of all functions of the form (3.3.5) in both variables g_1, g_2. It follows, however, from the assumption of uniform boundedness on $N(\mathbf{x}, E_2)$, analyticity of $U(g)x$ with respect to $U(G)$ and continuity of $U(g)$:

$$|(E_1 U(g_1)x, AE_2 U(g_2)x) - (E_1 x, AE_2 x)| \le$$
$$\|E_1(U(g_1) - I)x\| \cdot \|AE_2 U(g_2)x\| + \|E_2(U(g_2) - I)x\| \cdot \|AE_1 x\|. \tag{3.3.7}$$

This concludes the proof. $\square$

3.3.4 If the assumptions of the preceding lemma are valid for A, the explicit expressions for the partial derivatives $\partial_\xi f_A$ along the curves $t \mapsto exp(t\xi) \cdot \mathbf{x}$ have the form ($\|x\| = 1$, $\xi, \eta \in \mathfrak{g}$):

$$\partial_\xi f_A(\mathbf{x}) = 2\,\mathrm{Im}(x,\,AX_\xi x), \tag{3.3.8}$$

$$\partial_\eta \partial_\xi f_A(\mathbf{x}) = 2\,\mathrm{Re}[(X_\xi x,\,AX_\eta x) - (x,\,AX_\xi X_\eta x)], \tag{3.3.9}$$

and similarly for higher derivatives. For these expressions, we shall use also forms which are literally valid only if the set $\mathcal{E}(\mathfrak{g})U(G)x$ is mapped by A into $\mathcal{D}_G$:

$$\partial_\xi f_A(\mathbf{x}) =: i\,Tr(P_x[X_\xi,\,A]), \tag{3.3.10}$$

$$\partial_\eta \partial_\xi f_A(\mathbf{x}) =: i^2\,Tr(P_x[X_\eta,\,[X_\xi,\,A]]), \tag{3.3.11}$$

etc. Also in more general cases, we shall write symbolically

$$i\,Tr(P_x[X_\xi,\,A]) := f_{i[X_\xi,A]}(\mathbf{x}) := \partial_\xi f_A(\mathbf{x}). \tag{3.3.12}$$

The Definition 3.3.6 (ii) deals with such symbols.

3.3.5 Examples. *Assumptions of the Lemma 3.3.3 are satisfied, e.g. for*

(i) *all bounded operators* $A = A^* \in \mathcal{L}(\mathcal{H})$,
(ii) *all symmetric operators* $A \in \mathcal{E}(\mathfrak{g})$.

3.3.6 Definitions. (i) *Let A be a symmetric operator on $\mathcal{H}$ with $O_\mathbf{z} \subset D(A)$ for some $\mathbf{z} \in P\mathcal{A}_G$ and let $f_A : \mathbf{x} \mapsto f_A(\mathbf{x}) := Tr(P_x A)$ be infinitely differentiable on $O_\mathbf{z}$. Let $K_\mathbf{x}$ be the stability group of $F_\mathbf{x} \in \mathfrak{g}^*$, $F_\mathbf{x}(\xi) := Tr(P_x X_\xi)$, with respect to the coadjoint representation of G and $[\mathbf{x}] := K_\mathbf{x} \cdot \mathbf{x}$ $(\mathbf{x} \in O_\mathbf{z})$. If*

$$f_A([\mathbf{x}]) := f_A(\mathbf{x}) = f_A(h \cdot \mathbf{x}), \quad \forall h \in K_\mathbf{x}, \forall \mathbf{x} \in O_\mathbf{z}, \tag{3.3.13}$$

the operator A will be called a **U(G)-classical operator on $O_\mathbf{z}$** *or simply a* **z-classical operator**.
(ii) *Let $A := A_1 A_2 \dots A_n$ be formal product of some selfadjoint operators $A_j^* = A_j$, $j = 1, 2, \dots n$. Let $A_0 := I$. Suppose, that for some $j \in \{0, 1, 2, \dots n\}$ the products $A_{j+1} \dots A_n$ and $A_j A_{j-1} \dots A_1 A_0$ are well defined operators with $U(G)z$ $(0 \neq z \in \mathbf{z})$ lying in the intersection of their domains. Denote then (with $x \in \mathbf{x}$, $\|x\| = 1$, $\mathbf{x} \in O_\mathbf{z}$)*

$$f_A(\mathbf{x}) := f_{A_1 A_2 \dots A_n}(\mathbf{x}) := (A_j A_{j-1} \dots A_1 x,\; A_{j+1} A_{j+2} \dots A_n x). \tag{3.3.14}$$

For any other $j \in \{1, \dots n\}$ satisfying these conditions the values in (3.3.14) will be the same. If $f_A \in C^\infty(O_\mathbf{z})$ and if (3.3.13) is valid (with $A \hookrightarrow A$) for f_A, then A will be called a **generalized z-classical operator**. *The same name will be given to any formal complex finite linear combination B of generalized z-classical operators $A^\tau := A_1^\tau A_2^\tau \dots A_{n_\tau}^\tau$:*

$$B := \sum_\tau \lambda_\tau A^\tau, \tag{3.3.15}$$

and we shall set

$$f_B([\mathbf{x}]) := f_B(\mathbf{x}) := \sum_\tau \lambda_\tau f_{A^\tau}(\mathbf{x}). \tag{3.3.16}$$

The adjective 'generalized' will be sometimes omitted.

3.3.7 Examples.

(i) *All the generators X_ξ ($\xi \in \mathfrak{g}$) are $\mathbf{z}$-classical for all $\mathbf{z} \in P\mathcal{A}_G$.*

(ii) *If, for some $\mathbf{z} \in P(\mathcal{H}) : K_\mathbf{z} = K_\mathbf{z}^\circ$, (cf. 3.1.1) and $f_A \in C^\infty(O_\mathbf{z})$, then A is $\mathbf{z}$-classical.*

(iii) *If A is $\mathbf{z}$-classical and $X_\xi, \dots X_\chi \in U(\mathfrak{g})$, then all the symbols $[X_\xi, [X_\eta, \dots [X_\chi, A] \dots]]$ represent generalized $\mathbf{z}$-classical operators. We can see this from 3.3.4 and (3.2.11):*

$$f_A(g \cdot \mathbf{z}) = f_A([g \cdot \mathbf{z}]) = f_A([gh \cdot \mathbf{z}])$$

and differentiations and induction give the result.

(iv) *Let $f_A \in C^\infty(O_\mathbf{z})$ and all the $K_\mathbf{x}$ be symmetry groups of the observable $A : U(h^{-1})AU(h) = A$ for all $h \in K_\mathbf{x}$ and all $\mathbf{x} \in O_\mathbf{z}$ (e.g. if $K_\mathbf{z}$ is a normal subgroup of G and $K_\mathbf{z}$ is a symmetry group of A). Then A is $\mathbf{z}$-classical.*

3.3.8 If A is $\mathbf{z}$-*classical*, then the function f_A can be considered as a function on $M_\mathbf{z}$ according to (3.3.13) and then $f_A \in C^\infty(M)$. Denote by σ_A^M the *Hamiltonian vector field on M* corresponding to the Hamiltonian function $f_A : m \mapsto f_A(m)$, $m \in M$. Choose a system σ_j ($j = 1, \dots dim M$) of vector fields on M forming a basis of $T_m M$ for all m in a neighbourhood of $m_0 \in M$. Since the symplectic form Ω^M is nondegenerate, the inverse matrix to $\Omega_m^M(\sigma_j, \sigma_k)$ with elements $\Omega_M^{jk}(m)$ ($j, k = 1, 2, \dots \dim M$) exists:

$$\sum_i \Omega_M^{ji}(m)\Omega_m^M(\sigma_i, \sigma_k) = \sum_i \Omega_m^M(\sigma_k, \sigma_i)\Omega_M^{ij}(m) = \delta_{jk}. \tag{3.3.17}$$

From the connection between Hamiltonian vector fields and corresponding Hamiltonian functions, we obtain:

$$\sigma_A^M(m) = \sum_{j,k} \Omega_M^{jk}(m)d_m f_A(\sigma_k)\sigma_j(m). \tag{3.3.18}$$

For Poisson brackets of functions f_A and f_B on M corresponding to $\mathbf{z}$-classical operators A and B, we obtain with a help of (3.3.17):

$$\{f_A, f_B\}(m) := \Omega_m^M(\sigma_A^M, \sigma_B^M) = -\sum_{j,k} d_m f_A(\sigma_j)\Omega_M^{jk}(m)d_m f_B(\sigma_k). \tag{3.3.19}$$

If σ_j are Hamiltonian vector fields corresponding to generators $X_j \in U(\mathfrak{g})$, then we obtain according to (3.3.12)

$$d_m f_A(\sigma_j) = f_{i[X_j,A]}(m) \tag{3.3.20}$$

and the Poisson bracket (3.3.19) has the form

$$\{f_A, f_B\}(m) = - \sum_{j,k} f_{i[X_j,A]}(m)\Omega_M^{jk}(m) f_{i[X_k,B]}(m). \tag{3.3.21}$$

If the operator B is one of the generators of $U(G)$, $B := X \in U(\mathfrak{g})$, then the Poisson bracket (3.3.21) has the expression:

$$\{f_A, f_X\}(m) = i\, Tr(P_x[A, X]) = f_{i[A,X]}(m), \tag{3.3.22}$$

where $x \in \mathbf{x} \in [\mathbf{x}] := m \in M$. The results (3.3.21) and (3.3.22) have to be compared with 3.2.14. If the orbit O_z coincides with the manifold $M := M_z$, then the vector field σ_A^M in (3.3.18) is the skew-orthogonal projection of σ_A (from (2.3.8)) onto M, the skew-orthogonality being defined by the form Ω on $P(\mathcal{H})$, see Sect. 2.2.

3.3.9 The unitary group $U_A : t \mapsto U_A(t) := \exp(-itA)$ does not leave the orbit O_z invariant for a general selfadjoint $\mathbf{z}$-classical operator A. Then we would like to compare the classical Hamiltonian evolution on M_z generated by f_A (with the flow F_t^A) and the quantal evolution on $P(\mathcal{H})$ described by the flow $U_A(t)$. From the point of view of this work, the 'quantities of interest' are generators of the representation $U(G)$. The evolutions of the corresponding functions f_X ($X = X^* \in U(\mathfrak{g})$) are described by

$$\frac{d}{dt} f_X^t = \{f_A, f_X^t\} = f_{i[A,X]}^t \tag{3.3.23}$$

in both cases of the classical flow F_t^A as well as of the quantal evolution $U_A(t)$, compare 3.2.12, (2.3.27) and (3.3.22). The difference is between the two cases in the meaning of f^t:

(i) In the case of the flow F_t^A on M for any $f \in C^\infty(M)$, we define

$$f^t(m) := f(F_t^A m), \quad m \in M, \tag{3.3.24}$$

and the flow F_t^A has to be determined from (3.3.23) ($\forall X^* = X \in U(\mathfrak{g})$).

(ii) In the quantal case, we have given the flow U_A on $P(\mathcal{H})$ and for functions f on (the dense U_A-invariant subset of) $P(\mathcal{H})$ we set

$$f^t(\mathbf{x}) := f(U_A(t)\mathbf{x}). \tag{3.3.25}$$

The functions f_B for any $\mathbf{z}$-classical B are defined in the both cases by the formula

$$f_B(\mathbf{x}) := Tr(P_x B). \tag{3.3.26}$$

The 'classical f_B' is the restriction of the 'quantal f_B' to the manifold $M := M_\mathbf{z}$. The classical flow is the specific kind of restriction of the flow U_A onto M (compare (3.3.18) and the note in the last sentence of 3.3.8).

Although the rules for computation of the functions

$$t \mapsto f_X(F_t^A[\mathbf{x}]), \quad [\mathbf{x}] \in M_\mathbf{z}, \quad X = X^* \in U(\mathfrak{g}), \tag{3.3.27}$$

and

$$t \mapsto f_X(U_A(t)\mathbf{x}), \quad \mathbf{x} \in [\mathbf{x}] \in M_\mathbf{z}, \quad X = X^* \in U(\mathfrak{g}), \tag{3.3.28}$$

seem to be very similar, the mutually corresponding functions from (3.3.27) and (3.3.28) might be radically different for an abstractly defined selfadjoint ($\mathbf{z}$-classical) operator A. We shall give in the next chapter an example, in which both the functions from (3.3.27) and (3.3.28) (given by the same $X \in U(\mathfrak{g})$ and with the same initial condition $\mathbf{x} \in O_\mathbf{z}$) are periodic with different periods (and, moreover, with mutually different dependence of these periods on the initial condition $\mathbf{x}$); the corresponding orbits in $\mathfrak{g}^*$:

$$\{F_{[\mathbf{x}]}^{cl}(t) : t \in \mathbb{R}\} \subset \mathfrak{g}^*, \text{ with } F_{[\mathbf{x}]}^{cl}(t) : \xi \mapsto f_{X_\xi}(F_t^A[\mathbf{x}]), \ \xi \in \mathfrak{g}, \tag{3.3.29}$$

and the orbit

$$\{F_\mathbf{x}^q(t) : t \in \mathbb{R}\} \subset \mathfrak{g}^*, \text{ where } F_\mathbf{x}^q(t) : \mathfrak{g} \to \mathbb{R}, \ \xi \mapsto F_\mathbf{x}^q(t)(\xi) := f_{X_\xi}(U_A(t)\mathbf{x}), \tag{3.3.30}$$

are mutually different closed curves in $\mathfrak{g}^*$, see 4.1.10.

3.3.10 We expect, contrary to the above mentioned example, that in certain situations the parametrized curves in $\mathfrak{g}^*$ defined in (3.3.27) and (3.3.28) will be in some sense close one to another, at least for not too large times $t \in \mathbb{R}$. We mean namely such situations, in which A is the Hamiltonian operator of a 'realistic' quantal model and the initial condition $\mathbf{x}$ leads to subsequent evolution $U_A(t)\mathbf{x}$, which is sufficiently well approximated by laws of CM. For some estimates in these directions, they might be useful Taylor expansions of the functions in (3.3.27) and (3.3.28) in the initial point $t = 0$. Set, as usual,

$$\{f_A, f_X\}^{(n)} := \{f_A, \{f_A, f_X\}^{(n-1)}\}, \quad \{f_A, f_X\}^{(0)} := f_X, \text{ for } n \in \mathbb{Z}_+, \tag{3.3.31}$$

and also the corresponding notation for multiple commutators for operators. Then we have expressions for derivatives

$$\left.\frac{d^n}{dt^n}\right|_{t=0} f_X(F_t^A[\mathbf{x}]) = \{f_A, f_X\}^{(n)}(\mathbf{x}),$$ (3.3.32)

and

$$\left.\frac{d^n}{dt^n}\right|_{t=0} f_X(\mathbf{U}_A(t)\mathbf{x}) = i^n \, Tr(P_x[A, X]^{(n)}) =: f_{i^n[A,X]^{(n)}}(\mathbf{x}).$$ (3.3.33)

The right hand side of (3.3.32) can also be expressed as a polynomial in expectation values of quantal observables in the initial state $\mathbf{x}$ by multiple application of (3.3.21). To make these formulae clearly applicable it is necessary to have some assumptions on the domain of A, e.g. let A be $\mathbf{z}$-classical with $O_{\mathbf{z}}$ in its invariant analytic domain, $\mathbf{x} \in O_{\mathbf{z}}$ and $A^n x \in \mathcal{A}_G$ (:= the analytic domain of $U(\mathfrak{g})$) for all $n \in \mathbb{Z}_+$. If these assumptions are fulfilled, then the identity of functions (3.3.27) and (3.3.28) (for given $X \in U(\mathfrak{g})$ and $\mathbf{x} \in O_{\mathbf{z}}$) is equivalent to the equality of the right hand sides in (3.3.32) and (3.3.33) for all $n \in \mathbb{Z}_+$. This equality holds for any such A for $n = 0, 1$. The equality in higher orders is essentially dependent on the choice of A.

Content of this subsection is closely related to the investigation of $\hbar \to 0$ limit of quantal correlation functions in the work by Hepp [154], cf. also 4.1.8– 4.1.10.

3.3.11 Extended phase spaces: If the one-parameter group of time evolution is included into G as a subgroup, the reduction of the orbits $O_{\mathbf{z}}$ to the symplectic manifolds $M_{\mathbf{z}}$ can be sometimes replaced by a natural procedure of a reduction of $O_{\mathbf{z}}$ to odd dimensional manifolds of the dimension $2n + 1$, if the dimension of the corresponding classical phase space is equal to $2n$. In this case, the restriction of the form Ω° to such a manifold is degenerate, of the rank $2n$. Such odd dimensional manifolds with a given closed two-form of the maximal rank are called **contact manifolds**. Usage of contact manifolds in CM is convenient for a natural possibility of passing to moving reference frames. Another situation, in which they are useful is that of time dependent Hamiltonians, cf. [1, Chap. 5], and also [111, Sect. 18.5].

Sometimes it is useful to describe mechanical systems in CM by symplectic manifolds which are of the dimension higher by 2 than the usual ones. Any symplectic manifold can be extended to a **contact manifold** and any contact manifold can be extended to a symplectic manifold, each time increasing the dimension by one.

We shall not try to give here the theory of these situations. For generalities on such structures cf. e.g. [1, 7]. Some cases will be mentioned in the following chapter.

Chapter 4
Examples of Classical Mechanical Projections

4.1 The Heisenberg Group (CCR)

4.1.1 A physical system consisting of the finite number N of nonrelativistic (apriori mutually distinguishable) point particles is described in the conventional QM by an infinite dimensional unitary irreducible representation of the $2n + 1$ - dimensional **Heisenberg group** G ($n := N\nu$, ν is the dimension of the one-particle configuration space); cf. also [37, Sect. 3.3-b]. The Heisenberg group G is a *central extension* by $\mathbb{R}$ of the commutative group $\mathbb{R}^{2n}$ (which can be identified with the classical flat phase space $\mathbb{R}^{2n} = T^*\mathbb{R}^n$), compare [321] and [346]. The (scalar multiples of the) selfadjoint generators X_j, $j = 1, 2, \ldots 2n$, of the representation correspond to basic 'kinematical' observables of the system. The choice of X_j's is conveniently made in such a way, that on corresponding domains (e.g. on $\mathcal{D}_G$) the commutation relations (CCR) are fulfilled:

$$[X_j, X_k] = i\, S_{jk} X_0 \quad \text{for} \quad j, k = 1, 2, \ldots 2n; \tag{4.1.1a}$$

$$[X_j, X_0] = 0, \quad j = 1, 2, \ldots 2n. \tag{4.1.1b}$$

Here the elements S_{jk} of the $2n \times 2n$ real matrix S are defined:

$$S_{j\ j+n} = -S_{j+n\ j} = 1, \quad j = 1, 2, \ldots n, \quad S_{jk} = 0 \quad \text{otherwise.} \tag{4.1.2}$$

Hence $S^{-1} = S^T = -S$ where S^T is the transposed matrix to S. From (4.1.1b) we see, that

$$X_0 = \hbar I, \quad (I \text{ is the identity of } \mathcal{L}(\mathcal{H})). \tag{4.1.3}$$

The parameter $\hbar \in \mathbb{R}$, $\hbar \neq 0$, ($\hbar :=$ the '*Planck constant*', if its value is chosen properly) classifies all infinite-dimensional unitary irreducible representations of G; representations corresponding to various values of $\hbar$ are mutually inequivalent, [346].

© Springer Nature Switzerland AG 2020, corrected publication 2020
P. Bóna, *Classical Systems in Quantum Mechanics*,
https://doi.org/10.1007/978-3-030-45070-0_4

Setting

$$Q_j := X_j, \; P_j := X_{j+n} \quad \text{for } j = 1, 2, \ldots n \tag{4.1.4}$$

we obtain from (4.1.1a) the usual form of the **canonical commutation relations** (**CCR**). There is only one physically admissible choice of the constant $\hbar$: it is the Planck constant divided by 2π (its numerical value depends on a choice of physical units for determination of which it is necessary to consider also dynamics). Operators Q_j (resp. P_j) are interpreted to correspond to observables called 'coordinates of the configuration' (resp. 'coordinates of the linear momentum'), in a cartesian basis. Note, that this representation of G can be considered as a projective representation of $\mathbb{R}^{2n}$, as it was described in 1.2.7.

4.1.2 The Schrödinger form of the above mentioned representation of G consists of the realization of the Hilbert space $\mathcal{H}$ of the representation as $L^2(\mathbb{R}^n, \mathrm{d}^n q)$ ($\mathrm{d}^n q$ is the Lebesgue measure) and the action of X_j's can be defined on such $\varphi \in L^2(\mathbb{R}^n, \mathrm{d}^n q)$, which belong to Schwartz test functions:

$$(X_j \varphi)(q_1, q_2, \ldots q_n) := q_j \, \varphi(q_1, q_2, \ldots q_n) \tag{4.1.5a}$$

and

$$(X_{j+n}\varphi)(q_1, q_2, \ldots q_n) := -i\hbar \frac{\partial}{\partial q_j} \varphi(q_1, q_2, \ldots q_n) \tag{4.1.5b}$$

for $j = 1, 2, \ldots n$. An equivalent realization of CCR is obtained by an arbitrary unitary transformation U of $\mathcal{H}$ onto itself, e.g. by the scaling $U = U_\lambda$ ($\lambda \in \mathbb{R}_+ \setminus \{0\}$):

$$(U_\lambda \varphi)(\boldsymbol{q}) := \lambda^{n/2} \varphi(\lambda \boldsymbol{q}). \tag{4.1.6}$$

It is $U_\lambda^{-1} = U_{1/\lambda}$ and we have:

$$X_j' := U_\lambda X_j U_\lambda^{-1} = \lambda X_j, \; X_{j+n}' := U_\lambda X_{j+n} U_\lambda^{-1} = \frac{1}{\lambda} X_{j+n}, \quad j = 1, 2, \ldots n. \tag{4.1.7}$$

These transformations are useful for taking limits $\hbar \to 0$, compare [154] and also our 4.1.8.

4.1.3 Let $X \cdot S \cdot x := X_j S_{jk} x_k$ with summation over $j, k = 1, 2, \ldots 2n$, where $x_k \in \mathbb{R}$ for all k. Let W_x ($x \in \mathbb{R}^{2n}$) be unitary operators of the above mentioned projective representation (cf. 1.2.7) :

$$W_x := \exp\left(\frac{i}{\hbar} X \cdot S \cdot x \right). \tag{4.1.8}$$

From (4.1.1a) we obtain

$$W_x^{-1} X_j W_x = X_j + x_j I, \tag{4.1.9}$$

$$W_{x+x'} = \exp\left(\frac{i}{\hbar} x \cdot S \cdot x'\right) W_x W_{x'}. \tag{4.1.10}$$

Let us mention here, that the multiplier in (4.1.10) is determined by the standard symplectic form Ω^{cl} on the classical flat phase space $\mathbb{R}^{2n}$; setting $q_j := x_j$, $p_j := x_{j+n}$ for $j = 1, 2, \ldots n$, it is

$$\Omega^{cl} := \sum_{j=1}^{n} \mathrm{d} p_j \wedge \mathrm{d} q_j, \tag{4.1.11}$$

$$x' \cdot S \cdot x = \Omega^{cl}(x, x'). \tag{4.1.12}$$

4.1.4 Let $\varphi \in \mathcal{A}_G :=$ the analytic domain of $U(G)$, $\|\varphi\| = 1$, $\varphi_x := W_x \varphi$ ($x \in \mathbb{R}^{2n}$). Let $P_x^\varphi \in P(\mathcal{H})$ be the corresponding projectors, $Tr(P_x^\varphi A) := (\varphi_x, A\varphi_x)$ ($A \in \mathcal{L}(\mathcal{H})$) and $P_0^\varphi := P_\varphi$. From (4.1.9) one has

$$Tr(P_x^\varphi X_j) = Tr(P_\varphi X_j) + x_j. \tag{4.1.13}$$

Hence the **mapping** $P^\varphi : x \mapsto P_x^\varphi$ is a bijection of $\mathbb{R}^{2n}$ **onto the orbit** $O_\varphi := \{P_x^\varphi : x \in \mathbb{R}^{2n}\}$ and it is continuous if O_φ is taken in the relative topology from $P(\mathcal{H})$. Due to absolute continuity of spectra of all X_j ($j = 1, 2, \ldots 2n$) with respect to the Lebesgue measure on $\mathbb{R}$ the function $x \mapsto (\varphi, W_x \varphi)$ converges to zero with $|x| \to \infty$ and $|(\varphi, W_x \varphi)| = 1$ iff $x = 0$. Consequently, the mapping P^φ is also open (i.e. any open set is mapped to an open set), hence it is a regular C^∞-embedding of $\mathbb{R}^{2n}$ into $P(\mathcal{H})$; with our choice of $\varphi \in \mathcal{A}_G$, P^φ is even an analytic embedding into $P(\mathcal{H})$.

4.1.5 Let σ_j denote the vector field on O_φ corresponding to the generator $\frac{1}{\hbar} X_j$ ($j = 1, 2, \ldots 2n$). We shall denote by Ω^φ the restriction of the symplectic form Ω on $P(\mathcal{H})$, 2.2.1, onto O_φ. The form Ω^φ is nondegenerate, since for the values Ω_x^φ of Ω^φ in any point $\varphi_x \in O_\varphi$ we have:

$$\Omega_x^\varphi(\sigma_j, \sigma_k) = \frac{i}{\hbar^2} Tr(P_x^\varphi [X_j, X_k]) = -\frac{1}{\hbar} S_{jk} \tag{4.1.14}$$

and $\det S = 1$. Hence $M_\varphi = O_\varphi$ in this case. Let $f_{X_j}(x) := f_{X_j}(\varphi_x) := Tr(P_x^\varphi X_j)$ ($x \in \mathbb{R}^{2n}$) be the classical observable corresponding to X_j. From (4.1.13) we see, that a unique $\varphi_0 \in O_\varphi$ can be chosen such, that

$$Tr(P_{\varphi_0} X_j) = 0 \text{ for all } j = 1, 2, \ldots 2n. \tag{4.1.15}$$

In the following, we shall take $\varphi := \varphi_0$ according to (4.1.15). Then

$$f_{X_j}(x) = x_j, \quad j = 1.2.\dots.2n. \tag{4.1.16}$$

From (4.1.14) and (4.1.12) we see, that in the coordinates (4.1.16) the form $\Omega^{cl} := \hbar\Omega^{\varphi}$ is identical with Ω^{cl} defined earlier. Hence the brackets

$$\{f_{X_j}, f_{X_k}\}(x) := \hbar\,\Omega^{\varphi}_x(\sigma_j, \sigma_k) = -S_{jk} \tag{4.1.17}$$

are exactly the **classical Poisson brackets on** $\mathbb{R}^{2n}$, cf. also [31]. The Hamiltonian vector fields on O_{φ} corresponding to the Hamiltonian functions f_{X_j} are σ_j with flows $\exp(-\frac{i}{\hbar}tX_j)$. This recovers on O_{φ} the standard classical kinematics from the geometry of $P(\mathcal{H})$ and the CCR.

4.1.6 Let us look now on the dynamics on O_{φ} generated by the Hamiltonian operator

$$A := A_V := \frac{1}{2}\sum_{jk=1}^{n} a_{jk}P_j P_k + V(Q) \tag{4.1.18}$$

from the point of view of the Sect. 3.1 (see (4.1.4) for the notation). Here $a \equiv \{a_{jk}\}$ is a real symmetric positive matrix and V is a real distribution on $\mathbb{R}^n$ chosen such, that the operator A is φ-classical, Def. 3.3.6. The quantal dynamical group is $\exp(-\frac{i}{\hbar}tA)$ and the corresponding classical projection (= classical mechanical projection) F_t^A on O_{φ} is given by the Hamiltonian function

$$f_A(x) := Tr(P_x^{\varphi}A). \tag{4.1.19}$$

From (4.1.13) we obtain (with $(q; p) := x$):

$$f_A(q, p) = \frac{1}{2}\sum_{jk=1}^{n} a_{jk}p_j p_k + Tr(P_{\varphi}V(Q+q)) + \frac{1}{2}\sum_{jk=1}^{n} a_{jk}Tr(P_{\varphi}P_j P_k), \tag{4.1.20}$$

where we write $V(Q+q) := W_x^{-1}V(Q)W_x$. The potential term in the realization (4.1.5) is rewritten as

$$V_{\varphi}(q) := Tr(P_{\varphi}V(Q+q)) = \int_{\mathbb{R}^n} |\varphi(q')|^2\, V(q+q')\,\mathrm{d}^n q', \tag{4.1.21a}$$

or as a *convolution* $(\tilde{\varphi}(q) := \varphi(-q))$:

$$V_{\varphi}(q) = |\tilde{\varphi}|^2 * V(q) =: \varrho_{\varphi} * V(q). \tag{4.1.21b}$$

This 'smearing' of the potential energy by a density ϱ_{φ} is the only difference between the classical projections in the case of $G :=$(the Heisenberg group) and the usual classical limit with the 'unsmeared' potential energy $V(q)$ (up to the unessential additive constant term in (4.1.20)).

4.1.7 Notes. (i) *The quantal correlation functions are constant on the orbits O_φ;* e.g.

$$Tr(P_x^\varphi(X_j - x_j)(X_k - x_k)) = Tr(P_\varphi X_j X_k),$$
$$\text{for all } j, k, \text{ and for all } x \in \mathbb{R}^{2n}, \tag{4.1.22}$$

(ii) *If the Hamiltonian operator A is quadratic in all the generators X_j:*

$$A := \frac{1}{2} h^{jk} X_j X_k \Rightarrow f_A(x) = \frac{1}{2} h^{jk} x_j x_k + const, \tag{4.1.23}$$

i.e. in this case the usual classical limit coincides with the classical projections. This situation is analyzed in Sect. 4.2.

4.1.8 On the *limit* $\hbar \to 0$.

All the previous results and considerations are equally valid for any nonvanishing value of the parameter $\hbar$. Any change of the value of the parameter $\hbar$ might be interpreted from the point of view of mathematics, either as a change of the representation $U_\hbar(G)$ of the Heisenberg group G to an inequivalent one leaving the correspondence of the generators $\frac{1}{\hbar} X_j \in U_\hbar(\mathfrak{g})$ to fixed elements $\xi_j \in \mathfrak{g}$ of the Lie algebra unchanged, or as a change of the basis $\{\xi_j\}$ in $\mathfrak{g}$ into $\{\lambda\xi_j\}$ (corresponding to a 'reinterpretation' (i.e. change of units) of parameters x occurring in (4.1.8)), leaving the choice of the representation fixed.

Let a physical interpretation of the generators X_j be fixed (compare Sect. 1.2), leaving the value of $\hbar$ unspecified. If some empirical system is adequately described by QM with the given interpretation of X_j's, for some value of $\hbar$, then this value $\hbar$ is for the system unique (independently on any choices of generators of the evolution in time— consider, e.g. the occurrence of $\hbar$ in uncertainty relations). If two such systems could form one composite system the mutually noninteracting parts of which they are, then the value of $\hbar$ for both systems is the same (interpretation of X_j's fixed!), since each of the subsystems taken separately determines $\hbar$ for the whole system (we have now a $2(n_1 + n_2) + 1$—dimensional Heisenberg group, if the subsystems have n_1, resp. n_2 degrees of freedom).

These considerations show, that any *change of the value of $\hbar$* —if physically interpreted—has to be connected with a *change of interpretation of the generators* $X_j \in U_\hbar(\mathfrak{g})$. We obtain an example of such a reinterpretation, if we describe a system consisting of a large number of particles: in a description of the center of mass motion we can deal instead of with center of mass coordinates and total linear momenta (which satisfy CCR with the experimental value of Planck constant) rather with center of mass coordinates and averaged momenta per a particle.

If we keep the interpretation of X_j's fixed, then for various values of $\hbar$ we obtain different theories. We shall describe a transition of $\hbar \to 0$ in the context of the classical projections of QM. Let us write $\lambda^2 \hbar$ ($\lambda \in (0, 1]$) instead of $\hbar$ in all formulas of the subsections 4.1.1–4.1.7. Let $X_j(\lambda)$ be the Schrödinger realizations of the CCR-generators in $L^2(\mathbb{R}^n) =: \mathcal{H}$ and let us apply to them the transformation U_λ from (4.1.7) for each value of λ. **Let us denote** $X_j^\lambda := U_\lambda X_j(\lambda) U_\lambda^{-1}$. We obtain:

$$Q_j^\lambda \varphi(q) = \lambda q_j \varphi(q), \ P_j^\lambda \varphi(q) = -i \, \lambda \hbar \frac{\partial}{\partial q_j} \varphi(q), \qquad (4.1.24)$$

where

$$Q_j^\lambda := X_j^\lambda, \ P_j^\lambda := X_{j+n}^\lambda \ (j = 1, 2, \ldots n).$$

Let us fix $\varphi = \varphi_0 \in \mathcal{H}$ according to (4.1.15), which will be held unchanged for all the values of λ. Let W^λ be the unitary representation from 4.1.3:

$$W_x^\lambda := \exp\left(\frac{i}{\lambda^2 \hbar} X^\lambda \cdot S \cdot x \right). \qquad (4.1.25)$$

Let $\varphi_x^\lambda := W_x^\lambda \varphi$, i.e. for $x := (q; p) \in \mathbb{R}^{2n}$ we have

$$\varphi_x^\lambda(q') = \exp\left(-\frac{i}{2\lambda^2 \hbar} q \cdot p \right) \exp\left(\frac{i}{\lambda \hbar} q' \cdot p \right) \varphi(q' - \frac{q}{\lambda}); \ q, p, q' \in \mathbb{R}^n. \qquad (4.1.26)$$

Let
$$P_x^{(\lambda)} \text{ be the projector onto } \varphi_x^\lambda, \ P_0^{(\lambda)} = P_\varphi \equiv P_\varphi \text{ for all } \lambda.$$

The correlations of all orders are for any λ independent of x:

$$Tr\left(P_x^{(\lambda)}(X_j^\lambda - x_j)(X_k^\lambda - x_k) \ldots (X_r^\lambda - x_r) \right) = Tr(P_\varphi X_j^\lambda X_k^\lambda \ldots X_r^\lambda). \quad (4.1.27)$$

The right hand side of (4.1.27) is proportional to λ^s, where s is the number of X^λ in the right hand side of (4.1.27). From this we see that the

algebra $\mathcal{E}(\mathfrak{g})^\lambda$ of quantal observables consisting of polynomials in X^λ

is mapped onto a set of functions on $O_\varphi^\lambda := W^\lambda(G)\varphi$:

$$f^\lambda : E \mapsto f_E^\lambda(x) := Tr(P_x^{(\lambda)}E), \ E \in \mathcal{E}(\mathfrak{g})^\lambda; \ F_{X_j}^\lambda(x) = x_j, \qquad (4.1.28)$$

and this mapping f^λ becomes in the limit $\lambda \to 0$ a homomorphism of associative algebras.

For the generator f_A^λ of the 'projected' evolution in time, corresponding to the quantal generator (4.1.18), i.e., for each λ, to the operator

$$A^\lambda := \frac{1}{2} \sum_{jk=1}^{n} a_{jk} P_j^\lambda P_k^\lambda + V(Q^\lambda), \qquad (4.1.29)$$

we obtain:

$$f_A^\lambda(q, p) := Tr(P_x^{(\lambda)} A^\lambda) = \frac{1}{2} \sum_{jk=1}^{n} a_{jk} p_j p_k + \varrho^\lambda * V(q) + b_\lambda. \qquad (4.1.30)$$

Here b_λ is a constant depending on λ as $O(\lambda^2)$ and

$$\varrho^\lambda(q) := \lambda^{-n} |\varphi(-\frac{q}{\lambda})|^2 \qquad (4.1.31)$$

is a normalized density on $\mathbb{R}^n$, which weakly converges to the Dirac δ-function with $\lambda \to 0$. A comparison of flows on $O_\varphi^\lambda (= \mathbb{R}^{2n})$ generated by f_A^λ for various λ is not easy for given V and φ in general.

4.1.9 Let $U_A^\lambda(t) := \exp(-\frac{it}{\lambda^2} A^\lambda)$ be the time evolution group corresponding to the generator (4.1.29) (we set $\hbar = 1$). Let

$$x_j^\lambda(t, x) := Tr(U_A^\lambda(t) P_x^{(\lambda)} U_A^\lambda(-t) X_j^\lambda) = f_{X_j}^\lambda(U_A^\lambda(t)\varphi_x^\lambda) \qquad (4.1.32)$$

be time-evolved quantal expectations of the 'canonical' observables X_j^λ with initial values $\varphi_x^\lambda \in O_\varphi^\lambda$ (the mapping f^λ from (4.1.28) is here extended to a mapping into functions on $P(\mathcal{H})$). The well-known Ehrenfest's equations are certain equalities including the functions (4.1.32) and their time-derivatives, which have an analogous form to that of equations of motion of CM, being in the same time exact consequences of QM. We can write them in the form (with $x^\lambda := (q_1^\lambda, \ldots q_n^\lambda, p_1^\lambda, \ldots p_n^\lambda)$ and summation is over $1, 2, \ldots n$):

$$\frac{d}{dt} q_j^\lambda(t, x) = a_{jk} p_k^\lambda(t, x) = \frac{\partial}{\partial p_j} f_A^\lambda(x^\lambda(t, x)), \qquad (4.1.33)$$

$$\frac{d}{dt} p_j^\lambda(t, x) = -Tr\left(U_A^\lambda(t) P_x^{(\lambda)} U_A^\lambda(-t) \frac{\partial V}{\partial q_j}(Q^\lambda)\right). \qquad (4.1.34)$$

Here f_A^λ is the classical Hamiltonian function corresponding to the quantal generator A^λ. The corresponding equations for the classical projection on O_φ^λ are of the form:

$$\frac{d}{dt} q_j^\lambda(t, x)_{cl} = a_{jk} p_k^\lambda(t, x)_{cl} = \frac{\partial}{\partial p_j} f_A^\lambda(x^\lambda(t, x)_{cl}), \qquad (4.1.35)$$

$$\frac{d}{dt} p_j^\lambda(t, x)_{cl} = -\frac{\partial}{\partial q_j}\left(\varrho^\lambda * V(q^\lambda(t, x)_{cl})\right) = -\frac{\partial}{\partial q_j} f_A^\lambda(x^\lambda(t, x)_{cl}), \qquad (4.1.36)$$

with f_A^λ from (4.1.30). We shall rewrite (4.1.34) into a form similar to (4.1.36).[1] Let $y \in \mathbb{R}^{2n}$ and W_y^λ as in (4.1.25). Inserting $W_{\pm y}^\lambda$ into the trace in (4.1.34) we obtain:

$$\frac{d}{dt} p_j^\lambda(t, x) = -\varrho^\lambda(y, t, x) * \frac{\partial V}{\partial q_j}(q^y) = -\left(\frac{\partial}{\partial q_j} \varrho^\lambda(y, t, x)\right) * V(q^y), \quad (4.1.37)$$

where $q^y := (y_1, \ldots y_n)$ and

$$\varrho^\lambda(y, t, x)(q) := \frac{1}{\lambda^n} \left| (W_{-y}^\lambda U_A^\lambda(t) W_x^\lambda \varphi)(-\frac{q}{\lambda}) \right|^2. \quad (4.1.38)$$

Since the right hand side of (4.1.37) is independent of $y \in \mathbb{R}^{2n}$, we can insert there $y := x^\lambda(t, x)$ and obtain a formal analogy with (4.1.36). We expect, that the difference $\varrho^\lambda(y, t, x) - \varrho_\varphi^\lambda$ (compare (4.1.21b)) will converge to zero with $\lambda \to 0$ in the sense of distributions uniformly on compacts in t, if $y := x^\lambda(t, x)$, and also for $y := x^\lambda(t, x,)_{cl}$, with some reasonable choice of V. This conjecture is based on the results of [154]. (The convergence holds for each fixed $t \in \mathbb{R}$ for $y = x^\lambda(t, x))$.[2])

4.1.10 Example. We shall give here an elementary example showing possible differences between a quantal time-evolution and its classical projection. We shall notice also the behaviour of these evolutions in the limit of vanishing λ. In the formalism introduced above, let $\varphi \in L^2(\mathbb{R}, dq)$ represents a *'minimal wave packet'*:

$$\varphi(q) := \pi^{-\frac{1}{4}} \exp\left(-\frac{1}{2} q^2\right), \quad (4.1.39)$$

and choose $\varphi_z^\lambda := W_z^\lambda \varphi$ with $z := q - ip$, $W_z^\lambda := \exp[i\lambda^{-2}(Q^\lambda p - P^\lambda q)]$. Let the generator A^λ of quantal time-evolution be

$$A^\lambda := a(\lambda) P_\varphi, \quad (4.1.40)$$

where $a(\lambda)$ is some real function. Then the classical Hamiltonian function on the orbit O_φ^λ of the Heisenberg group in $L^2(\mathbb{R})$ is

$$f_A^\lambda(z) := Tr(P_z^{(\lambda)} A^\lambda) = a(\lambda) \exp\left(-\frac{\bar{z}z}{2\lambda^2}\right) \quad (4.1.41)$$

with $\bar{z}$ being the complex conjugate of $z \in \mathbb{C}$. We are interested in the comparison of solutions of classical Hamiltonian equations on O_φ, $z^\lambda(t, z)_{cl}$, and the corresponding quantal expectations:

$$z^\lambda(t, z) := Tr(U_A^\lambda(t) P_z^{(\lambda)} U_A^\lambda(-t) Z^\lambda), \quad Z^\lambda := Q^\lambda - iP^\lambda, \quad (4.1.42)$$

---

[1]It is left to the reader's assessment, whether the forthcoming reformulation could be helpful for better understanding of the "classical limit $\hbar \to 0$" of the dynamics.

[2]This fact was kindly announced to the author by Prof. Klaus Hepp (in 1985).

with the same initial values $z = q - ip$. Elementary calculations give:

$$z^\lambda(t, z) = \left[1 - \frac{1}{a(\lambda)} f_A^\lambda(z)\right] z + \frac{1}{a(\lambda)} f_A^\lambda(z) \exp\left(-\frac{it}{\lambda^2} a(\lambda)\right) z, \qquad (4.1.43)$$

$$z^\lambda(t, z)_{cl} = \exp\left(-\frac{it}{\lambda^2} f_A^\lambda(z)\right) z. \qquad (4.1.44)$$

We see that (4.1.43) and (4.1.44) describe motions on mutually tangent circles in $\mathbb{C}$ with different radii and different dependence of frequencies on initial conditions as well as on the parameter λ. For $\lambda \to 0$ the quantum evolution vanishes independently on the 'renormalization' $a(\lambda)$. For slowly varying $a(\lambda)$, the classical evolution vanishes too, but the way of this vanishing looks qualitatively differently. If, e.g. $a(\lambda) = \lambda^2 \exp(b/2\lambda^2)$, $b > 0$, then $\bar{z}z = b$ is a critical value for $\lambda \to 0$.

4.2 Extension of CCR by a Quadratic Generator

4.2.1 All the orbits O_φ occurring in Sect. 4.1 were mutually homeomorphic (and homeomorphic to $\mathbb{R}^{2n}$). In this section, we shall give examples of irreducible representations $U(G)$ of some Lie groups G in a Hilbert space $\mathcal{H}$ containing various mutually nonhomeomorphic orbits $O_j := U(G)\varphi_j$ in $P(\mathcal{H})$, $(j = 1, 2, \dots)$. Let G be a connected Lie group containing the $2n + 1$—dimensional Heisenberg group G_n as an invariant (i.e. normal) subgroup (G will be specified later). Let U be such a unitary continuous representation of G, the restriction of which to G_n coincides with the irreducible representation described in Sect. 4.1 with $\hbar = 1 = \lambda$. With the notation of the previous section, for $m = 1, 2, \dots K$, $A_m \in U(\mathfrak{g})$, set

$$A_m := \frac{1}{2} h_m^{jk} X_j X_k, \quad \text{(summation over} j, k = 1, 2, \dots 2n), \qquad (4.2.1)$$

with any h_m a real symmetric $2n \times 2n$—matrix; the formally defined operator A_m is symmetric on the Gårding domain of $U(G_n)$. From (4.1.1a) we have commutation relations (cf. also 4.1.3):

$$[X_j, X_k] = i\, S_{jk} I, \quad [X_j, A_m] = i\, S_{jk} h_m^{kl} X_l =: i\, (S \cdot h_m \cdot X)_j, \qquad (4.2.2)$$

$$[A_m, A_k] = \frac{i}{2} X \cdot (h_m \cdot S \cdot h_k - h_k \cdot S \cdot h_m) \cdot X, \quad m, k = 1, 2, \dots K. \qquad (4.2.3)$$

Assume that for any m, k there are reals c_{mk}^j such, that

$$h_m \cdot S \cdot h_k - h_k \cdot S \cdot h_m = \sum_{j=1}^{K} c_{mk}^j h_j, \quad m, k = 1, 2, \dots K. \qquad (4.2.4)$$

then the linear hull of the operators X_j $(j = 1, 2, \ldots 2n)$, A_m $(m = 1, 2, \ldots K)$ and $I := I_{\mathcal{H}} := id_{\mathcal{H}}$ forms the Lie algebra $U(\mathfrak{g})$. We have (cf. also [37, Proposition 3.3.12]):

4.2.2 Proposition. *Let $U(\mathfrak{g})$ be the above defined representation of a Lie algebra $\mathfrak{g}$ in $\mathcal{H}$ and let G be the corresponding simply connected Lie group with the Lie algebra g. Then the representation $U(G_n)$ of the Heisenberg group G_n has a unique extension to the representation $U(G)$ of G in $\mathcal{H}$ such, that the closures of the operators X_j $(j = 1, 2, \ldots 2n)$, A_m $(m = 1, 2, \ldots K)$ and $I_{\mathcal{H}}$ are selfadjoint generators of $U(G)$ corresponding to basis vectors in $\mathfrak{g}$ according to (2.3.20). In particular the operators A_m $(m = 1, 2, \ldots K)$ are essentially selfadjoint on the Gårding domain of $U(G_n)$.*

Proof. The **Gårding domain of $U(G_n)$** is a common dense invariant domain of all the operators in $U(\mathfrak{g})$. According to a Nelson's theorem (see [13, Theorem 11.5.2.]) it suffices to prove essential selfadjointness of the operator Δ,

$$\Delta := \sum_{j=1}^{2n} X_j^2 + \sum_{m=1}^{K} A_m^2, \tag{4.2.5}$$

on the invariant domain. First we shall choose $m := (j; k)$ with $j, k = 1, 2, \ldots 2n$ and

$$A_m := A_{(j;k)} := \frac{1}{2}(X_j X_k + X_k X_j). \tag{4.2.6}$$

In this case the operator Δ in (4.2.5) can be expressed in the form

$$\Delta = \frac{3}{2} n I_{\mathcal{H}} + \sum_{j=1}^{n} (P_j^2 + Q_j^2)(I_{\mathcal{H}} + \sum_{k=1}^{n} (P_k^2 + Q_k^2)), \tag{4.2.7}$$

where we used CCR and the notation (4.1.4). From the known properties of the Hamiltonians $P_j^2 + Q_j^2$ of independent linear oscillators, we conclude (with a help, e.g., of [262, Theorem VIII.33]) that Δ is essentially selfadjoint.

  Denote the Lie algebra generated by X_j's and $A_{(j;k)}$ $(j, k = 1, 2, \ldots 2n)$ by $\mathfrak{g}_{max}$ and the corresponding **simply connected group** by G_{max}. Any A_m of the form (4.2.1) is a linear combination of $A_{(j;k)}$'s. Consequently, any Lie algebra $U(\mathfrak{g})$ from 4.2.1 is a subalgebra of $U(\mathfrak{g}_{max})$ and the corresponding group G is a subgroup of G_{max}. From this just proved integrability of $U(\mathfrak{g}_{max})$ to a unitary representation $U(G_{max})$, it follows integrability of $U(\mathfrak{g})$ for any $\mathfrak{g}$ introduced in 4.2.1. This implies the selfadjointness of (4.2.5) with arbitrary A_m of the form (4.2.1) and this, in turn, implies uniqueness of $U(G)$. $\square$

4.2.3 In this section we shall restrict our attention to the cases of representations $U(G)$ obtained from $U(G_n)$ by addition of only one generator $A := A_m$ of the form (4.2.1) in the manner described above. Let h be any nonzero real symmetric matrix with elements h_{jk} $(j, k = 1, 2, \ldots 2n)$ and let

$$A := \frac{1}{2} h^{jk} X_j X_k \qquad (4.2.8)$$

denote here the selfadjoint operator corresponding to the right hand side of (4.2.8). According to (4.2.6) the operators X_j $(j = 1, 2, \ldots 2n)$, A and $I_\mathcal{H}$ are selfadjoint generators of an irreducible unitary representation $U(G)$ of a $2n + 2$—dimensional connected Lie group G containing G_n as a normal subgroup. The restriction $U(G_n)$ of $U(G)$ is irreducible, too. Let $\boldsymbol{U}(G)$ be the realization of G in $P(\mathcal{H})$ obtained by the natural projection of $U(G)$, 2.3.9. We shall investigate infinitely differentiable orbits of $\boldsymbol{U}(G)$ in $P(\mathcal{H})$.

4.2.4 Lemma. *Let $\boldsymbol{U}(G)$ be as in 4.2.3 and $\mathcal{D}_G$ be a dense invariant subset in $\mathcal{H}$ consisting of infinitely differentiable vectors of $U(G)$, e.g. $\mathcal{D}_G$ is the Gårding subspace for $U(G)$([13, 11.1.8]). Let $\varphi \in \mathcal{D}_G$, $\|\varphi\| = 1$ and $O_\varphi := \boldsymbol{U}(G)\varphi$ be the immersed submanifold of $P(\mathcal{H})$ according to 3.1.2. The orbit O_φ is $2n$-dimensional iff there is an element $C_\varphi \in U(\mathfrak{g})$,*

$$C_\varphi := c_\varphi^j X_j - A, \qquad (4.2.9)$$

such, that φ is its eigenvector:

$$C_\varphi \varphi = \lambda \varphi \quad \text{for some } \lambda \in \mathbb{R}. \qquad (4.2.10)$$

If (4.2.9) with (4.2.10) is the case, then $O_\varphi = \boldsymbol{U}(G_n)\varphi$.

Proof. The tangent space to O_φ at φ is the linear hull of vectors $\sigma_j(\varphi)$ $(j = 1, 2, \ldots 2n)$ (see 4.1.5) and $\sigma_A(\varphi)$ (see e.g. 2.3.5 and Sect. 3.2). According to 4.1.5, all the σ_j's are linearly independent. Hence O_φ is $2n$-dimensional iff

$$\sigma_A(\varphi) = c_\varphi^j \sigma_j(\varphi) \qquad (4.2.11)$$

for some reals c_φ^j. According to (2.3.8), the equation (4.2.10) implies (4.2.11). Assuming (4.2.11), we have in the standard identification of $T_\varphi P(\mathcal{H})$ with $[\varphi]^\perp$ by the help of Ψ_φ (see 2.1.7 and (2.1.16)):

$$(I - P_\varphi)(c_\varphi^j X_j - A)\varphi = 0 \Rightarrow (c_\varphi^j X_j - A)\varphi = \lambda \varphi \qquad (4.2.12)$$

with $\lambda := \lambda(\varphi) := Tr(P_\varphi(c_\varphi^j X_j - A))$. $\qquad \square$

4.2.5 Let $C := C_\varphi$ have the form (4.2.9) and let (4.2.10) be fulfilled. Then C satisfies the system of linear equations:

$$Tr(P_\varphi[C, X_j]) = 0, \quad j = 1, 2, \ldots 2n, \tag{4.2.13}$$

$$Tr(P_\varphi[C, A]) = 0, \tag{4.2.14}$$

where (4.2.14) follows from (4.2.13). The equations (4.2.13) have unique solution C of the form (4.2.9) for any $\varphi \in \mathcal{D}_G$, even if the relation (4.2.10) is not fulfilled:

$$c_\varphi^j = h^{jk} \, Tr(P_\varphi X_k). \tag{4.2.15}$$

The corresponding operator C_φ represents the generator of the isotropy subgroup $K_\varphi \subset G$ at the point $F_\varphi \in \mathfrak{g}^*$ in the $Ad^*(G)$-representation; here $F_\varphi(\xi) := Tr(P_\varphi X_\xi)$ for $\xi \in \mathfrak{g}$, compare 3.2.3 and 3.2.6. From (4.2.15) and (4.2.9), we have immediately:

4.2.6 Lemma. *If* $\varphi \in O_\varphi$ *is chosen such that* $Tr(P_\varphi X_j) = 0$ *for all* $j = 1, 2, \ldots 2n$, *then* $C_\varphi = -A$.

4.2.7 Proposition. *The orbit* O_φ *of* $U(G)$ *is 2n-dimensional iff it contains an eigen-projector* P_φ *of* A, *i.e. iff for some* $\varphi \in O_\varphi$ *it is*

$$Tr(P_\varphi A^2) = \left(Tr(P_\varphi A)\right)^2. \tag{4.2.16}$$

Proof. In any orbit lying in $\mathcal{D}_G$ there is a point φ satisfying the conditions of the Lemma 4.2.6, compare 4.1.5. The assertion is an immediate consequence of the Lemmas 4.2.6 and 4.2.4. $\qquad\square$

4.2.8 Corollaries. *(i) Let* $\varphi_0 \in U(G)\varphi$ *satisfy* (4.1.15). *If* φ *is an eigenvector of* A, *then also* φ_0 *is an eigenvector of* A, *if* $\varphi \in \mathcal{D}_G$.

 (ii) If $\varphi_0 \in \mathcal{D}_{G_{max}}$ *satisfies* (4.1.15), *then those relations are satisfied by all the vectors*

$$\varphi_t^A := \exp(-itA)\varphi_0 \tag{4.2.17}$$

for all $t \in \mathbb{R}$ *and all the choices of* A; (4.2.8).

4.2.9 Proposition. *For any choice of* A *in 4.2.3 there is in* $P(\mathcal{H})$ *a 2n+1 - dimensional orbit of the corresponding representation* $U(G)$ *(defined in 4.2.3), which is an infinitely differentiable immersed submanifold of* $P(\mathcal{H})$.

Proof. Remember that any A is an unbounded selfadjoint operator. Let $U_\pi := P_+ - P_-$ be the 'parity operator' defined by [3]

$$U_\pi^* = U_\pi^{-1} = U_\pi, \quad U_\pi X_j U_\pi = -X_j \quad (j = 1, 2, \ldots 2n), \tag{4.2.18}$$

---

[3] In $\mathcal{H} \equiv L^2(\mathbb{R}^n, d^n x)$, it is defined as $[U_\pi \psi](x) := \psi(-x), \, \forall \psi \in \mathcal{H}, \, x \in \mathbb{R}^n$.

and P_+ (resp. P_-) are corresponding orthogonal projectors,

$$P_+ + P_- = I_{\mathcal{H}}. \tag{4.2.19}$$

Choose a dense invariant linear subset $\mathcal{D}_G$ of $\mathcal{H}$ consisting of infinitely differentiable vectors of $U(G)$ such, that (as usually)

$$U_\pi \mathcal{D}_G \subset \mathcal{D}_G, \text{ hence } P_\pm \mathcal{D}_G \subset \mathcal{D}_G. \tag{4.2.20}$$

This condition implies, that $P_+ \mathcal{D}_G$ (resp. $P_- \mathcal{D}_G$) is dense in $\mathcal{H}_+ := P_+ \mathcal{H}$ (resp. in $\mathcal{H}_- := P_- \mathcal{H}$). For any $\varphi \in \mathcal{D}_G^+ \cup \mathcal{D}_G^-$ (with $\mathcal{D}_G^\pm := P_\pm \mathcal{D}_G$), the assumption of 4.2.6 is fulfilled due to (4.2.18). If φ is not an eigenvector of A, then the orbit O_φ is 2n+1-dimensional. Assume, that $A\varphi = \lambda\varphi$. Let $\varphi \in \mathcal{D}_G^+$, for definiteness. Since A is U_π-invariant: $[A, U_\pi] = 0$, its spectral measure E_A commutes with projectors $P_\pm$. Denote for any Borel set $B \subset \mathbb{R}$

$$E_A^\pm(B) := P_\pm E_A(B), \text{ hence } E_A = E_A^+ + E_A^-, \tag{4.2.21}$$

and E_A^+ is the spectral measure of the restriction of A to the $U(G)$-invariant (infinite dimensional) subspace $\mathcal{H}_+$ of $\mathcal{H}$. Due to unboundedness of A, we can assume that the subspace $(P_+ - E_A^+(\{\lambda\}))\mathcal{H}$ of $\mathcal{H}_+$ is nonempty; here $E_A^+(\{\lambda\})$ is the eigenprojector of $P_+ A$ corresponding to the eigenvalue λ. Choose a nonzero vector

$$\varphi' \in (P_+ - E_A^+(\{\lambda\}))\mathcal{H} \tag{4.2.22}$$

and assume the normalization $\|\varphi\| = \|\varphi'\| = 1$. Let $\chi := \frac{1}{\sqrt{2}}(\varphi' + \varphi)$. Since $\mathcal{D}_G^+$ is dense in $\mathcal{H}_+$, we can find for arbitrarily small $\delta > 0$ a vector φ_0:

$$\varphi_0 \in \mathcal{D}_G^+ : \|\varphi_0 - \chi\|^2 < \delta, \|\varphi_0\| = 1. \tag{4.2.23}$$

With $\delta < 2 - \sqrt{2}$, the vector φ_0 cannot be an eigenvector of A and, moreover, it satisfies (4.1.15). Hence the corresponding orbit O_{φ_0} is $2n + 1$-dimensional. The manifold structure was proved in 3.1.2. $\qquad\square$

4.2.10 Let O_φ (with $\varphi \in \mathcal{D}_G$) be a 2n+1-dimensional orbit of $U(G)$ and let Ω° be the restriction of the standard symplectic form Ω on $P(\mathcal{H})$ onto O_φ, compare 3.2.2. According to the previous results (Sects. 3.2 and 4.1), Ω° is a closed two-form of the maximal rank 2n, hence it is a contact two-form on O_φ (see, e.g. [1, Chap. 5.1.]). The equations (4.2.13) determine the characteristic line-bundle of Ω° in terms of operators $C = C_\varphi$ corresponding to generators of stability groups of $F_\varphi \in \mathfrak{g}^*$ (see 4.2.5) with respect to $Ad^*(G)$. The characteristic line bundle of Ω° is integrable, determining a regular foliation of O_φ. The factorization of O_φ with respect to this foliation is the symplectic manifold M_φ (symplectomorphic to the classical phase space $T^*\mathbb{R}^n$) as it was constructed in Sect. 3.2 (for definition of the **cotangent bundle** $T^*(M)$ of a general manifold M see e.g. [37, A.3.6 Definitions (v)]).

4.2.11 The quantal and classical evolutions corresponding to the generator A, cf. 4.2.3, (resp. to the Hamiltonian function f_A) coincide in our examples in the sense of $3.3.9 + 3.3.10$, independently of the dimension ($= 2n$ or $2n+1$) of the orbit O_φ. The time-evolved quantal states remain all the time on the orbit O_φ. We might be interested also in time evolution of other quantities than (the expectations of) X_j in the quantal interpretation. According to 4.1.7(i), in the case of dim $O_\varphi = 2n = \dim M_\varphi$, any *'spreading of the wave packet'* does not occur. The situation is different, however, on $2n+1$ - dimensional orbits. For various φ_j ($j = 1, 2$) corresponding to distinct quantal states in the same leaf $[\varphi] \in M_\varphi$ we have in general (cf. 4.1.4 for notation)

$$Tr(P_x^{\varphi_1}(X_j - x_j)(X_k - x_k)) \neq Tr(P_x^{\varphi_2}(X_j - x_j)(X_k - x_k)). \qquad (4.2.24)$$

This is the case of e.g., free particle motions. This fact makes a certain difference between classical and quantal interpretations of the 'extended phase spaces' O_φ (dim $O_\varphi = 2n + 1$). This will be briefly discussed later on, in 4.3.5.

4.3   Notes on Other Examples

4.3.1 By the method developed in our Chap. 3, we can construct from an arbitrary continuous unitary representation $U(G)$ of a Lie group G 'classical phase spaces', which are diffeomorphic (and even symplectomorphic) to orbits of $Ad^*(G)$. It can be shown, [174, 15.2], that any symplectic homogenous space of any connected Lie group G is a *covering symplectic space* of either an orbit of Ad*(G), or an orbit of $Ad^*(G_1)$, where G_1 is a *central extension of G by* $\mathbb{R}$—see also 1.3.7. On the other hand, unitary continuous representations of G can be constructed from orbits of $Ad^*(G)$, [174]. Considerations in Sects. 1.2 and 1.3 show reasons for modeling state spaces of CM-systems as homogeneous symplectic spaces of some Lie groups, at least for 'basic' or 'elementary' physical systems. In this section we shall outline further examples of obtaining CM-systems from unitary group representations which suggest, that all generally accepted models of 'elementary' finite dimensional CM-systems could be obtained in this way.

4.3.2   Classical spin from SO(3): Let U be a (projective) irreducible representation of the compact Lie group SO(3)—the connected component of the 3-dimensional orthogonal group O(3) of orthogonal transformations of a 3-dimensional Euclidean space E_3. The representation space $\mathcal{H} = \mathbb{C}^{2J+1}$ ($J = \frac{n}{2}$, $n \in \mathbb{Z}_+$) is finite dimensional. Generators Y_k ($k = 1, 2, 3$) of U corresponding to rotations around orthogonal axes in E_3 satisfy the commutation relations (with $\epsilon_{jkm} = -\epsilon_{kjm} = -\epsilon_{jmk}$, $\epsilon_{123} = 1$):

$$[Y_k, Y_m] = i\, \epsilon_{kmj} Y_j. \qquad (4.3.1)$$

Choose any nonzero $\varphi \in \mathcal{H}$ and form the orbit $O_\varphi := \{U(g)\varphi : g \in SO(3)\}$. Let us denote by Y_ξ the generator of $t \mapsto U(\exp(t\xi))$ corresponding to an element ξ of the Lie algebra $\mathfrak{g} := so(3)$. We are interested in the $Ad^*(SO(3))$-action onto $F_\varphi \in so(3)^*$, where

$$F_\varphi : \xi \mapsto F_\varphi(\xi) := Tr(P_\varphi Y_\xi), \ \xi \in so(3). \tag{4.3.2}$$

Generators $C_\varphi := c_\varphi^j Y_j$ of one-parameter subgroups of the isotropy group of F_φ are just all nonzero solutions of equations

$$Tr(P_\varphi[Y_k, C_\varphi]) = 0, \quad k = 1, 2, 3. \tag{4.3.3}$$

With $y^k := y^k(\varphi) := Tr(P_\varphi Y_k)$, the only linearly independent solution C_φ of (4.3.3) can be written:

$$C_\varphi = y^k(\varphi)Y_k. \tag{4.3.4}$$

One could easily check that $C_\varphi = 0$ in (4.3.4) for some φ, iff $C_\varphi = 0$ for all $\varphi \in \mathcal{H}$, iff $J = 0$ (i.e. $\dim_\mathbb{C} \mathcal{H} = 1$), iff the matrix of the homogeneous equations (4.3.3) is identically zero. In all other cases the rank of the matrix of the system (4.3.3) equals to 2. For $J = 0$ the corresponding classical phase space degenerates to a point: this corresponds to the traditional point of view according to of which spin does not occur in classical mechanics.

For orbits O_φ in representations with $J \neq 0$ we have two possibilities:

(i) The vector φ is an eigenvector of C_φ and the orbit O_φ is two-dimensional (any generator $Y \in U(so(3))$ which is linearly independent of C_φ cannot be a solution of (4.3.3): $Tr(P_\varphi[Y_k, Y]) = 0$ for $k = 1, 2, 3$ implies $Y = \lambda C_\varphi$; hence Y linearly independent of C_φ generate two-dimensional tangent space to O_φ at φ).

(ii) If φ is not an eigenvector of C_φ, then the generator C_φ generates a one-dimensional submanifold of O_φ diffeomorphic to a circle S^1 (C_φ generates the *isotropy subgroup* of $SO(3)$ at F_φ, which is closed, hence compact). In this case O_φ is 3-dimensional.

Note that for $J = \frac{1}{2}$ only the possibility (i) occurs, since $\mathcal{H} = \mathbb{C}^2$ and $\dim_\mathbb{R} P(\mathcal{H}) = 2$.

It can be easily shown that in the both cases the corresponding classical phase space M_φ (in the case (i) identical with O_φ) is diffeomorphic to the sphere S^2 in $so(3)^*$ with coordinates

$$y^k : F_\varphi \mapsto y^k(\varphi) := Tr(P_\varphi Y_k), \ k = 1, 2, 3; \quad \varphi \in O_\varphi. \tag{4.3.5}$$

Let $t \in \mathbb{R}$ and let $\tau \in \mathbb{R}^3$ be any unit vector: $\sum_k (\tau^k)^2 = 1$. Let $y(\varphi) \in \mathbb{R}^3$ be given by coordinates y^k in (4.3.5) and $\tau \cdot y := \sum_k \tau^k y^k$. Using (4.3.1) we obtain:

$$y^k(\exp(-it\tau^j Y_j)\varphi) = y^k(\varphi)\cos t + \epsilon_{kjm}\tau^j y^m(\varphi)\sin t$$
$$+2\tau^k\tau\cdot y(\varphi)\sin^2\frac{t}{2}, \tag{4.3.6}$$

what gives an explicit expression for the sphere $S^2 \subset so(3)^*$. The $r_\varphi :=$ *radius of the sphere* is equal to the length of $y(\varphi)$,

$$|y(\varphi)|^2 = y(\varphi)\cdot y(\varphi) = r_\varphi^2. \tag{4.3.7}$$

In the case (i) the values of (4.3.7) might be only the numbers J^2, $(J-1)^2$, $(J-2)^2, \ldots$, i.e. the orbits $O_\varphi \subset P(\mathcal{H})$ are mapped by the association $\varphi \mapsto F_\varphi$ (cf. (4.3.2)) onto a finite-number of $[J+1]$ distinct spheres in the three-dimensional linear space $so(3)^*$ (here $[k]$ is the integer part of $k \in \mathbb{R}_+$; if $J \in \mathbb{Z}_+$ one of the spheres degenerates into a point). But $P(\mathcal{H})$ is a connected manifold and the mapping $\varphi \mapsto F_\varphi$ is continuous, hence for $J \geq 1$ also the cases (ii) occur and the numbers (4.3.7) acquire values from a whole interval of $\mathbb{R}_+$, if φ runs over $P(\mathcal{H})$.

Let us write explicitly the symplectic form Ω^M on the phase space $M_\varphi = S^2$. In terms of coordinate functions y_k from (4.3.5), we obtain in the region where $y_3(\varphi) \neq 0$ (indices are written down for convenience):

$$\Omega^M = -\frac{1}{y_3}dy_1 \wedge dy_2, \quad y_3^2 := r_\varphi^2 - y_1^2 - y_2^2. \tag{4.3.8}$$

The Poisson bracket of these coordinate functions is

$$\{y_k, y_m\} = -\epsilon_{kmj}y_j. \tag{4.3.9}$$

The sphere S^2 with this symplectic structure is interpreted as the phase space of an (isolated) *classical spin*. It is an example of a compact symplectic manifold.

4.3.3 We can construct now certain combinations of the previous example with those of Sects. 4.1 and 4.2. Let us distinguish generators X_j of the representation of $6N + 1$—dimensional Heisenberg group corresponding to coordinates of positions and momenta of N individual particles. Denote them Q_j^a, P_j^a ($a = 1, 2, \ldots N$; $j = 1, 2, 3$) with CCR in the form

$$[Q_j^a, P_k^b] = i\,I\delta_{ab}\delta_{jk}, \quad [Q_j^a, Q_k^b] = [P_j^a, P_k^b] = 0, \tag{4.3.10}$$

for all $a, b = 1, 2, \ldots N$; $j, k = 1, 2, 3$. Now we define operators of orbital momenta (no summation over repeated indices a, b):

$$Y_j^a := \epsilon_{jkm}Q_k^a P_m^a, \quad Y_j := Y_j^{tot} := \sum_a Y_j^a \tag{4.3.11}$$

satisfying (4.3.1) (up to domain specifications) for any upper index (a, or *tot*). Relations (4.2.2) have now the form:

$$[Y_j^a, Q_k^b] = i\,\delta_{ab}\epsilon_{jkm}Q_m^a, \quad [Y_j^a, P_k^b] = i\,\delta_{ab}\epsilon_{jkm}P_m^a, \qquad (4.3.12)$$

$$[Y_j, Q_k^a] = i\,\epsilon_{jkm}Q_m^a, \quad [Y_j, P_k^a] = i\,\epsilon_{jkm}P_m^a, \quad [Y_j, Y_k^a] = i\,\epsilon_{jkm}Y_m^a. \qquad (4.3.13)$$

Let us first consider the Lie algebra $\mathfrak{g}_0$ represented by generators Q_j^a, P_j^a and Y_j $(j = 1, 2, 3;\ a = 1, 2, \ldots N)$ of the representation $U(G_0)$ of the corresponding group G_0, compare Proposition 4.2.2. We see that G_0 is a semidirect product of of $SU(2)$ with the Heisenberg group G_{3N} (with the notation from 4.2.1), $G_0 = SU(2) \ltimes G_{3N}$, where the Heisenberg group is a normal subgroup. Let us investigate the orbits $O_\varphi := U(G_0)\varphi \subset \mathcal{D}_{G_0}$ in $P(\mathcal{H})$ and the corresponding classical phase spaces M_φ. Since any O_φ is a homogeneous space of G_0, it can be generated from a point φ satisfying (see 4.1.5)

$$Tr(P_\varphi X_j) = 0, \text{ for all } j = 1, 2, \ldots 6N. \qquad (4.3.14)$$

The local structure of O_φ is most easily seen in a neighbourhood of such φ. The isotropy group of $F_\varphi \in \mathfrak{g}_0^*$ (see (3.1.5)) with respect to $Ad^*(G_0)$ has the Lie algebra generated by such $C \in U(\mathfrak{g}_0)$, which are solutions of the system

$$Tr(P_\varphi[C, X_j]) = 0\ (j = 1, 2, \ldots 6N), \quad Tr(P_\varphi[C, Y_k]) = 0\ (k = 1, 2, 3). \qquad (4.3.15)$$

The corank of the matrix of this homogeneous system is:

(i) equal to 3 iff $Tr(P_\varphi Y_k) = 0$ for all $k = 1, 2, 3$; in this situation there might occur cases with dim $O_\varphi = 6N$, $6N + 2$, $6N + 3$ corresponding to such φ, for which $Y\varphi = 0$ for all $Y \in U(so(3))$, (resp. $Y\varphi = 0$ for just one linearly independent $Y \in U(so(3))$, resp. $Y\varphi \neq 0$ for all $Y \neq 0$); as an example of the case of dim $O_\varphi = 6N + 3$ we can take φ for $N = 1$ in Schrödinger realization of CCR:

$$\varphi(q) := \varphi(q_1, q_2, q_3) := c\,q_1 q_2 q_3\,\exp(-q_1^2 - q_2^2 - q_3^2), \ c := \left(\frac{2^5}{\pi}\right)^{\frac{3}{4}}, \quad (4.3.16)$$

corresponding to the value $J = 3$ of the total momentum. In all these cases of various values of dim O_φ the corresponding symplectic spaces M_φ are homeomorphic to $T^*\mathbb{R}^{3N} = \mathbb{R}^{6N}$.

(ii) equal to 1 in all other cases; now all the solutions C of (4.3.15) are proportional to C_φ of the form (4.3.4). If φ is an eigenvector of C_φ, then dim $O_\varphi = $ dim $M_\varphi = 6N + 2$. In the remaining case it is dim $O_\varphi = 6N + 3$ and dim $M_\varphi = 6N + 2$. If φ is proportional to $C_\varphi\varphi$, the orbit O_φ is the fiber-bundle with base $\mathbb{R}^{6N}$ and typical fiber S^2; if φ is not an eigenvector of C_φ, then the fiber on R^{6N} is the whole group $SO(3)$. In the both cases the phase space is $T^*\mathbb{R}^{3N}$ fibered by two dimensional spheres S^2 with the canonical symplectic form from $P(\mathcal{H})$ being the sum of the canonical form on $T^*\mathbb{R}^{3N}$ and that on S^2 described in (4.3.8).

Let us take now all the operators Q_j^a, P_j^a, Y_j^a ($a = 1, 2, \ldots N; j = 1, 2, 3$) as generators of the considered representation $U(G)$ (now G is semi direct product of the Heisenberg group G_{3N} and of the direct product of N copies of the group $SU(2)$). The orbits and corresponding phase spaces arising from the action of this group G on $P(\mathcal{H})$ with $\mathcal{H} = L^2(\mathbb{R}^{3N}) = L^2(\mathbb{R}^3) \otimes L^2(\mathbb{R}^3) \otimes \ldots L^2(\mathbb{R}^3)$ (N-tuple tensor product) can be constructed as N-tuple direct product manifolds; each of the multipled manifolds can be obtained by the above described procedure with N = 1.

Examples of classical systems obtained in this subsection include systems of several nonrelativistic spinning particles. Here the '*classical spin*' was obtained from quantal **orbital** momentum.

4.3.4 The groups which are, perhaps, physically most important ones, are Galilean and Poincaré groups. Because of relative complexity of any complete exposition of these important examples, we shall restrict our present exposition to several notes and remarks. For more detailed nice exposition see e.g. in [321].

(i) **The *Galilean group*.**
This group realizes the nonrelativistic (better: Galilean relativistic) conception of relative positions and motions of mechanical systems (particles, bodies etc.). It is a ten parameter Lie group, the parameters of which can be chosen to describe time and space translations (4 parameters), space rotations (3 parameters) and transition to uniformly moving systems (3 coordinates of a velocity). Any unitary (vector) representation of this group cannot be, however, interpreted in terms of really observed physical systems, see e.g. [321, Sect. XII.8]. Physically interpreted projective representations correspond to multipliers m_τ of the Galilean group characterized by a real parameter τ—the mass of the system. Let us denote by G the *central extension* (cf. [174, 321], resp. also [37, Note 3.3.6]) of (the covering group of) the Galilean group by $\mathbb{R}$ corresponding to a multiplier m_τ with $\tau \neq 0$ (all such groups are mutually isomorphic). Orbits of $Ad^*(G)$ (described e.g. in [5]) are just one particle phase spaces obtained in our subsection 4.3.3. Unitary representations of G, in which the central subgroup $\mathbb{R}$ acts by a multiplication by constants, correspond to physically interesting projective representations of the Galilean group. Irreducible representations of G describe one-particle systems. The projected orbits O_φ of these representations are either seven or nine or ten dimensional (this is a consequence of 4.3.3, 4.2.7 and absolute continuity of the spectrum of the time-evolution generator $P_1^2 + P_2^2 + P_3^2$ of $U(G)$). In the cases dim $O_\varphi = 7$ or 9 the manifolds O_φ with the two-form Ω° (cf. 3.2.2) are just contact manifolds of the extended phase spaces, dim $O_\varphi = $ dim $M_\varphi + 1$.

(ii) **The *Poincaré group*.**
Let now G be the ten-parameter covering group of the Poincaré group. Physical interpretation of the parameters is the same as that of the corresponding parameters of the Galilean group. In the present case of G, however, the conception of Galilean relativity is replaced by the conception of Einstein relativity of mechanical motions. Since the second cohomology group of G is now trivial, we have to deal with unitary (vector) representations of G only. The orbits of the coadjoint action of G

corresponding to phase spaces of particles with nonvanishing masses have the same topological and symplectic structure as in the case (i). The action $Ad^*(G)$ is, however, different from that of the Galilean case; with this are connected also different interpretations of coordinates determined by the mutually corresponding generators in cases (i) and (ii). The dimensionality of orbits O_φ of unitary irreducible representations $U(G)$ corresponding to nonzero masses is the same as in (i). Also here, we obtain 7- and 9-dimensional contact manifolds the contact two-form $\Omega°$ on them coincides with the standard two-form of classical relativistic mechanics (which, in the case of dim $O_\varphi = 7$, comes from the restriction of $d p_\mu \wedge d q^\mu$ defined on $T^*\mathbb{R}^4$ onto the submanifold $p_0^2 - \sum_j p_j^2 = (mass)^2$).

4.3.5 Remark. Any symplectic manifold can be trivially extended to a contact manifold by taking the direct product with $\mathbb{R}$. If M is a symplectic phase space of some physical system, then the added dimension in $\mathbb{R} \times M$ can be interpreted as the 'time variable' t. Let Ω be the symplectic form on M, $\pi : \mathbb{R} \times M \to M$ be the canonical projection and σ_A the Hamiltonian vector field on M with Hamiltonian function f_A, i.e. $i(\sigma_A)\Omega = -d f_A$. The contact two-forms $\Omega° := \pi^*\Omega$, resp. $\Omega^A := \Omega° - d f_A \wedge dt$ on the manifold $\mathbb{R} \times M$ have characteristic vector fields δ_t (defined by $dt(\delta_t) = 1$ and $df(\delta_t) = 0$ for any function f of the form $f := \pi^* f'$, where $f' \in C^\infty(M)$), resp. $\sigma_A° := \pi^*\sigma_A + \delta_t$ (with the identification $T(\mathbb{R} \times M) = T\mathbb{R} \times TM$ in the sense of vector bundle isomorphisms). Clearly $\pi_*\sigma_A° = \sigma_A$. For a time-independent vector field σ_A this procedure is trivial, if we have no possibility to distinguish various points of the fibres $\mathbb{R} = \pi^{-1}(x)$ ($x \in M$) by some measurements, i.e. if time is homogeneous with respect to the considered physical system. This is the case of classical mechanics determined by $(M; \Omega)$ and $f_A \in C^\infty(M)$.

The situation is different for contact orbits $O_\varphi \subset P(\mathcal{H})$. Each point of O_φ corresponds to a quantum mechanically clearly distinguishable physical state: by measuring of also quantities other than expectations of generators of $U(G)$, we can empirically distinguish various points of the same fibre, on which all the expectations of the generators in $U(\mathfrak{g})$ are constant. This fact breaks, in a certain sense, the homogeneity of time on contact orbits of the representations, which contain also time evolution of the system as a one parameter subgroup.

4.3.6 Identical particles.

If the physical system consists of N mutually distinguishable, but otherwise equal subsystems, it is described in QM by the N-fold tensor product Hilbert space $\mathcal{H}_N := \mathcal{H} \otimes \mathcal{H} \otimes \cdots \otimes \mathcal{H}$ with the Hilbert space $\mathcal{H}$ describing a single subsystem. If the 'basic observables' of a single subsystem are determined by a representation $U(G)$ in $\mathcal{H}$, observables of the whole compound system might be determined by the representation U_N of the N-fold direct product group $G_N := G \times G \times \cdots \times G$, i.e. for $\varphi := \varphi_1 \otimes \varphi_2 \otimes \cdots \otimes \varphi_N \in \mathcal{H}_N$, $\varphi_j \in \mathcal{H}$, we set $U_N(g_1 \times g_2 \times \cdots \times g_N)\varphi := U(g_1)\varphi_1 \otimes U(g_2)\varphi_2 \otimes \ldots U(g_N)\varphi_N$ for all $g_j \in G$, and extend U_N onto $\mathcal{H}_N$ by linearity and continuity. This is the case, e.g. of the example in Sect. 4.1. Then we can construct in the usual way orbits $O_\varphi := U_N(G_N)\varphi$ in $P(\mathcal{H}_N)$ and corresponding symplectic manifolds M_φ. We shall write also $U(G_N) := U_N(G_N)$.

In physics, however, 'equal (micro-)subsystems' are *indistinguishable*. If the N subsystems are indistinguishable (identical), then for any permutation $\pi \in \Pi_N$ (:= the permutation group of N elements) the product-vectors $\varphi := \varphi_1 \otimes \varphi_2 \otimes \cdots \otimes \varphi_N$ and $\pi \cdot \varphi := \varphi_{\pi(1)} \otimes \varphi_{\pi(2)} \otimes \cdots \otimes \varphi_{\pi(N)}$, as well as their linear combinations (the permutations $\pi \in \Pi_N$ act here also as linear operators on $\mathcal{H}_N$) are physically indistinguishable. There were discovered in the particle and statistical physics two kinds of particles: **Bose particles**—*bosons* (e.g. photons, mesons) specified by their integer particle spin, and **Fermi particles**—*fermions* (e.g. electrons, protons, neutrinos) having half-integer spins. Collections of N *identical particles* of each of these kinds behave according of their own specific 'statistics': Bose, resp. Fermi statistics. The two 'statistics' are formalized by two different symmetry properties of multiparticle wave functions of corresponding collections of particles. In the case of Bose (resp. Fermi) statistics the only physically realizable states correspond to *totally symmetric* (resp. *totally antisymmetric*) vectors $\varphi \in \mathcal{H}_N$:

$$\pi \cdot \varphi = \epsilon_+(\pi)\varphi, \quad \epsilon_+(\pi) := 1, \quad \text{for all } \pi \in \Pi_N, \tag{4.3.17a}$$

in the case of Bose statistics, resp.

$$\pi \cdot \varphi = \epsilon_-(\pi)\varphi, \quad \epsilon_-(\pi) := \pm 1 := \text{parity of } \pi \in \Pi_N. \tag{4.3.17b}$$

in the case of Fermi statistics.[4]

Let P_+ (resp. P_-) be the orthogonal projector in $\mathcal{H}_N$ onto the subspace $\mathcal{H}_N^+$ (resp. $\mathcal{H}_N^-$) of the totally symmetric (4.3.17a) (resp. totally antisymmetric (4.3.17b)) vectors. Now we intend to project the above mentioned orbits $O_\varphi \subset P(\mathcal{H}_N)$ into $P(\mathcal{H}_N^+)$, resp. into $P(\mathcal{H}_N^-)$. To make the procedure more transparent we shall divide it to more steps then it is, perhaps, necessary. For a $U(G_N)$-analytic vector $\varphi \in \mathcal{H}_N$ ($\varphi \neq 0$) let $\tilde{O}_\varphi := U(G_N)\varphi$, so that $O_\varphi := P\tilde{O}_\varphi$. We shall denote by $P : \mathcal{H}' \to P(\mathcal{H}')$, $\varphi \mapsto P_\varphi$, the natural projection in all the cases of $\mathcal{H}' := \mathcal{H}_N, \mathcal{H}_N^+, \mathcal{H}_N^-$. Let

$$\tilde{O}_\varphi^+ := P_+\tilde{O}_\varphi, \quad \tilde{O}_\varphi^- := P_-\tilde{O}_\varphi \text{ be subsets of } \mathcal{H}_N^+ \text{ (resp. } \mathcal{H}_N^-). \tag{4.3.18}$$

Assume, for definiteness, that $P_+\varphi \neq 0$, and *concentrate ourselves to the Bosonic case* (the formal procedures are similar with the fermions). Let K^φ be the stability group of φ with respect to $U(G_N)$. Considerations similar to those of Sect. 3.1 show that $\tilde{O}_\varphi$, as an immersed submanifold of $\mathcal{H}_N$, is diffeomorphic to G_N/K^φ. We shall consider $\tilde{O}_\varphi$ with the differentiable manifold structure of G_N/K^φ. The restricted mapping of P_+:

$$P_+^\varphi : \tilde{O}_\varphi \to \mathcal{H}_N^+, \ \psi \mapsto P_+\psi, \ \psi \in \tilde{O}_\varphi, \tag{4.3.19}$$

is (infinitely) differentiable. Hence the set

---

[4] This relation between spin and statistics can be obtained as a consequence of mathematical axiomatics of relativistic quantum field theory, cf. e.g. [301].

$$\tilde{O}^\circ_{\varphi+} := (P^\varphi_+)^{-1}(0) \subset \tilde{O}_\varphi \tag{4.3.20}$$

is closed in $\tilde{O}_\varphi$, and

$$\tilde{O}_{\varphi+} := \tilde{O}_\varphi \setminus \tilde{O}^\circ_{\varphi+} \text{ is a submanifold of } \tilde{O}_\varphi.$$

Each point of $P^\varphi_+ \tilde{O}_{\varphi+}$ has a well defined projection into $P(\mathcal{H}^+_N)$ and the mapping PP^φ_+,

$$PP^\varphi_+ : \tilde{O}_{\varphi+} \to P(\mathcal{H}^+_N), \ \varphi' \mapsto PP^\varphi_+\varphi' := \{\lambda P^\varphi_+\varphi' : \lambda \in \mathbb{C}\} \in P(\mathcal{H}^+_N), \tag{4.3.21}$$

is real analytic. The number $rg(\varphi') \in \mathbb{Z}_+ \ (\varphi' \in \tilde{O}_{\varphi+})$:

$$rg(\varphi') := rank \ T_{\varphi'}(PP^\varphi_+), \tag{4.3.22}$$

where $T_{\varphi'}$ is the tangent mapping in an arbitrarily chosen point $\varphi' \in \tilde{O}_{\varphi+}$, is given in some charts on $\tilde{O}_{\varphi+}$ around φ' and on $P(\mathcal{H}^+_N)$ around $PP^\varphi_+\varphi'$ as the dimension of the vector space[5] $T_{\varphi'}(PP^\varphi_+)[T_{\varphi'}\tilde{O}_{\varphi+}]$ (which is, roughly speaking, the maximal rank of submatrices of the mapping $T_{\varphi'}(PP^\varphi_+)$ in these charts with nonvanishing determinants). The function $\varphi' \mapsto rg(\varphi')$ is lower semicontinuous, and possesses only finite number of values. Hence for $m_\varphi := \max\{rg(\varphi') : \varphi' \in \tilde{O}_{\varphi+}\}$ the subset $\tilde{O}^m_{\varphi+}$ of $\tilde{O}_\varphi$ defined by:

$$\tilde{O}^m_{\varphi+} := rg^{-1}(m_\varphi) := \{\varphi' \in \tilde{O}_{\varphi+} : rg(\varphi') = m_\varphi\}, \tag{4.3.23}$$

is open, hence it is a submanifold of $\tilde{O}_\varphi$. We can assume that φ was chosen such, that $\varphi \in \tilde{O}^m_{\varphi+}$. Let, for any $\psi \in \tilde{O}^m_{\varphi+}$, the $\mathfrak{k}^\psi_0 \subset \mathfrak{g}_N$ (:= the **Lie algebra of** G_N) be the linear space consisting of those generators $\xi \in \mathfrak{g}_N$, for which

$$T_\psi(PP^\varphi_+)X_\xi\psi := i \left.\frac{d}{dt}\right|_{t=0} PP^\varphi_+ \exp(-itX_\xi)\psi = 0. \tag{4.3.24}$$

Clearly, $\dim \mathfrak{k}^\psi_0 = \dim G_N - m_\varphi$ is constant for all $\psi \in \tilde{O}^m_{\varphi+}$. The equation (4.3.24) is equivalent to the equation

$$(I_\mathcal{H} - P_{\psi^{(+)}})P_+X_\xi\psi = 0, \text{ with } \psi^{(+)} := P^\varphi_+\psi. \tag{4.3.25}$$

By the relation $\psi^{(\pm)} \in \mathcal{H}^\pm_N$ is **defined** the completely symmetric (resp. antisymmetric) part of the vector $\psi \in \mathcal{H}_N$. Let

---

[5]This vector space is, as could be seen from the formula, the image of the tangent space $T_{\varphi'}\tilde{O}_{\varphi+}$ by the tangent map of the mapping PP^φ_+.

$\mathfrak{m}_0^\varphi$ be a complementary subspace in $\mathfrak{g}_N$ to $\mathfrak{k}_0^\varphi$.

Since the mapping $P P_+^\varphi$ restricted to $\widetilde{O}_{\varphi+}^m$ is smooth and of constant rank m_φ, it is a *subimmersion* (compare [51, 5.10.6.]), hence **there is a manifold** $Z_{\varphi+}^m$ of dimension m_φ and a *submersion* $s_+^\varphi : \widetilde{O}_{\varphi+}^m \to Z_{\varphi+}^m$ as well as an *immersion* $i_+^\varphi : Z_{\varphi+}^m \to P(\mathcal{H}_N^+)$ such, that

$$P P_+^\varphi = i_+^\varphi \circ s_+^\varphi \text{ on } \widetilde{O}_{\varphi+}^m. \tag{4.3.26a}$$

This means, that the image $P P_+^\varphi(\widetilde{O}_{\varphi+}^m) \subset P(\mathcal{H}_N^+)$ can be considered as an *immersed submanifold* (with possible *selfintersections*) of $P(\mathcal{H}_N^+)$:

$$P P_+^\varphi(\widetilde{O}_{\varphi+}^m) = i_+^\varphi(Z_{\varphi+}^m). \tag{4.3.26b}$$

A basis of the tangent space to $Z_{\varphi+}^m$ is generated in the point $\nu := s_+^\varphi(\varphi)$ by curves $t \mapsto s_+^\varphi(\exp(-it X_\xi)\varphi$ with $\xi \in \mathfrak{m}_0^\varphi$. The image by $T_\nu i_+^\varphi$ of this tangent space in $T_{\varphi^{(+)}} P(\mathcal{H}_N^+)$ is generated by vectors which, in the chart $\Psi_{\varphi^{(+)}}$ (see 2.1.5, 2.1.8), have the form

$$T_{\varphi^{(+)}} \Psi_{\varphi^{(+)}}(v_\xi) := -i (I - P_{\varphi^{(+)}}) P_+ X_\xi \varphi, \quad \xi \in \mathfrak{m}_0^\varphi. \tag{4.3.27}$$

The values of the symplectic form Ω on $P(\mathcal{H}_N)$ on these vectors are:

$$\Omega_{\varphi^{(+)}}(v_\eta, v_\xi) = -2\|\varphi^{(+)}\|^{-2}\mathrm{Im}(P_+ X_\eta \varphi, (I - P_{\varphi^{(+)}})P_+ X_\xi \varphi). \tag{4.3.28}$$

The pull-back of Ω by i_+^φ makes $Z_{\varphi+}^m$ a manifold endowed with a canonical two-form. It is known, that the factorization of the subimmersion $P P_+^\varphi$ (together with the choice of the manifold $Z_{\varphi+}^m$) can be chosen in a canonical way, see [51, 5.10.7]. We assume here, that the mapping s_+^φ is onto (i.e. surjective), what is possible, because any submersion is an open mapping. The form $i_+^{\varphi*}\Omega$ on $Z_{\varphi+}^m$ is closed. The subset of $Z_{\varphi+}^m$ on which the form $i_+^{\varphi*}\Omega$ has its maximal rank is an open set, hence a submanifold $Z_{\varphi+}$ of $Z_{\varphi+}^m$. Denote by Ω_+° the restriction of $i_+^{\varphi*}\Omega$ onto $Z_{\varphi+}$. Since $d\Omega_+^\circ = 0$, the *characteristic bundle* of Ω_+° (consisting of vector fields on $Z_{\varphi+}$ annihilating the form Ω_+°) is an *integrable subbundle* of $T Z_{\varphi+}$, see e.g. [1, 5.1.2], determining a *natural foliation* of $Z_{\varphi+}$; any *leaf of this foliation* is an immersed connected submanifold of $Z_{\varphi+}$. Let M_φ^+ be the factor space obtained from $Z_{\varphi+}$ by its decomposition into the leaves of this foliation and let $p_M^+ : Z_{\varphi+} \to M_\varphi^+$ be the natural projection. If the equivalence relation on $Z_{\varphi+}$ given by classes identical with leaves $[p_M^+]^{-1}(x)$ $(x \in M_\varphi^+)$ is regular (see [51, 5.9.5]), then there is unique manifold structure on M_φ^+ such that p_M^+ is a submersion. In this case there is, on the malnifold M_φ^+, a unique symplectic form Ω_+^M satisfying

$$p_M^{+*}\Omega_+^M = \Omega_+^\circ. \tag{4.3.29}$$

The Proposition 3.2.10 is a special case of this assertion.

Note: In the above presented construction of the symplectic manifold $(M_\varphi^+, \Omega_+^M)$, we did not use any specific properties of the projector P_+ and of the group action $U(G_N)$. These properties enter in constructions of specific orbits.

4.3.7 We shall specify here the previous construction to the case of $G_N := N$-fold direct product of $2n + 1$—dimensional Heisenberg group G with infinite-dimensional unitary irreducible representations U in $\mathcal{H}$. The linear space $U_N(\mathfrak{g}_N)$ is spanned by elements[6]

$$X_\xi := \sum_{j=1}^N X_\xi^j \quad \text{with any } X_\xi^j \in U(\mathfrak{g}), \ \xi \in \mathfrak{g}_N, \tag{4.3.30}$$

where the index j has the following meaning: If $\varphi \in \mathcal{H}_N$ has the form

$$\varphi := \varphi_1 \otimes \varphi_2 \otimes \cdots \otimes \varphi_N, \tag{4.3.31}$$

then the linear operator X^j on $\mathcal{H}_N$ corresponds to an (equally denoted) operator on $\mathcal{H}$ by:

$$X^j \varphi := \varphi_1 \otimes \varphi_2 \otimes \cdots \otimes X^j \varphi_j \otimes \varphi_{j+1} \otimes \cdots \otimes \varphi_N. \tag{4.3.32}$$

(No summation! In this subsection all sums are explicitly indicated.)

Let us work in the Schrödinger realization of CCR, i.e. $\mathcal{H} = L^2(\mathbb{R}^n)$, $\mathcal{H}_N = L^2(\mathbb{R}^{Nn})$ and operators X_j^k ($k = 1, 2, \ldots N$; $j = 1, 2, \ldots 2n$) acting on the k-th copy of $L^2(\mathbb{R}^n)$ are chosen as in (4.1.5). Let $\varphi \in \mathcal{H}_N$ be given by (4.3.31) with $\varphi_j \in L^2(\mathbb{R}^n)$, supp $\varphi_j \cap$ supp $\varphi_k = \emptyset$ ($j \neq k$) and such, that there is a neighbourhood of unity $e \in G$ so that for any g_j ($j = 1, 2, \ldots N$) in this neighbourhood also $U(g_j)\varphi_j$ and $U(g_k)\varphi_k$ ($j \neq k$) have disjoint supports. We assume, moreover, that φ is a smooth function on $\mathbb{R}^{Nn}$. With these assumptions, we obtain from (4.3.28) in a neighbourhood of the point $s_\pm^\varphi(\varphi)$ on $Z_{\varphi\pm}^m$ (the following result shows that the mappings $P P_\pm^\varphi$ have at φ the maximal rank):

$$\Omega_{\varphi^{(\pm)}}(v_\eta, v_\xi) = i \sum_{j=1}^N (\varphi_j, [X_\eta^j, X_\xi^j]\varphi_j), \tag{4.3.33}$$

where we assumed for all the j : $\|\varphi_j\| = 1$, and X_η, X_ξ in (4.3.28) are of the form (4.3.30). The expression (4.3.33) shows, that $Z_{\varphi\pm} = M_\varphi^\pm$ is a $2Nn$-dimensional sym-

---

[6]For $\mathfrak{g}_N = \bigoplus_{j=1}^N \mathfrak{g}^{(j)}$, $\mathfrak{g}^{(j)}$ are copies of $\mathfrak{g}$, one has $\xi := \sum_{j=1}^N \xi_j$ with $\xi_j \in \mathfrak{g}^{(j)}$, $X_\xi^j := X_{\xi_j} \in U(\mathfrak{g})$.

plectic manifold. This means, that $Z_{\varphi\pm}$ for both signs are locally diffeomorphic (and symplectomorphic) to $M_\varphi = O_\varphi = \mathbb{R}^{2Nn} \subset P(\mathcal{H}_N)$ (Sect. 4.1). In a neighbourhood of $\varphi' \in O_\varphi$, the functions $\varphi' \mapsto Tr(P_{\varphi'} X_k^j)$ ($j = 1, 2, \ldots N$; $k = 1, 2, \ldots 2n$) are symplectic coordinates. Similarly, in a neighbourhood of $s_\pm^\varphi(\varphi)$ the functions

$$f_k^j : s_\pm^\varphi(\varphi') \mapsto Tr(P_{\varphi'} X_k^j) = (\varphi_j', X_k^j \varphi_j'), \; j = 1, 2, \ldots N; \; k = 1, 2, \ldots 2n, \tag{4.3.34}$$

are symplectic coordinate functions on $Z_{\varphi\pm}$.

Let us assume now, that φ_j's in (4.3.31) have the form

$$\varphi_j := W_{x^{(j)}} \varphi_0 \text{ for some } \varphi_0 \in L^2(\mathbb{R}^n), \; x^{(j)} \in \mathbb{R}^n, \tag{4.3.35}$$

assuming φ_0 to be smooth with compact support, and $x^{(j)} \neq x^{(k)}$ ($j \neq k$) such that φ_j, φ_k have mutually disjoint supports, see 4.1.3 for the notation. On the orbit $\widetilde{O}_\varphi$ in $\mathcal{H}_N$, there is also the point

$$(\otimes\varphi_0)^N := \varphi_0 \otimes \varphi_0 \otimes \cdots \otimes \varphi_0. \tag{4.3.36}$$

Choose now φ equal to (4.3.36) and calculate the values of (4.3.28) in the points $\varphi_\pm \in P(\mathcal{H}_N^\pm)$. In the antisymmetric case we obtain zero, since $P_-\varphi =: \varphi_- = 0$ (hence $\varphi \in \widetilde{O}_{\varphi_-}^\circ$, (4.3.20), and $PP_-\varphi$ is not defined).

In the case of Bose statistics we have:

$$\Omega_{\varphi^{(+)}}(v_\eta, v_\xi) = \frac{i}{N} \sum_{j=1}^{N} \sum_{k=1}^{N} (\varphi_0, [X_\eta^k, X_\xi^j]\varphi_0), \tag{4.3.37}$$

where X_η^j ($j = 1, 2, \ldots N$) should be considered as operators in $L^2(\mathbb{R}^n)$, ignoring the definition (4.3.32): they act on $L^2(\mathbb{R}^n)$ regardless of its order in the tensor product forming the whole Hilbert space $\mathcal{H}_N$. The rank of the form (4.3.37) equals to $2n$ and the point $PP_+^\varphi(\varphi)$ does not belong to $i_+^\varphi(Z_{\varphi+})$ for $N \geq 2$, i.e. φ is not mapped by s_+^φ into the symplectic manifold $Z_{\varphi+}$. We see that, although locally symplectomorhic to R^{2Nn}, the both classical phase spaces $Z_{\varphi-}$ and $Z_{\varphi+}$ of identical particles are globally different from the standard cotangent bundle $T^*\mathbb{R}^{Nn}$: in classical projections the Pauli exclusion principle holds for identical particles: regardless to the kind of their statistics.

4.3.8 With the notation from 4.3.6, let $V_N(G)$ be the unitary representation of G in $\mathcal{H}_N$ (reducible for $N \geq 2$) defined as the diagonal part of U_N:

$$V_N(g) := U_N(g \times g \times \cdots \times g), \text{ for all } g \in G. \tag{4.3.38}$$

The **Lie algebra** $V_N(\mathfrak{g})$ is generated by the basis of the form (4.3.30) with $X_\xi^j = X_\xi^k$ (considered as operators in $\mathcal{H}$) for all $j, k = 1, 2, \ldots N$, $\xi \in \mathfrak{g}$. Such operators

$X_\xi \in V_N(\mathfrak{g})$ commute with projectors $P_\pm$. Hence V_N leaves the subspaces $\mathcal{H}_N^+$ and $\mathcal{H}_N^-$ invariant, and we can obtain the classical projections of this 'macroscopic' (for large N) subsystem in the standard way, (Sect. 3.2); the obtained classical phase spaces are orbits of $Ad^*(G)$ with their canonical symplectic structure—there is no difference in the kinds of statistics, from the point of view of kinematics.

In trying to extend our constructions to systems consisting of infinite number $N \to \infty$ of equal (or identical) subsystems, we meet the problems of divergence of 'global (or collective) observables' $X_\xi^N := X_\xi$ and of discontinuity of the resulting representation V_∞ of G. We give a formalization of this 'large N limit' in the next Sect. 5.1, and in the Sect. 5.2 we outline a possible generalization of obtaining classical subsystems of collective observables from infinite quantal systems. We shall not take any care of statistics of subsystems, what could be motivated by results of the last two subsections: the statistics seems to have no essential influence upon the classical phase spaces of systems of identical particles.

Chapter 5
Macroscopic Limits

5.1 Multiple Systems

5.1.1 We shall construct in this chapter classical subsystems of a large quantal system. We shall assume here that the large system consists of infinite number of copies of a finite subsystem of the type dealt with in preceding chapters. The infinite "macroscopic" system is obtained as an inductive limit of a net of systems consisting of an increasing number of copies of the mentioned finite systems. The symmetry group G of a single finite subsystem is then also a symmetry group of the large system. An essential formal difference with respect to the systems discussed in preceding sections is that the action of G on the large system is not described by a continuous unitary representation, hence we cannot introduce generators corresponding to one-parameter subgroups of G as operators in some Hilbert space.

5.1.2 To make the following considerations more intuitive, let us come back for a while to finite systems consisting of N equal subsystems. Let the unitary representation $V_N(G)$ and its generators $X_\xi^N := X_\xi$ ($\xi \in \mathfrak{g}$) be defined as in 4.3.7 and 4.3.8, esp. in (4.3.30). Then

$$[X_\xi^N, X_\eta^N] = i\, X_{[\xi,\eta]}^N \quad (\xi, \eta \in \mathfrak{g}) \tag{5.1.1}$$

and the restriction to the orbit $O_\varphi^N := V_N(G)\varphi$ ($\varphi \in \mathcal{H}_N$) of the canonical symplectic form Ω^N on $P(\mathcal{H}_N)$ is determined by

$$\Omega_\varphi^N(\sigma_\xi, \sigma_\eta) = i\, Tr(P_\varphi[X_\xi^N, X_\eta^N]), \ (\xi, \eta \in \mathfrak{g}). \tag{5.1.2}$$

Here σ_ξ is the vector field on $P(\mathcal{H}_N)$ corresponding to the unitary flow

$$(t; \varphi) \mapsto \exp(-itX_\xi^N)\varphi, \quad \varphi \in \mathcal{H}_N, t \in \mathbb{R}. \tag{5.1.3}$$

For $N \to \infty$, the operators X_ξ^N diverge and $V_N(G)$ does not converge to any continuous unitary representation—compare the next subsection. Let

© Springer Nature Switzerland AG 2020, corrected publication 2020
P. Bóna, *Classical Systems in Quantum Mechanics*,
https://doi.org/10.1007/978-3-030-45070-0_5

$$X_{\xi N} := \frac{1}{N} X_{\xi}^N, \quad \xi \in \mathfrak{g}, \ N = 1, 2, \dots. \tag{5.1.4}$$

In terms of [155] X_{ξ}^N (resp. $X_{\xi N}$) are 'extensive (resp. intensive) observables' but, contrary to [155], they can be unbounded in our case. The limits for large N of $X_{\xi N}$'s could exist in some convenient sense, but they are not generators of any unitary representation of the group G. Due to the commutation relations

$$[X_{\xi N}, X_{\eta N}] = \frac{i}{N} X_{[\xi, \eta] N}, \tag{5.1.5}$$

the limits of $X_{\xi N}$ ($\xi \in \mathfrak{g}$) will be mutually commuting operators. To obtain correct classical commutation relations (i.e. the Poisson brackets, see 1.3.5) for functions $f_{\xi N}$ on the orbits O_x^N ($x \in \mathcal{H}_N$),

$$f_{\xi N} : x \mapsto f_{\xi N}(x) := Tr(P_x X_{\xi N}), \tag{5.1.6}$$

in the limit $N \to \infty$, the two-form Ω^N from (5.1.2) should be 'renormalized'. We define

$$\Omega_N := \frac{1}{N} \Omega^N. \tag{5.1.7}$$

The form Ω_N (if restricted onto the symplectic manifold M_x^N obtained from O_x^N as in Sect. 3.2) associates with the Hamiltonian function $f_{\xi N}$ the vector field σ_ξ (restricted to M_x^N) given by the flow (5.1.3). It is

$$\Omega_{N\bullet}(\sigma_\xi, \sigma_\eta) = i \, Tr(P_\bullet[X_\xi^N, X_{\eta N}]) = -Tr(P_\bullet X_{[\xi, \eta] N}). \tag{5.1.8}$$

We intend to develop a corresponding formalism for infinite systems, i.e. a suitable one for the work in the limit $N = $ 'actual infinity'.

5.1.3 Let $U(G)$ be a continuous unitary representation of a connected Lie group G on a separable Hilbert space $\mathcal{H}$. We shall use notation of Chap. 4 for concepts related to $U(G)$. Let Π be an index set (of arbitrary cardinality) and $\mathcal{H}_j$ ($j \in \Pi$) be copies of $\mathcal{H}$. Let us fix unitary maps

$$u_j : \mathcal{H} \to \mathcal{H}_j, \quad j \in \Pi, \tag{5.1.9}$$

of $\mathcal{H}$ onto $\mathcal{H}_j$'s. Let

$$\mathcal{H}_\Pi := \bigotimes_{j \in \Pi} \mathcal{H}_j \tag{5.1.10}$$

be the tensor product defined according to von Neumann [227] and known as *CTPS* (:= **complete tensor product space**—see also notes in the text in 5.1.4 below and [35, 106, 274]). For $\varphi_j \in \mathcal{H}_j$ let

$$\Phi := \bigotimes_{j \in \Pi} \varphi_j \tag{5.1.11}$$

be a *product-vector in* $\mathcal{H}_\Pi$. For any linear densely defined operator A on $\mathcal{H}$ (with domain $D(A) \subset \mathcal{H}$) and for $\varphi_j \in \mathcal{H}_j$ such that $u_j^{-1}\varphi_j \in D(A)$ let $\pi_j(A)$ be the operator on $\mathcal{H}_\Pi$ determined by

$$\pi_j(A)\Phi := \left(\bigotimes_{k \in \Pi \setminus \{j\}} \varphi_k \right) \otimes (u_j A u_j^{-1} \varphi_j). \tag{5.1.12}$$

Symbolically: $\pi_j(A) := I_1 \otimes I_2 \otimes \cdots \otimes I_{j-1} \otimes A \otimes I_{j+1} \otimes \ldots$, if $\Pi = \mathbb{Z}_+ \setminus \{0\}$.
Unitary group action U_Π of G on $\mathcal{H}_\Pi$ is determined by

$$U_\Pi(g)\Phi := \bigotimes_{j \in \Pi} (u_j U(g) u_j^{-1} \varphi_j). \tag{5.1.13}$$

For $|\Pi|$ ($:=$ the cardinality of Π) finite, the representation U_Π is strongly continuous with generators

$$X_\xi^\Pi := \sum_{j \in \Pi} \pi_j(X_\xi), \ \xi \in \mathfrak{g}. \tag{5.1.14}$$

U_Π is not weakly continuous in the case of infinite Π : If $\varphi \in \mathcal{H}$ is not an eigenvector of X_ξ, $\varphi_j := u_j \varphi$ for all $j \in \Pi$, $\|\varphi\| = 1$ and Φ is the corresponding product-vector (5.1.11) in $\mathcal{H}_\Pi$, then $\|\Phi\| = 1$ and

$$(\Phi, \ U_\Pi(\exp(t\xi))\Phi) = 0 \tag{5.1.15}$$

for all sufficiently small $|t| \neq 0, t \in \mathbb{R}$, since

$$|(\varphi_j, u_j \exp(-itX_\xi) u_j^{-1} \varphi_j)| = |(\varphi, \exp(-itX_\xi)\varphi)| < 1 \quad \text{if} \quad e^{-itX_\xi}\varphi \neq \lambda\varphi, \tag{5.1.16}$$

for any $\lambda \in \mathbb{C}$, i.e. the function in (5.1.15) is discontinuous at $t = 0$.

5.1.4 Notes on the structure of CTPS.

We shall not give here a thorough definition of **CTPS**. We shall assume that the definitions of (convergence and quasiconvergence of) infinite products and sums of complex numbers as well as of the scalar product in $\mathcal{H}_\Pi$ according to [227] are known to the reader. Let $z \in \mathbb{C}^\Pi$, i.e. z is a function

$$z : \Pi \to \mathbb{C}, \quad j \mapsto z_j. \tag{5.1.17}$$

Assume that $|z_j| = 1$ for all $j \in \Pi$ and define a unitary operator U_z on $\mathcal{H}_\Pi$ by its linear action on product vectors (5.1.11) (the set of which is total in $\mathcal{H}_\Pi$) given by

$$U_z \Phi := \bigotimes_{j \in \Pi} (z_j \varphi_j). \qquad (5.1.18)$$

Let $\{\varphi^n : n \in \mathbb{Z}_+\}$ be an orthonormal basis in $\mathcal{H}$. Let $a, b \in \mathbb{Z}_+^\Pi$ with components $a_j, b_j \in \mathbb{Z}_+$ ($j \in \Pi$), and set

$$\varphi_j^a := u_j(\varphi^{a_j}) \in \mathcal{H}_j, \ \Phi^a := \bigotimes_{j \in \Pi} \varphi_j^a. \qquad (5.1.19)$$

For $a \neq b$, the vectors Φ^a and Φ^b are mutually orthogonal: $(\Phi^a, \Phi^b) = 0$. Let $\Phi := \Phi^a$ for some a (this can be done so for any normalized product-vector $\Phi \in \mathcal{H}_\Pi$ by a choice of the identifications u_j, $j \in \Pi$, of $\mathcal{H}_j$ with $\mathcal{H}$). The vectors Φ^b, for which $b_j = a_j$ for all $j \in \Pi \backslash J_b$, $b_j \in \mathbb{Z}_+$ for all $j \in J_b$, where J_b runs over all *finite* subsets of Π,[1] form an orthonormal basis in a closed subspace of $\mathcal{H}_\Pi$ **denoted by $\mathcal{H}_\Pi^\Phi$ and called ITPS (incomplete tensor product space)**. Let P_Φ be the orthogonal projector in $\mathcal{H}_\Pi$ onto $\mathcal{H}_\Pi^\Phi$. For two arbitrary product vectors $\Phi, \Psi \in \mathcal{H}_\Pi$ the projectors P_Φ and P_Ψ are either orthogonal or equal. For any U_z from (5.1.18) we have

$$U_z P_\Psi U_z^* = P_{U_z \Psi}, \qquad (5.1.20)$$

and the product vectors Ψ and $U_z \Psi$ are **weakly equivalent**, cf. [227]. If $P_\Phi \Psi = \Psi$ (hence $P_\Psi \Phi = \Phi$), then Φ and Ψ are **(strongly) equivalent**. The set of all product vectors Φ weakly equivalent to a product vector Ψ form a total set in a closed subspace of $\mathcal{H}_\Pi$ with the **orthogonal projector** P_Ψ^w. Clearly, P_Ψ^w is the sum of all such P_Φ, which correspond to mutually *strongly inequivalent* product vectors Φ, all of them being weakly equivalent to Ψ. The sum of all mutually strongly inequivalent P_Ψ (we use an obvious licence in language) is the unit operator in $\mathcal{H}_\Pi$.[2]

Let $\mathfrak{A}^\Pi$ denotes the C^*-subalgebra of the algebra of all bounded operators on $\mathcal{H}_\Pi$ (denoted by $\mathcal{L}(\mathcal{H}_\Pi)$), generated by the elements

$$\{\pi_j(A) \in \mathcal{L}(\mathcal{H}_\Pi) : A \in \mathcal{L}(\mathcal{H}), j \in \Pi\}, \qquad (5.1.21)$$

where $\mathcal{L}(\mathcal{H})$ is the algebra of all bounded operators on the Hilbert space $\mathcal{H}$. For any $x \in \mathfrak{A}^\Pi$, the following relations are valid, [227]:

$$[x, P_\Psi] = [x, U_z] = 0 \text{ for all } U_z, \text{ and for all } P_\Psi, \qquad (5.1.22)$$

with U_z from (5.1.18). If p is another orthogonal projector in $\mathcal{L}(\mathcal{H}_\Pi)$, and for some product-vector Ψ it is $pP_\Psi = p$, then

$$\text{if } [x, p] = 0 \text{ for all } x \in \mathfrak{A}^\Pi \Rightarrow p = P_\Psi \text{ or } p = 0, \qquad (5.1.23)$$

---

[1] i.e. all the vectors Φ^b for which $b_j \neq a_j$ for finite number of indices $j \in \Pi$ only.

[2] We shall use sometimes projectors instead of the corresponding subspaces.

i.e. irreducibility of the action of $\mathfrak{A}^{\Pi}$ in each $\mathcal{H}_{\Pi}^{\Psi}$. The weak closure of $\mathfrak{A}^{\Pi}$ in $\mathcal{L}(\mathcal{H}_{\Pi})$ consists of all elements $x \in \mathcal{L}(\mathcal{H}_{\Pi})$ satisfying (5.1.22). The action of $\mathfrak{A}^{\Pi}$ in $\mathcal{H}_{\Pi}^{\Psi}$ is a representation of this C^*-algebra. Such representations (all irreducible and faithful) for two product vectors are unitarily equivalent iff these vectors are weakly equivalent. The *center* of the weak closure of $\mathfrak{A}^{\Pi}$ in $\mathcal{L}(\mathcal{H}_{\Pi})$ is generated by the projectors P_{Ψ}^{w}. **Denote** this weak closure by $\mathfrak{B}^{\#}$ and by $\mathfrak{Z}^{\#}$ its center: $x \in \mathfrak{Z}^{\#} \subset \mathfrak{B}^{\#}$ iff $[x, y] = 0$ for all $y \in \mathfrak{B}^{\#}$.

5.1.5 Proposition. *The mapping*

$$\sigma : G \to {}^*\text{-}Aut\,\mathfrak{A}^{\Pi}, \ g \mapsto \sigma_g, \tag{5.1.24}$$

defined by (see (5.1.13))

$$\sigma_g(x) := U_{\Pi}(g)xU_{\Pi}(g^{-1}), \ \forall x \in \mathfrak{A}^{\Pi}, \ g \in G, \tag{5.1.25}$$

is a group homomorphism of G into the group *-$Aut\,\mathfrak{A}^{\Pi}$ of *-automorphisms of the C^*-algebra $\mathfrak{A}^{\Pi}$. For any normalized vector $\Psi \in \mathcal{H}_{\Pi}$ define the vector state ω^{Ψ} on $\mathfrak{A}^{\Pi}$ by

$$\omega^{\Psi} : x \mapsto \omega^{\Psi}(x) := (\Psi, x\Psi). \tag{5.1.26}$$

The functions

$$g \mapsto \omega^{\Psi}(\sigma_g(x)) \tag{5.1.27}$$

for any $x \in \mathfrak{A}^{\Pi}$ and any $\Psi \in \mathcal{H}_{\Pi}$ are continuous functions from G to $\mathbb{C}$.

Proof. The mapping $A \mapsto U(g)AU(g^{-1})$ is a *-automorphism of $\mathcal{L}(\mathcal{H})$, $A \in \mathcal{L}(\mathcal{H})$. Since $\mathfrak{A}^{\Pi}$ is generated by elements $x := \pi_j(A)$ ($j \in \Pi$, $A \in \mathcal{L}(\mathcal{H})$) defined in (5.1.12) (i.e. $\mathfrak{A}^{\Pi}$ is the norm-closure of finite linear combinations of finite products of such elements), the first statement follows from the definition (5.1.13) of U_{Π}. The functions (5.1.27) are continuous for all $x = \pi_j(A)$ and all product states ω^{Ψ} (i.e. states corresponding via (5.1.26) to product vectors Ψ of the form (5.1.11)). The set of product vectors is total in $\mathcal{H}_{\Pi}$ and any *-automorphism of a C^*-algebra is norm-continuous. These facts imply by standard considerations validity of the last statement. $\qquad\square$

5.1.6 Note. *Due to weak discontinuity of U_{Π}, the second statement of 5.1.5 is not valid if $\mathfrak{A}^{\Pi}$ would be replaced by its weak closure $\mathfrak{B}^{\#}$ in $\mathcal{L}(\mathcal{H}_{\Pi})$. This can be seen by setting $\Psi := \Phi^a$ from (5.1.19) with $a_j := 0$ (for all $j \in \Pi$) and with a choice $\varphi^0 \in \mathcal{H}$ such that it is not an eigenvector of the generator X_{ξ} of $U(G)$ for some $\xi \in \mathfrak{g}$. Then, setting $x := P_{\Psi}^{w} \in \mathfrak{B}^{\#}$ in (5.1.26), the function*

$$t \mapsto \omega^{\Psi}(\sigma_{\exp(t\xi)}(P_{\Psi}^{w})) \tag{5.1.28}$$

is discontinuous at $t = 0$: For $t = 0$ its value equals to 1, but for arbitrarily small nonzero values of $t \in \mathbb{R}$ the values of (5.1.28) are found to be zero.

5.1.7 To simplify notations, we shall set $\Pi := \mathbb{Z}_+ \backslash \{0\}$ for the rest of the present section. For a densely defined linear operator A on $\mathcal{H}$ with domain $D(A)$, let

$$D^\Pi(A) := \bigotimes_{j \in \Pi} u_j D(A) \tag{5.1.29}$$

be the linear subset of $\mathcal{H}_\Pi$ consisting of finite linear combinations of product vectors Φ, (5.1.11), with $\varphi_j \in u_j D(A)$ ($j \in \Pi$). $D^\Pi(A)$ is not, in general, dense in $\mathcal{H}_\Pi$. Let

$$A_N := \frac{1}{N} \sum_{j=1}^N \pi_j(A), \quad (N \in \Pi), \tag{5.1.30}$$

be (densely defined) operators on $\mathcal{H}_\Pi$, a common domain of which contains $D^\Pi(A)$. Let $D_\Pi(A)$ be the set of vectors $\Psi \in \mathcal{H}_\Pi$ such, that

$$A_\Pi \Psi := norm\text{-} \lim_{N \to \infty} A_N \Psi \tag{5.1.31}$$

exists in $\mathcal{H}_\Pi$. The set $D_\Pi(A)$ is a nonzero linear subset of $\mathcal{H}_\Pi$: for $\varphi \in D(A)$ and $\varphi_j := u_j \varphi$ ($j \in \Pi$), the product vector Φ from (5.1.11) belongs to $D_\Pi(A)$. Let $\{\varphi^n : n \in \mathbb{Z}_+\} \subset D(A)$ be an orthonormal basis in $\mathcal{H}$ and, for some $a \in \mathbb{Z}_+^\Pi$, let Φ^a defined according to (5.1.19) belongs to $D_\Pi(A)$. Then, for $b \in \mathbb{Z}_+^\Pi$ differing from a in at most finite number of components, it is $\Phi^b \in D_\Pi(A)$. With $\Psi := \Phi^a$, such vectors Φ^b form an orthonormal basis in $\mathcal{H}_\Pi^\Psi$, hence $P_\Psi D_\Pi(A)$ is dense in $\mathcal{H}_\Pi^\Psi$, and (5.1.31) give a densely defined operator on $\mathcal{H}_\Pi^\Psi$. For any product vector $\Psi \in D_\Pi(A)$, let us define a densely defined operator on $\mathcal{H}_\Pi^\Psi$:

$$A^\Psi := P_\Psi A_\Pi P_\Psi = P_\Psi A_\Pi. \tag{5.1.32}$$

The second equality is a consequence of the obvious commutativity of A_Π with P_Ψ for any product vector $\Psi \in D_\Pi(A)$. The restriction of A_Π to the subspace $\mathcal{H}_\Pi^\Psi$ (which clearly is a linear, not densely defined operator on $\mathcal{H}_\Pi$) will be denoted by A_Π^Ψ, or simply A^Ψ ($\Psi \in D_\Pi(A)$). Now it is easy to prove

5.1.8 Lemma. *For a densely defined operator A on $\mathcal{H}$, let $\Psi \in D_\Pi(A)$ be a product vector in $\mathcal{H}_\Pi$. Then $A^\Psi = \lambda P_\Psi$ for some $\lambda \in \mathbb{C}$, on $D_\Pi(A)$.*

Proof. Since $\Psi \in D_\Pi(A)$ is a product vector, it is also $\Psi \in D^\Pi(A)$. We shall assume that Ψ is normalized. Then it can be written in the form

$$\Psi = \bigotimes_{j=1}^\infty \varphi_j, \quad \text{with } u_j^{-1} \varphi_j \in D(A) \text{ for } j = 1, 2, \ldots, \tag{5.1.33}$$

where each φ_j ($j \in \Pi$) is normalized in $\mathcal{H}_j$: $\|\varphi_j\|^2 = (\varphi_j, \varphi_j) = 1$. Let $\Psi_k \in D^\Pi(A)$ ($k = 1, 2$) be such product vectors in $\mathcal{H}_\Pi^\Psi$ which differ from (5.1.33) at most in the first n factors φ_j. Such vectors Ψ_k, with $n \in \Pi$, form a total set in $\mathcal{H}_\Pi^\Psi$. We have

$$(\Psi_1, A^\Psi \Psi_2) = \lim_{N \to \infty} \frac{1}{N} \left(\sum_{j=1}^n (\Psi_1, \pi_j(A)\Psi_2) + \sum_{j=n+1}^N (\Psi_1, \pi_j(A)\Psi_2) \right) =$$

$$= \lim_{N \to \infty} \frac{1}{N} \sum_{j=n+1}^N (\varphi_j, u_j A u_j^{-1} \varphi_j)(\Psi_1, \Psi_2) =$$

$$= \lim_{N \to \infty} \frac{1}{N} \sum_{j=1}^N (\Psi, \pi_j(A)\Psi)(\Psi_1, \Psi_2) = (\Psi, A^\Psi \Psi)(\Psi_1, \Psi_2). \quad (5.1.34)$$

By linearity, the obtained relation extends to all $\Psi_k \in P_\Psi D_\Pi(A)$. On that domain, we obtain

$$A^\Psi = Tr(P_\Psi^\circ A^\Psi)P_\Psi = Tr(P_\Psi^\circ A_\Pi)P_\Psi, \quad (5.1.35)$$

where P_Ψ° is the projector onto the one-dimensional subspace of $\mathcal{H}_\Pi^\Psi$ spanned by the vector Ψ. □

Note: Since A^Ψ is bounded on $\mathcal{H}_\Pi$ (if $\Psi \in D_\Pi(A)$ is a product-vector), we shall extend this operator to the whole $\mathcal{H}_\Pi$ by continuity and we shall denote this extension by the same symbol, hence: $A^\Psi \in \mathcal{L}(\mathcal{H}_\Pi)$.

5.1.9 Proposition. *Let $\Psi \in D_\Pi(\mathfrak{g})$ be an arbitrary vector from*

$$D_\Pi(\mathfrak{g}) := \bigcap_{\xi \in \mathfrak{g}} D_\Pi(X_\xi), \quad (5.1.36)$$

in the notation of 5.1.3 and 5.1.7. Then $U_\Pi(g)\Psi \in D_\Pi(\mathfrak{g})$, for all $g \in G$. In particular, with $g \cdot \Psi := U_\Pi(g)\Psi$, we have for product-vectors $\Psi \in D_\Pi(\mathfrak{g})$:

$$X_\xi^{g \cdot \Psi} = Tr(P_{g \cdot \Psi}^\circ X_{\xi\Pi})P_{g \cdot \Psi} = Tr(P_\Psi^\circ X_{Ad(g^{-1})\xi\Pi})P_{g \cdot \Psi}. \quad (5.1.37)$$

Proof. According to Lemma 3.1.4, $U(g)X_\xi U(g^{-1}) = X_{Ad(g)\xi}$ for any $\xi \in \mathfrak{g}$. Then, according to 5.1.3, we have also

$$U_\Pi(g^{-1})\pi_j(X_\xi)U_\Pi(g) = \pi_j(X_{Ad(g^{-1})\xi}). \quad (5.1.38)$$

For $\Psi \in D_\Pi(\mathfrak{g})$ there exist X_ξ^Ψ for all $\xi \in \mathfrak{g}$. Because of continuity of unitary operators $U_\Pi(g)$ for any fixed $g \in G$, there exist also the limits

$$\lim_{N \to \infty} U_\Pi(g)X_{Ad(g^{-1})\xi N}\Psi = U_\Pi(g)X_{Ad(g^{-1})\xi}^\Psi \Psi \quad (5.1.39)$$

for all $\xi \in \mathfrak{g}$. Rewriting the expression on the left hand side of (5.1.39) we get

$$U_\Pi(g)X_{Ad(g^{-1})\xi N}\Psi = \frac{1}{N}\sum_{j=1}^{N}U_\Pi(g)\pi_j(X_{Ad(g^{-1})\xi})\Psi = X_{\xi N}U_\Pi(g)\Psi. \quad (5.1.40)$$

This shows that the limit of the right hand side of (5.1.40) for large N exists for any $\xi \in \mathfrak{g}$, what proves the first assertion. The proof of the second assertion is a corollary of the proof of the first one for the case of a product vector $\Psi \in D_\Pi(\mathfrak{g})$, obtained from (5.1.35). $\qquad\qquad\qquad\qquad\qquad\qquad\qquad\qquad\qquad\qquad\qquad\qquad\qquad\qquad \square$

5.1.10 For a product vector $\Psi \in D_\Pi(\mathfrak{g})$, let ω^Ψ be the corresponding state on $\mathfrak{A}^\Pi$ defined in (5.1.26). We shall denote the obvious extension of this state to the unbounded observables $X_{\xi N}$ ($N \in \Pi$) by the same symbol. Then we have

$$\lim_{N\to\infty}\omega^\Psi(X_{\xi N}) = Tr(P_\psi^\circ X_{\xi\Pi}) =: \omega^\Psi(X_{\xi\Pi}). \quad (5.1.41)$$

We see that the value of expressions in (5.1.41) can be interpreted as the value of the intensive (unbounded) observable $X_{\xi\Pi}$ in the state ω^Ψ. Define the linear functional $F_\Psi \in \mathfrak{g}^*$ by

$$F_\Psi : \xi \mapsto F_\Psi(\xi) := Tr(P_\psi^\circ X_{\xi\Pi}), \text{ for product vectors } \Psi \in D_\Pi(\mathfrak{g}). \quad (5.1.42)$$

According to (5.1.37), the action $g \cdot F_\Psi := F_{g\cdot\Psi}$ of G coincides with the $Ad^*(G)$-action:

$$(g \cdot F_\Psi)(\xi) = F_{g\cdot\Psi}(\xi) = F_\Psi(Ad(g^{-1})\xi) = (Ad^*(g)F_\Psi)(\xi). \quad (5.1.43)$$

According to 5.1.9, the set of product vectors in $D_\Pi(\mathfrak{g})$ is $U_\Pi(G)$-invariant, hence any point of the orbit $G \cdot F_\Psi$ has the form (5.1.42).

Define the group homomorphism σ^* of G into the group of affine transformations of the state-space $\mathcal{S}(\mathfrak{A}^\Pi)$:

$$\sigma^* : G \to \sigma_G^*, \ g \mapsto \sigma_g^*, \text{ where } (\sigma_g^*\omega)(x) := \omega(\sigma_{g^{-1}}(x)) \quad (5.1.44)$$

for all $g \in G$, $\omega \in \mathcal{S}(\mathfrak{A}^\Pi)$ and $x \in \mathfrak{A}^\Pi$ with σ_g defined in (5.1.25). Let $\Psi \in D_\Pi(\mathfrak{g})$ be a product vector and

$$O_\Psi := \{\sigma_g^*\omega^\Psi : g \in G\} \subset \mathcal{S}(\mathfrak{A}^\Pi) \quad (5.1.45)$$

be the orbit through ω^Ψ of the action σ_G^*. For $\omega \in O_\Psi$ let

$$F_\omega \in \mathfrak{g}^* : F_\omega(\xi) := \omega(X_{\xi\Pi}). \quad (5.1.46)$$

Let us write also $g \cdot \omega := \sigma_g^*\omega$. Clearly $g \cdot \omega^\Psi := \omega^{g\cdot\Psi}$. According to (5.1.43), the mapping F from the state space into the dual $\mathfrak{g}^*$ of the Lie algebra:

$$\boldsymbol{F} : O_\Psi \to \mathfrak{g}^*, \ \omega \mapsto \boldsymbol{F}(\omega) := F_\omega, \qquad (5.1.47)$$

maps the orbit O_Ψ onto an orbit of $Ad^*(G)$. Let

$$[\omega] := \boldsymbol{F}^{-1}(F_\omega), \ \text{for } \omega \in O_\Psi, \ \text{be equivalence classes in } O_\Psi.$$

The corresponding factor space M_Ψ is mapped by $\boldsymbol{F}$ (which is constant on classes $[\omega]$) bijectively onto the orbit $G \cdot F_\Psi$. The last orbit is endowed by the *Kirillov-Kostant symplectic structure*. The functions f_ξ on M_Ψ:

$$[\omega] \mapsto f_\xi(\omega) := \omega(X_{\xi\Pi}), \ \omega \in O_\Psi, \ \xi \in \mathfrak{g}, \qquad (5.1.48)$$

are the Hamiltonian functions generating the flows

$$(t; [\omega]) \mapsto [\exp(t\xi) \cdot \omega]. \qquad (5.1.49)$$

Corresponding Poisson brackets are:

$$\{f_\xi, f_\eta\}([\omega]) = -F_\omega([\xi, \eta]), \ \xi, \eta \in \mathfrak{g}, \qquad (5.1.50)$$

compare e.g. (3.2.2). Here it is assumed that M_Ψ is endowed by the manifold structure of the $Ad^*(G)$-orbit $\boldsymbol{F}(M_\Psi)$. We have obtained here classical phase spaces from equivalence classes of states in $\mathcal{S}(\mathfrak{A}^\Pi)$ determined by the group action σ_G^*. Although the construction is formally parallel to that in the case of finite systems, there are certain physically significant differences in the interpretation, as mentioned in 1.1.6.

5.1.11 Let P_G be the orthogonal projector in $\mathcal{L}(\mathcal{H}_\Pi)$ onto the subspace of $\mathcal{H}_\Pi$ spanned by all product vectors $\Psi \in D_\Pi(\mathfrak{g})$. The operator P_G is equal to the sum of all mutually orthogonal projectors P_Ψ^w corresponding to the product vectors $\Psi \in D_\Pi(\mathfrak{g})$, as is seen from (5.1.35) and obvious commutativity of any A_Π with all the U_z, (5.1.18). Hence

$$P_G \in \mathfrak{Z}^\# := \text{ the center of } \mathfrak{B}^\#,$$
$$\mathfrak{B}^\# := (\mathfrak{A}^\Pi)'' := \text{ the weak operator closure of } \mathfrak{A}^\Pi \text{ in } \mathcal{L}(\mathcal{H}_\Pi)$$

(commas denote here the double commutant). The mapping

$$\rho : \mathfrak{A}^\Pi \to P_G \mathfrak{B}^\#, \ x \mapsto P_G x, \qquad (5.1.51)$$

is a *-representation of the C^*-algebra $\mathfrak{A}^\Pi$ in the Hilbert space $P_G \mathcal{H}_\Pi$.

The *representation* ρ can be uniquely extended to a W^*-**representation** of the W^*-algebra (i.e. abstract von Neumann algebra) $(\mathfrak{A}^\Pi)^{**} := \text{the double dual of } \mathfrak{A}^\Pi$, see [274, 1.21.13]. (The unique extensions of mappings from a C^*-algebra to mappings from its double dual will be usually denoted by the same symbols used for the original mappings.) The image of $(\mathfrak{A}^\Pi)^{**}$ under ρ is $P_G \mathfrak{B}^\#$. Let

$$s_G \in \mathfrak{Z} := \text{the center of } (\mathfrak{A}^\Pi)^{**},$$

be the support of ρ, i.e. $(I - s_G)(\mathfrak{A}^\Pi)^{**}$ is the kernel of ρ (I is here the identity of $(\mathfrak{A}^\Pi)^{**}$). The restriction ρ_G to $s_G(\mathfrak{A}^\Pi)^{**}$ of ρ is an isomorphism of W^*-algebras (which is σ - σ continuous, see [274, 1.21.13+4.1.23]). **Let $\mathcal{S}_\mathfrak{g} \subset \mathcal{S}(\mathfrak{A}^\Pi)$ consists of such states ω, the central supports $s_\omega \in \mathfrak{Z}$ of which are contained in s_G,** i.e. $s_\omega s_G = s_\omega$ (the **central support** of a state is defined as the central support, equiv. **central cover**—cf. [235, 3.8.1], [306], [274, 1.14.2], of the extension to $(\mathfrak{A}^\Pi)^{**}$ of the corresponding cyclic representation of $\mathfrak{A}^\Pi$). The set $\mathcal{S}_\mathfrak{g}$ will play an important role in the following.

The automorphisms σ_g ($g \in G$), (5.1.25), have unique extensions to automorphisms of the W^*-algebra $(\mathfrak{A}^\Pi)^{**}$, which are σ-σ and also norm–norm continuous, [274, 1.21.13]. The σ_g can be understood also as an (uniquely defined) automorphism of the von Neumann algebra $\mathfrak{B}^\#$. Due to Proposition 5.1.9, it is

$$\sigma_g(P_G) = P_G \quad \text{for all } g \in G, \tag{5.1.52}$$

hence also

$$\sigma_g(s_G) = s_G, \quad g \in G. \tag{5.1.53}$$

Let us keep the **notation** $X_{\xi\Pi}$ ($\xi \in \mathfrak{g}$) for the closures of the restrictions to $P_G \mathcal{H}_\Pi$ of operators denoted previously by the same symbols. According to (5.1.37), all the $X_{\xi\Pi}$'s have in $P_G \mathcal{H}_\Pi$ a common complete orthonormal set (a basis) of eigenvectors consisting of product vectors $\Psi \in D_\Pi(G)$, with real eigenvalues. Hence, they form a set of mutually commuting selfadjoint operators on $P_G \mathcal{H}_\Pi$. Let $E_{\xi\Pi}^\#(B)$ ($B :=$ any Borel subset of $\mathbb{R}$) be projectors forming their spectral measures $E_{\xi\Pi}^\#$. All these projectors belong to $P_G \mathfrak{Z}^\#$, since any $X_{\xi\Pi}$ ($\xi \in \mathfrak{g}$) is a constant on each $P_\Psi^w < P_G$. Define

$$E_{\xi\Pi}(B) := \rho_G^{-1}[E_{\xi\Pi}^\#(B)] \in s_G \mathfrak{Z} \quad \text{for all } \xi \in \mathfrak{g} \text{ and Borel } B \subset \mathbb{R}. \tag{5.1.54}$$

Any $E_{\xi\Pi}$ ($\xi \in \mathfrak{g}$) is a resolution of identity in the W^*-algebra $s_G \mathfrak{Z}$. Let us define also

$$E_{\xi\Pi}'(B) := E_{\xi\Pi}(B), \quad \text{if } B \text{ does not contain the zero } 0 \in \mathbb{R}, \tag{5.1.55}$$
$$:= E_{\xi\Pi}(B) + I - s_G, \quad \text{if } 0 \in B.$$

Here I is the identity of $\mathfrak{Z}$. Then $E_{\xi\Pi}'$ ($\xi \in \mathfrak{g}$) is a resolution of identity in $\mathfrak{Z}$.

5.1.12 Definition. *Let $\mathfrak{M}_G$ be the W^*-subalgebra of $\mathfrak{Z}$ generated by projectors $E_{\xi\Pi}'(B)$ ($\xi \in \mathfrak{g}$, B - Borel in $\mathbb{R}$) and by I. $\mathfrak{M}_G$ is called the **algebra of G-macroscopic observables** of the system $(\mathfrak{A}^\Pi, \sigma_G)$, or simply the $(G\text{-})$**macroscopic algebra**. Let $\mathfrak{N}_G := s_G \mathfrak{M}_G$ be the W^*-subalgebra of $\mathfrak{M}_G$ generated by projectors $E_{\xi\Pi}(B)$ and called the **algebra of G-definiteness** of $(\mathfrak{A}^\Pi, \sigma_G)$, or sometimes also the $(G\text{-})$macroscopic algebra, if there will be no confusion possible.*

5.1.13 Lemma. *Let ξ_j $(j = 1, 2, \ldots n := \dim G)$ form a basis in $\mathfrak{g}$. For $\lambda \in \mathbb{R}^n$ let $F := \sum_j \lambda_j F_j \in \mathfrak{g}^*$ expressed in the corresponding dual basis $\{F_j\} \subset \mathfrak{g}^*$. Let*

$$E_\mathfrak{g}(F) := E_{\xi_1\Pi}(\lambda_1) E_{\xi_2\Pi}(\lambda_2) \ldots E_{\xi_n\Pi}(\lambda_n) \in \mathfrak{N}_G. \qquad (5.1.56)$$

The projectors $E_\mathfrak{g}(F)$ $(F \in \mathfrak{g}^*)$ do not depend on a specific choice of the basis in $\mathfrak{g}$ and they are all minimal projectors in $\mathfrak{N}_G$. Here $E_{\xi\Pi}(\lambda) := E_{\xi\Pi}(\{\lambda\})$, and $E_g(F) := E_g(\{F\})$.

Proof. The restriction of the mapping ρ_G to $\mathfrak{N}_G$ is a W^*-isomorphism of $\mathfrak{N}_G$ into $P_G 3^\# \subset \mathfrak{B}^\#$. Let $\Psi \in \rho_G(E_\mathfrak{g}(F)) \mathcal{H}_\Pi$. From linearity of the mapping $\xi \mapsto X_{\xi\Pi}$ for $\xi = \sum \tau_j \xi_j$, we have

$$X_{\xi\Pi}\Psi = \sum_j \tau_j X_{\xi_j\Pi}\Psi = \sum_j \tau_j \lambda_j \Psi = F(\xi)\Psi. \qquad (5.1.57)$$

The second equality is due to the definition of $E_{\xi\Pi}(\lambda_j)$ as the projector corresponding to the eigenvalue $\lambda_j \in \mathbb{R}$ of $X_{\xi\Pi}$ (we write λ_j in the place of the one-point set $\{\lambda_j\}$ for simplicity). The last equality in (5.1.57) is due to definition of the dual basis and shows the stated independence of $E_\mathfrak{g}(F)$ on the choice of a basis.

Let

$$E_\mathfrak{g}^\#(F) := \rho_G(E_\mathfrak{g}(F)) \ .$$

Any projector $E_\mathfrak{g}^\#(B)$ is a sum (uncountable—in general, see also [274, 1.13.4]) of projectors $E_\mathfrak{g}^\#(F)$ $(F(\xi) \in B)$. The algebra $\rho_G(\mathfrak{N}_G)$ is the double commutant of the set

$$\{E_\mathfrak{g}^\#(F) \ : \ F \in \mathfrak{g}^*\}, \qquad (5.1.58)$$

according to the bicommutant theorem by von Neumann taken in the algebra $\mathcal{L}(P_G \mathcal{H}_\Pi)$ of bounded operators on $P_G \mathcal{H}_\Pi$. All the projectors $E_\mathfrak{g}^\#(F)$ in (5.1.58) are mutually orthogonal. The commutant of (5.1.58) contains all the orthogonal projectors $p \leq E_\mathfrak{g}^\#(F)$. But any nonzero orthogonal projector $q < E_\mathfrak{g}^\#(F)$ (strict inequality!) cannot commute with all such p's. Hence $E_\mathfrak{g}^\#(F)$ is minimal in $\rho_G(\mathfrak{N}_G)$ and $E_G(F)$ is minimal in $\mathfrak{N}_G$ for any $F \in \mathfrak{g}^*$. Since $E_{\xi\Pi}(\mathbb{R}) = s_G$ (= the identity of $\mathfrak{N}_G$) is a sum of $E_\mathfrak{g}(F)$'s and $\mathfrak{N}_G$ is commutative, the set of all the $E_\mathfrak{g}(F)$'s exhausts the set of all the minimal projectors in $\mathfrak{N}_G$. $\qquad\square$

5.1.14 Any state $\omega \in \mathcal{S}(\mathfrak{A}^\Pi)$ on the algebra of bounded observables of our system has unique extension to a normal state on the algebra $(\mathfrak{A}^\Pi)^{**}$ and its restriction to $\mathfrak{M}_G$ is a normal state $\omega \in \mathcal{S}(\mathfrak{M}_G)$. Any normal state on $\mathfrak{M}_G$ can be obtained in this way, [274, 1.24.5]. **Let $\mathcal{M}$ be the spectrum space of $\mathfrak{M}_G$**, i.e. the compact set of all pure states on $\mathfrak{M}_G$ endowed with the induced topology from the w^*-topology of its dual $\mathfrak{M}_G^*$. Then $\mathfrak{M}_G$ is isomorphic (denoted by $\sim$) to the C^*-algebra $C(\mathcal{M})$ of all complex valued continuous functions on $\mathcal{M}$ (by a Gel'fand-Najmark theorem, cf. [223, 16.2 Thm.1],[53, Thm.2.1.11A]): x $(\in \mathfrak{M}_G) \leftrightarrow \hat{x}$ $(\in C(\mathcal{M}))$. An element $x \in \mathfrak{M}_G$ is an orthogonal projector iff the corresponding element $\hat{x} \in C(\mathcal{M})$ is characteristic function of some Borel subset B of $\mathcal{M}$, i.e.

$$\hat{x}(m) = \chi_B(m) \text{ for all } m \in \mathcal{M}.$$

A pure state $m \in \mathcal{M}$ is normal, iff the characteristic function $\chi_{\{m\}}$ of the one-point set $\{m\}$ is continuous, $\chi_{\{m\}} \in C(\mathcal{M})$. This means, that *normal pure states* on $\mathfrak{M}_G$ are just the isolated points of $\mathcal{M}$. The corresponding projectors $\chi_{\{m\}}$ are *minimal projectors* in $\mathfrak{M}_G \sim C(\mathcal{M})$. The spectrum space $\mathcal{M}$ is Hausdorff and the family of **clopen** (i.e. closed and open) **sets** forms a basis of the topology of $\mathcal{M}$, cf. [274]. Hence, any minimal projector in $C(\mathcal{M})$ is of the form $\chi_{\{m\}}$.

Any state $\omega \in \mathcal{S}(\mathfrak{M}_G)$ is represented by a probability Baire (i.e. regular Borel) measure on $\mathcal{M}$ and any such measure μ_ω represents a state on $\mathfrak{M}_G : \omega(x) = \mu_\omega(\hat{x})$, where x in the left hand side is an element of the abstract algebra $\mathfrak{M}_G$ and $\hat{x}$ in the right hand side denotes the corresponding function $\hat{x} \in C(\mathcal{M})$. Any pure state $m \in \mathcal{M}$ corresponds to the Dirac measure δ_m.

5.1.15 The algebra $\mathfrak{M}_G$ (and also $\mathfrak{N}_G$) is σ_G−invariant:

$$\sigma_g x \in \mathfrak{N}_G \text{ for all } g \in G \text{ and any } x \in \mathfrak{N}_G. \tag{5.1.59}$$

This is a consequence of the relation (compare the proof of 5.1.9)

$$U_\Pi(g) X_{\xi\Pi} U_\Pi(g^{-1}) = X_{Ad(g)\xi\Pi}, \quad (g \in G,\ \xi \in \mathfrak{g}), \tag{5.1.60}$$

what implies

$$\sigma_g[E_\mathfrak{g}(B)] = E_{Ad(g)\xi\Pi}(B) \quad (g \in G \text{ and Borel } B \subset \mathbb{R}), \tag{5.1.61}$$

due to uniqueness of spectral measures of selfadjoint operators and also due to continuity properties of the used mappings. From (5.1.61), we obtain immediately (by calculation of the eigenvalues of $X_{\xi\Pi}$):

$$\sigma_g[E_\mathfrak{g}(F)] = E_\mathfrak{g}(Ad^*(g)F), \quad (g \in G,\ F \in \mathfrak{g}^*). \tag{5.1.62}$$

This specifies, according to 5.1.13 and 5.1.14, the action of G on the set of all normal pure states on $\mathfrak{N}_G$. The remaining normal pure state on $\mathfrak{M}_G$ corresponds to the σ_G-invariant minimal projector $I - s_G$. Hence, σ_G acts on $\mathfrak{M}_G$ as a group of W^*−automorphisms and σ_G^* acts on $\mathcal{M}$ (resp. on $\mathcal{S}(\mathfrak{M}_G)$) as a group of homeomorphisms (resp. a group of continuous affine transformations). As a consequence, the orbits

$$O_\omega := \{\sigma_g^* \omega :\ g \in G\} \subset \mathcal{S}(\mathfrak{A}^\Pi) \tag{5.1.63}$$

are canonically mapped onto orbits of σ_G^* in $\mathcal{S}(\mathfrak{M}_G)$ consisting of normal states on $\mathfrak{M}_G$. By this mapping orbits consisting of vector states ω^{Ψ} are [3] mapped onto orbits in $\mathcal{M}$. The functions

$$\sigma_m^* : G \to \mathcal{M}, \ g \mapsto \sigma_m^*(g) := \sigma_g^* m, \ (m \in \mathcal{M}) \qquad (5.1.64)$$

are not continuous in the given topology on $\mathcal{M}$, 5.1.14. The orbits of σ_G^* consisting of normal pure states on $\mathfrak{M}_G$ are, due to (5.1.62), bijective images of (some) orbits of $Ad^*(G)$ in $\mathfrak{g}^*$. It is also clear that the normal pure states on $\mathfrak{M}_G$ form a G-**invariant** **subset** $\mathcal{M}_*$ of all states $\mathcal{S}(\mathfrak{M}_G)$ on $\mathfrak{M}_G$:

$$\sigma_G^* \mathcal{M}_* = \mathcal{M}_*, \ i.e. \ m \in \mathcal{M}_* \Rightarrow \sigma_g^* m \in \mathcal{M}_* \text{ for all } g \in G \ (\sigma_e^* m \equiv m). \quad (5.1.65)$$

5.1.16 Proposition. *Let $p = p^* = p^2 \in \mathfrak{M}_G$ be any projector and*

$$p\mathfrak{g}^* := \{ F \in \mathfrak{g}^* : 0 \neq E_{\mathfrak{g}}(F) \leq p \}. \qquad (5.1.66)$$

Let $J \subset \mathfrak{g}^*$ be a finite set and let by p_J be denoted

$$p_J := \sum_{F \in J} E_{\mathfrak{g}}(F), \ \text{for any finite } J \subset \mathfrak{g}^*. \qquad (5.1.67)$$

Denote further for any subset $K \subset \mathfrak{g}^*$:

$$c(K) := \text{l.u.b.}\{ p_J : \ J \subset K, \ J \text{ finite} \}. \qquad (5.1.68)$$

Assume $p s_G = p$.

Then the following assertions are fulfilled:

(i) $p = c(p\mathfrak{g}^*)$, and (ii) $\mathcal{M} = \overline{\mathcal{M}}_* :=$ the closure of $\mathcal{M}_*$.

Proof. The projector s_G is constructed in such a way that $\rho_G(s_G) = P_G$ and $P_G = E_{\xi\Pi}^{\#}(\mathbb{R})$ for any $\xi \in \mathfrak{g}$. Since ρ_G is an isomorphism of $\mathfrak{N}_G = s_G \mathfrak{M}_G$ into $\mathfrak{Z}^{\#}$, 5.1.11, it is $s_G = c(s_G \mathfrak{g}^*)$. Hence, for any projector $q = q s_G$ in $\mathfrak{M}_G$, there is a nonzero minimal projector $E_{\mathfrak{g}}(F_o) = E_{\mathfrak{g}}(F_o)q$, if q is nonzero. Let $q := p - c(p\mathfrak{g}^*)$ (≥ 0, according to the definition (5.1.68)) and assume that $q \neq 0$. Let $0 \neq E_{\mathfrak{g}}(F_o) = q E_{\mathfrak{g}}(F_o)$. But $E_{\mathfrak{g}}(F_o) \leq p$, hence $E_{\mathfrak{g}}(F_o))c(p\mathfrak{g}^*) = E_{\mathfrak{g}}(F_o)$. This ia a contradiction, since q is orthogonal to $c(p\mathfrak{g}^*)$. Hence $q = 0$, what proves (i).

Any projector in $\mathfrak{M}_G$ is represented in $C(\mathcal{M})$ by the characteristic function of a **clopen set**, and conversely, the characteristic function of a clopen set in $\mathcal{M}$ represents by Gel'fand isomorphism a projector in $\mathfrak{M}_G$, 5.1.14. The minimal projector $E_{\mathfrak{g}}(F)$ corresponds to the one-point clopen set $\{m_F\}$ containing $m_F \in \mathcal{M}_*$. The union of

---

[3] Where $\Psi \in \mathcal{H}_{\Pi}$ such that there is an $F \in \mathfrak{g}^*$ satisfying: $E_{\mathfrak{g}}^{\#}(F)\Psi = \Psi$.

all $\{m_F\}$ ($F \in \mathfrak{g}^*$) is an open subset the closure of which is clopen, since $\mathcal{M}$ is a Stonean space, see 5.1.14, and [274]. According to (i), it is the support of characteristic function corresponding to $s_G = c(s_G \mathfrak{g}^*) = c(\mathfrak{g}^*)$. The projector s_G is the unit element in $\mathfrak{N}_G$ and the projector $I - s_G$ is minimal. This shows that the sum of the characteristic functions corresponding to s_G and $I - s_G$ is the characteristic function of the whole $\mathcal{M}$, i.e. $\mathcal{M}$ is the union of a one-point set $\{m_\circ\}$ corresponding to $I - s_G$ and of the closure of

$$\mathcal{N}_* := \mathcal{M}_* \setminus \{m_\circ\} = \{m_F : F \in \mathfrak{g}^*\}, \qquad (5.1.69)$$

where we set $\{m_F\} := \emptyset :=$ the empty set, if $E_{\mathfrak{g}}(F) = 0$. This is (ii). $\qquad\square$

Notation: Let us introduce, for following usage, some further concepts. Let

$$\mu_\xi^\omega : B \rightarrow \mu_\xi^\omega(B) := \omega(E_{\xi\Pi}(B)), \text{ for any } \omega \in \mathcal{S}(\mathfrak{M}_G) \text{ and Borel } B \subset \mathbb{R}, \ (5.1.70)$$

be a *finitely additive Borel measure* on $\mathbb{R}$. For mutually dual bases $\{\xi_j : j = 1, 2, \ldots n\}$ in $\mathfrak{g}$ and $\{F_j : j = 1, 2, \ldots n\}$ in $\mathfrak{g}^*$ define $\mu_{\mathfrak{g}}^\omega$ on $\mathfrak{g}^*$ by:

$$\mu_{\mathfrak{g}}^\omega(\boldsymbol{B}) := \omega(E_{\xi_1\Pi}(B_1)E_{\xi_2\Pi}(B_2)\ldots E_{\xi_n\Pi}(B_n)) \text{ for } \boldsymbol{B} := \{F \in \mathfrak{g}^* : F(\xi_j) \in B_j\}.$$
$$(5.1.71)$$

If $\xi \in L^1(\mu_{\mathfrak{g}}^\omega, \mathfrak{g}^*)$ with $\xi \in (\mathfrak{g}^*)^* = \mathfrak{g}$, then

$$\omega(X_{\xi\Pi}) := \mu_{\mathfrak{g}}^\omega(\xi) = \int \lambda \mu_\xi^\omega(\mathrm{d}\lambda). \qquad (5.1.72)$$

5.1.17 Lemma. *The image by the natural map defined in 5.1.14 of any factor state $\omega \in \mathcal{S}(\mathfrak{A}^\Pi)$ into $\mathcal{S}(\mathfrak{M}_G)$ is an equally denoted pure state $\omega \in \mathcal{M}_*$ ($:=$ the set of all normal pure states on $\mathfrak{M}_G$).*

Proof. The *canonical cyclic representation* $\{\pi_\omega, \mathcal{H}_\omega, \varphi_\omega\}$ of $\mathfrak{A}^\Pi$ (here φ_ω is the cyclic vector in the Hilbert space $\mathcal{H}_\omega$ for the representation π_ω such, that

$$\omega(x) = (\varphi_\omega, \pi_\omega(x)\varphi_\omega) \qquad (5.1.73)$$

for all $x \in \mathfrak{A}_\Pi$) corresponding to a **factor state** $\omega \in \mathcal{S}(\mathfrak{A}^\Pi)$ has trivial center. Hence, any projector in the center of the commutant $\pi_\omega(\mathfrak{A}^\Pi)'$ is trivial. The canonical extension to $(\mathfrak{A}^\Pi)^{**}$ (i.e. unique W^*-continuous) of π_ω maps the bidual $(\mathfrak{A}^\Pi)^{**}$ onto the double commutant $\pi_\omega(\mathfrak{A}^\Pi)''$ by which $\mathfrak{M}_G \subset 3$ is mapped into the center $\pi_\omega(3)$ of this bicommutant. Since $\pi_\omega(3) \subset \pi_\omega(\mathfrak{A}^\Pi)'$, any projector in $\pi_\omega(\mathfrak{M}_G)$ is trivial. The corresponding $\omega \in \mathcal{S}(\mathfrak{M}_G)$ is expressed by (5.1.73) for $x \in \mathfrak{M}_G$. This ω is normal: $\omega \in \mathcal{S}_*(\mathfrak{M}_G)$, hence there exists a unique projector s_ω in (the center of) $\mathfrak{M}_G$ such, that

$$\omega(x) = \omega(xs_\omega), \text{ for all } \{x \in \mathfrak{M}_G : \omega(x^*x) = 0\} \Rightarrow x = x(I - s_\omega). \quad (5.1.74)$$

Hence for any nonzero projector $s \leq s_\omega$ one has $\omega(s) \neq 0$ and $\pi_\omega(s) = I_\omega :=$ the identity of $\mathcal{L}(\mathcal{H}_\omega)$. From this follows $\omega(s_\omega - s) = 0$ and $s_\omega - s = (s_\omega - s)(I - s_\omega) = 0$, so that s_ω is a minimal projector in $\mathfrak{M}_G$. This proves that $\omega \in \mathcal{M}_*$. $\qquad\square$

5.1.18 For any state $\omega \in \mathcal{S}_\mathfrak{g}$, the measures μ_ξ^ω ($\xi \in \mathfrak{g}$) are probability (σ-additive) regular Borel measures on $\mathbb{R}$, due to normality of $\omega \in \mathcal{S}(\mathfrak{M}_G)$, (5.1.70). Define the subset $\mathcal{S}_\mathfrak{g}^d \subset \mathcal{S}(\mathfrak{A}^\Pi)$:

$$\mathcal{S}_\mathfrak{g}^d := \{\omega \in \mathcal{S}_\mathfrak{g} : \omega(X_{\xi\Pi}) \text{ is finite for all } \xi \in \mathfrak{g}\}, \qquad (5.1.75)$$

where $\omega(X_{\xi\Pi})$ is defined in (5.1.72). Due to (5.1.61), the set $\mathcal{S}_\mathfrak{g}^d$ is σ_G^*−invariant. For any $f \in L^1(\mathbb{R}, \mu_\xi^\omega)$ define

$$\omega(f(X_{\xi\Pi})) := \int_\mathbb{R} f(\lambda) \, \omega(E_{\xi\Pi}(d\lambda)). \qquad (5.1.76)$$

Any state $\omega \in \mathcal{S}_\mathfrak{g}$ which is mapped into $\mathcal{M}_*$, e.g. any pure state $\omega \in \mathcal{S}_\mathfrak{g}$, belongs to $\mathcal{S}_\mathfrak{g}^d$ and, moreover,

$$\omega(X_{\xi\Pi}^2) = [\omega(X_{\xi\Pi})]^2 \text{ for all } \xi \in \mathfrak{g}. \qquad (5.1.77)$$

Denote $F_\omega(\xi) := \omega(X_{\xi\Pi})$ for $\omega \in \mathcal{S}_\mathfrak{g}^d$. The mapping

$$\boldsymbol{F} : \mathcal{S}_\mathfrak{g}^d \to \mathfrak{g}^*, \ \omega \mapsto \boldsymbol{F}(\omega) := F_\omega; \ F_\omega(\xi) := \omega(X_{\xi\Pi}), \quad \xi \in \mathfrak{g}, \qquad (5.1.78)$$

maps orbits of σ_G^* in $\mathcal{S}_\mathfrak{g}^d$ onto orbits of $Ad^*(G)$ in $\mathfrak{g}^*$. Let $\omega \in \mathcal{S}_\mathfrak{g}^d$ **and** $O_\omega := \sigma_G^* \omega$ **be the corresponding orbit**. If (5.1.77) is valid for ω then it is valid for all the states in O_ω, as it is seen from (5.1.61). We shall call orbits $O_\omega \subset \mathcal{S}_\mathfrak{g}^d$ satisfying (5.1.77) the **G-macroscopically pure orbits**, and similarly for single states; simply, we shall use also **(G-)pure orbits** (resp. **G-pure states**). The

set of all G-pure states will be denoted by $\mathcal{E}_\mathfrak{g}$ ($\subset \mathcal{S}_\mathfrak{g}^d$) .

The state $\omega \in \mathcal{E}_\mathfrak{g}$ need not be a (pure) state in $\mathcal{ES}(\mathfrak{A}^\Pi)$ or in $\mathcal{ES}_\mathfrak{g}$. But the following assertion is valid:

5.1.19 Proposition. *For $\omega \in \mathcal{S}_\mathfrak{g}^d$ and its canonical image $\omega \in \mathcal{S}(\mathfrak{M}_G)$ the following statements are equivalent:*

(i) $\omega \in \mathcal{E}_\mathfrak{g}$; (ii) $\omega \in \mathcal{M}_*$.

Proof. The implication (ii) $\Rightarrow$ (i) is clear. Let $\omega \in \mathcal{E}_\mathfrak{g}$ and let μ_ω be the Baire measure on $\mathcal{M}$ corresponding to $\omega \in \mathcal{S}_*(\mathfrak{M}_G)$. We shall prove that μ_ω is concentrated on a one point set $\{\omega\} \subset \mathcal{M}_*$. Let $B_n \subset \mathbb{R}$, $n \in \mathbb{Z}_+$, be an increasing absorbing sequence of Borel sets, i.e. for any bounded Borel $B \subset \mathbb{R}$ there is some $n_B \in \mathbb{Z}_+$ that for all

$n \geq n_B$ it is $B \subset B_n$, and $B_n \subset B_{n+1}$ for $n \in \mathbb{Z}_+$. If $f : \mathbb{R} \to \mathbb{C}$ is any Borel function which is uniformly bounded on each bounded Borel subset B of $\mathbb{R}$, then

$$E_{\xi\Pi}(B)f(X_{\xi\Pi}) := \int_B f(\lambda)E_{\xi\Pi}(d\lambda) = f(E_{\xi\Pi}(B)X_{\xi\Pi}) \qquad (5.1.79)$$

is a well defined element of $\mathfrak{N}_G$, [274, 1.11.3]. Since ω is normal, we can write for such 'locally finite' functions $f \in L^1(\mathbb{R}, \mu_\xi^\omega)$:

$$\omega(f(X_{\xi\Pi})) = \lim_{n \to \infty} \omega(E_{\xi\Pi}(B_n)f(X_{\xi\Pi})), \qquad (5.1.80)$$

$$\omega(E_{\xi\Pi}(B)f(X_{\xi\Pi})) = \int_{\mathcal{M}} m(E_{\xi\Pi}(B)f(X_{\xi\Pi}))\mu_\omega(dm), \qquad (5.1.81)$$

$$m(E_{\xi\Pi}(B)f(X_{\xi\Pi})) = m(E_{\xi\Pi}(B))m(E_{\xi\Pi}(B)f(X_{\xi\Pi})); \qquad (5.1.82)$$

in (5.1.82) we have used the character-property of $m \in \mathcal{M} := \mathcal{ES}(\mathfrak{M}_G)$.

For $n \in \mathbb{Z}_+$, the function $\chi_{\xi n} : m \mapsto m(E_{\xi\Pi}(B_n))$ is continuous characteristic function of a clopen set $\mathcal{M}_{\xi n} \subset \mathcal{M}$. From the monotonicity property of spectral measures, we have $\mathcal{M}_{\xi(n+1)} \supset \mathcal{M}_{\xi n}$. The union

$$\bigcup_{n \in \mathbb{Z}_+} \mathcal{M}_{\xi n} =: \mathcal{M}_\xi \qquad (5.1.83)$$

is open, hence measurable together with all the $\mathcal{M}_{\xi n}$. We see from (5.1.80), (5.1.81) and (5.1.82) that μ_ω is concentrated on $\mathcal{M}_\xi$:

$$\mu_\omega(\mathcal{M}_\xi) = \mu_\omega(\mathcal{M}) = 1, \quad \forall \xi \in \mathfrak{g}; \qquad (5.1.84)$$

it suffices to set for f a (nonzero) constant function. But

$$\mathcal{M}_{\mathfrak{g}} := \bigcap_{\xi \in \mathfrak{g}} \mathcal{M}_\xi = \bigcap_{j=1}^n \mathcal{M}_{\xi_j} \supset \mathcal{M}_* \backslash \{m_\circ\} = \mathcal{N}_*, \qquad (5.1.85)$$

where $\{\xi_j : j = 1, 2, \ldots n\}$ is a basis of $\mathfrak{g}$, 5.1.13, and μ_ω is concentrated on states in $\mathcal{M}_{\mathfrak{g}}$,

$$\mu_\omega(\mathcal{M}_{\mathfrak{g}}) = \mu_\omega(\mathcal{M}) = 1, \text{ for any } \omega \in \mathcal{S}_*(\mathfrak{M}_G). \qquad (5.1.86)$$

Let

$$F_\xi : \mathcal{M}_\xi \to \mathbb{R}, \ m \mapsto F_\xi(m) := F_m(\xi) := \lim_n m(E_{\xi\Pi}(B_n)X_{\xi\Pi}), \qquad (5.1.87)$$

what is a bounded continuous function on each $\mathcal{M}_{\xi n}$, and due to monotonicity it is continuous on the whole $\mathcal{M}_\xi$. For f in (5.1.80) we have:

$$m(E_{\xi\Pi}(B_n)f(X_{\xi\Pi})) = f(F_m(\xi)) \text{ for } m \in \mathcal{M}_{\xi n}, \tag{5.1.88}$$

if for $\lambda = F_m(\xi)$ the value $f(\lambda)$ is defined. From (5.1.81), one sees that the functions $m \mapsto \chi_{\xi n}(m)f(F_m(\xi))$ are in $L^1(\mathcal{M}, \mu_\omega)$. By an application of the Beppo-Levi theorem to their absolute values, we obtain:

The functions $F_\xi^* f \in L^1(\mathcal{M}, \mu_\omega)$; here it is

$$F_\xi^* f := f \circ F_\xi : \mathcal{M}_\xi \to \mathbb{R}, \quad m \mapsto f(F_\xi(m)) = f(F_m(\xi)). \tag{5.1.89}$$

We have used here (5.1.80) and (5.1.81). After a subsequent application of the Lebesgue dominated convergence theorem we arrive at:

$$\omega(f(X_{\xi\Pi})) = \mu_\omega(F_\xi^* f) := \int_{\mathcal{M}} f(F_m(\xi)) \, \mu_\omega(dm). \tag{5.1.90}$$

The relation (5.1.77) is valid due to (i). This means that the functions $f_1(\lambda) := \lambda$, $f_2(\lambda) := \lambda^2$, $(\lambda \in \mathbb{R})$, are both in $L^1(\mathbb{R}, \mu_\xi^\omega)$ for all $\xi \in \mathfrak{g}$ and for $f := f_j$ $(j = 1, 2)$ (5.1.90) is valid. Hence $F_\xi \in L^2(\mathcal{M}, \mu_\omega)$ for all $\xi \in \mathfrak{g}$ and, due to (5.1.77), we have

$$(F_\xi, F_\xi) = (F_\xi, \mathbf{1})(\mathbf{1}, F_\xi), \quad \text{for all } \xi \in \mathfrak{g}. \tag{5.1.91}$$

The brackets denote here here the scalar product in $L^2(\mathcal{M}, \mu_\omega)$ and $\mathbf{1} \in L^2(\mathcal{M}, \mu_\omega)$ is the function identically equal to one: $\mathbf{1}(m) := 1$ for all $m \in \mathcal{M}$. Applying the Schwarz inequality to (5.1.91), we obtain:

$$F_\xi = const. = (\mathbf{1}, F_\xi)\mathbf{1} = F_\omega(\xi)\mathbf{1}, \quad \mu_\omega\text{-a.e. for all } \xi \in \mathfrak{g}. \tag{5.1.92}$$

This means that **the function**

$$F_\mathfrak{g} : \mathcal{M}_\mathfrak{g} \to \mathfrak{g}^*, \quad m \mapsto F_m, \tag{5.1.93}$$

is constant μ_ω**-almost everywhere**, too. The restriction of $F_\mathfrak{g}$ to the set of normal states $\mathcal{N}_*$ separates points in $\mathcal{N}_*$ according to 5.1.13 and (5.1.87). Hence the set $F_\mathfrak{g}^{-1}(F_\omega) \subset \mathcal{M}_\mathfrak{g}$ contains at most one $m \in \mathcal{N}_*$. Due to continuity of $F_\mathfrak{g}$, the set $F_\mathfrak{g}^{-1}(F_\omega)$ is closed in $\mathcal{M}_\mathfrak{g} = \mathcal{M}_\mathfrak{g}^\circ$ (:= the interior of $\mathcal{M}_\mathfrak{g}$), what implies measurability of $F_\mathfrak{g}^{-1}(F_\omega)$. Due to (5.1.92):

$$\mu_\omega(F_\mathfrak{g}^{-1}(F_\omega)) = \mu_\omega(\mathcal{M}) = 1. \tag{5.1.94}$$

It is known, see e.g. [274], that for any $\omega \in \mathcal{S}_*(\mathfrak{M}_G)$ there is a unique projector $s_\omega \in \mathfrak{M}_G$ such that $\omega(x) = \omega(xs_\omega)$ for all $x \in \mathfrak{M}_G$ and $\omega(x^*x) = 0$ implies $xs_\omega = 0$. The characteristic function in $C(\mathcal{M})$ corresponding to s_ω is supported by the clopen set $\operatorname{supp}\mu_\omega \subset \mathcal{M}$. Since it is nonempty, it contains some $m \in \mathcal{M}_*\backslash\{m_\circ\} = \mathcal{N}_*$, and all these m's are contained in $F_{\mathfrak{g}}^{-1}(F_\omega)$ due to (5.1.85). Hence the clopen set supp μ_ω contains exactly one point of $\mathcal{M}_*$ which means, according to Proposition 5.1.16, that supp μ_ω is a one point subset of $\mathcal{M}_*$ and $s_\omega = E_{\mathfrak{g}}(F_\omega)$. This proves the implication (i) $\Rightarrow$ (ii). $\qquad\qquad\square$

5.1.20 Corollary. $\sigma_G^* \mathcal{E}_{\mathfrak{g}} = \mathcal{E}_{\mathfrak{g}}$, i.e. $\mathcal{E}_{\mathfrak{g}}$ is σ_G^*-invariant (:= 'G-invariant').

Proof. According to (5.1.62) and Lemma 5.1.13, the set $\mathcal{N}_*$ is G-invariant. The action of G (via σ_G^*) commutes with the mapping ω ($\in \mathcal{S}(\mathfrak{A}^\Pi)) \mapsto \omega$ ($\in \mathcal{S}_*(\mathfrak{M}_G)$). Then the result is immediate after an application of 5.1.19. $\qquad\qquad\square$

5.1.21 Proposition. *For any $\omega \in \mathcal{S}_*(\mathfrak{M}_G)$, the corresponding probability Radon measure μ_ω on $\mathcal{M}$ is supported by $\mathcal{M}_*$:*

$$\mu_\omega(\mathcal{M}_*) = \mu_\omega(\mathcal{M}) = 1. \qquad (5.1.95)$$

Proof. We can assume that $s_G s_\omega = s_\omega$ for the support projector s_ω of ω. We have, according to 5.1.16 (i), $s_\omega = c(s_\omega \mathfrak{g}^*)$. Due to normality of ω, it is

$$1 = \omega(s_\omega) = \mathrm{l.u.b.}\{\omega(p_J): \ p_J := \sum_{F \in J} E_{\mathfrak{g}}(F), \text{ finite } J \subset s_\omega \mathfrak{g}^*\}. \qquad (5.1.96)$$

Let $m_F \in \mathcal{M}_*$ ($F \in \mathfrak{g}^*$, $E_{\mathfrak{g}}(F) \neq 0$) be defined by $m_F(E_{\mathfrak{g}}(F)) = 1$. For any subset $K \subset \mathfrak{g}^*$, the open set (which is clopen for finite K)

$$\mathcal{M}(K) := \{m_F \in \mathcal{M}_* : \ F \in K\}, \ m_F \text{ is void if } E_{\mathfrak{g}}(F) = 0, \qquad (5.1.97)$$

is μ_ω–measurable. But $\omega(p_J) = \mu_\omega(\mathcal{M}(J))$, and μ_ω is regular. Hence,

$$1 = \mathrm{l.u.b.}\{\mu_\omega(\mathcal{M}(J)): \ J \subset s_\omega \mathfrak{g}^* \text{ finite}\} \leq \mu_\omega(\mathcal{M}(s_\omega \mathfrak{g}^*)) \leq \mu_\omega(\mathcal{M}_*) \leq 1,$$
$$(5.1.98)$$

what proves (5.1.95). $\qquad\qquad\square$

5.1.22 Lemma. *Any uniformly bounded function on $\mathcal{M}_*$ with values in $\mathbb{C}$ can be uniquely extended to a continuous function on $\mathcal{M}$, i.e. the spectrum space $\mathcal{M}$ of $\mathfrak{M}_G$ is the Stone-Čech compactification of the discrete space $\mathcal{M}_*$ of normal pure states on $\mathfrak{M}_G$.*

Proof. Since $\mathcal{M}_*$ is discrete, $C(\mathcal{M}_*)$ consists of all bounded complex valued functions on $\mathcal{M}_*$. The Stone-Čech compactification of a normal topological space $\mathcal{S}$ is a compact Hausdorff space $\mathcal{S}'$ and a homeomorphism τ of $\mathcal{S}$ into $\mathcal{S}'$ such, that $\tau(\mathcal{S})$ is dense in $\mathcal{S}'$ and any $f \in C(\mathcal{S})$ can be continued to some $\tilde{f} \in C(\mathcal{S}')$. It is clear, that the continuation $\tilde{f}$ is uniquely determined by f.

Let $f \in C(\mathcal{M}_*)$, $f \geq 0$. For any $\iota \in [0, \|f\|]$ (:= closed interval in $\mathbb{R}$) define (cf. (5.1.68)) $p_\iota := 0$ for $\iota \notin sp(f)$ and (let $f(m_\circ) = 0$):

$$p_\iota := c(\{F_m \in \mathfrak{g}^* : \ f(m) = \iota, \ m \in \mathcal{M}_*\}), \quad \iota \in sp(f), \qquad (5.1.99)$$

where $sp(f)$ denotes the spectrum of f. For any finite subset $J \subset sp(f)$ define

$$x_J := \sum_{\iota \in J} \iota \, p_\iota \in \mathfrak{M}_G. \tag{5.1.100}$$

The finite subsets J of $sp(f)$ are directed by inclusion and the net $\{x_J : \text{finite } J \subset sp(f)\}$ is increasing. Any increasing net of selfadjoint elements of a W^*-algebra $\mathfrak{M}$ converges to its least upper bound in $\mathfrak{M}$, [274, 1.7.4]. Let $x_f \in \mathfrak{M}_G$ be the limit of $\{x_J\}$. We claim that the function $\tilde{f} \in C(\mathcal{M})$, $\tilde{f}(m) := m(x_f)$ coincides with f on $\mathcal{M}_*$.

Let $m \in \mathcal{M}_*$. Then, due to normality of m,

$$\tilde{f}(m) = \text{l.u.b.}\{m(x_J) : \text{finite } J \subset sp(f)\} = \text{l.u.b.}\{\sum_{\iota \in J} \iota m(p_\iota) : \text{finite } J \subset sp(f)\}. \tag{5.1.101}$$

But $m \in \mathcal{M}_*$ lies in support of the characteristic function $m \mapsto m(p_\iota)$ iff $f(m) = \iota$, compare 5.1.16. Hence $\tilde{f}(m) = f(m)$, what we intended to prove. □

5.1.23 Lemma. *For a finitely additive probability measure μ on $\mathfrak{g}^*$ (without any specification of a Σ-algebra of measurable subsets in $\mathfrak{g}^*$) supported by $s_G \mathfrak{g}^*$, (5.1.66), the following assertions are equivalent:*

(i) $F_\mathfrak{g}^* \mu = \mu_\omega$ on $\mathcal{N}_*$ for some $\omega \in \mathcal{S}_*(\mathfrak{N}_G)$, i.e. $\mu = \mu_\omega \circ F_\mathfrak{g}^{-1}$.

(ii) μ is supported by a countable subset of $\mathfrak{g}^*$.

If these conditions are fulfilled, μ is σ-additive. Any μ_ω ($\omega \in \mathcal{S}_*(\mathfrak{N}_G)$) is of the form $F_\mathfrak{g}^* \mu$ for some σ-additive probability Borel measure μ on $\mathfrak{g}^*$ with at most countable supporting set in $s_G \mathfrak{g}^*$.

Proof. $\mathcal{N}_*$ is mapped bijectively by $F_\mathfrak{g}$ onto $s_G \mathfrak{g}^*$ and μ_ω is supported by $\mathcal{N}_*$ for all $\omega \in \mathcal{S}_*(\mathfrak{N}_G)$. Hence (i) is fulfilled for $\mu := \mu_\omega \circ F_\mathfrak{g}^{-1}$. Complete additivity of μ_ω (what is a consequence of normality of ω) leads then to the expression

$$\mu_\omega = \sum_{m \in \mathcal{N}_*} \omega(E_\mathfrak{g}(F_m)) \, \delta_m, \quad (\delta_m := \text{Dirac measure at } m). \tag{5.1.102}$$

Hence at most countable number of coefficients $\omega(E_\mathfrak{g}(F_m)) \neq 0$. This proves (i) $\Rightarrow$ (ii) as well as the last assertion of the Lemma. Let

$$\mu = \sum_{j \in \mathbb{Z}_+} \lambda_j \delta_{F_j}, \text{ with } \lambda_j \geq 0, \sum_j \lambda_j = 1, \ F_j \in s_G \mathfrak{g}^*. \tag{5.1.103}$$

Then $F_\mathfrak{g}^* \mu := \mu \circ F_\mathfrak{g}$ is a Baire measure on $\mathcal{N}_*$, hence represents a (normal) state ω on $\mathfrak{N}_G$. The σ-additivity is clear. □

5.1.24 Lemma. *Let, for $\omega \in \mathcal{S}(\mathfrak{N}_G)$, $\mu_\mathfrak{g}^\omega$ be the additive function of Borel subsets of $\mathfrak{g}^*$ defined in (5.1.71). Then $\mu_\mathfrak{g}^\omega$ has a unique extension to a finitely additive probability* *measure on the set of all subsets of $s_G \mathfrak{g}^*$. Conversely, any finitely additive probability measure on $s_G \mathfrak{g}^*$ is of the form $\mu_{\mathfrak{g}}^\omega$ for some $\omega \in \mathcal{S}(\mathfrak{N}_G)$.*

Proof. For any subset $K \subset \mathfrak{g}^*$ define, (5.1.68),

$$E_{\mathfrak{g}}(K) := c(K) := \sum_{F \in K} E_{\mathfrak{g}}(F). \qquad (5.1.104)$$

Then $E_{\mathfrak{g}}(K)$ is a projector in $\mathfrak{N}_G$ and we can define

$$\mu_{\mathfrak{g}}^\omega(K) := \omega(E_{\mathfrak{g}}(K)) \text{ for any } K \subset \mathfrak{g}^* \text{ and any } \omega \in \mathcal{S}(\mathfrak{N}_G). \qquad (5.1.105)$$

It is easily to see that $\mu_{\mathfrak{g}}^\omega$ in (5.1.105) is the desired unique extension. For the proof of the second assertion, choose any finitely additive probability measure μ on $s_G \mathfrak{g}^*$, μ defined on all subsets K of $s_G \mathfrak{g}^*$. Define a positive linear functional on $\mathfrak{N}_G$, ω_μ, by its values on all projectors:

$$\omega_\mu(E_{\mathfrak{g}}(K)) := \mu(K), \qquad (5.1.106)$$

compare 5.1.16. The von Neumann algebra $\mathfrak{N}_G$ is generated by the set of all its projectors and (5.1.106) defines uniquely a state on $\mathfrak{N}_G$. $\qquad \square$

5.1.25 Let us look what measures $\mu_{\mathfrak{g}}^\omega$ correspond to pure states $\omega \in \mathcal{M}$, which are not normal. From the character property of pure states we have $\omega(E_{\mathfrak{g}}(K_1 \cap K_2)) = \omega(E_{\mathfrak{g}}(K_1)E_{\mathfrak{g}}(K_2)) = \omega(E_{\mathfrak{g}}(K_1))\omega(E_{\mathfrak{g}}(K_2))$ what together with finite additivity gives:

$$K \subset \mathfrak{g}^* \Rightarrow \mu_{\mathfrak{g}}^\omega(K) \in \{0, 1\}. \qquad (5.1.107)$$

Remember that supp $\mu_{\mathfrak{g}}^\omega \subset s_G \mathfrak{g}^*$. Any finitely additive measure μ on $s_G \mathfrak{g}^*$ satisfying (5.1.107) corresponds to a pure state $\omega_\mu \in \mathcal{M}$. It determines also an *ultrafilter* on $s_G \mathfrak{g}^*$ consisting of all subsets K for which it is $\mu(K) = 1$. This is clearly a bijection between the set of all ultrafilters on $s_G \mathfrak{g}^*$ and the set of pure states

$$\mathcal{ES}(\mathfrak{N}_G) = \mathcal{N}.$$

Remember that to any $m \in \mathcal{M}$ corresponds the Dirac measure δ_m on $\mathcal{M}$ which is concentrated at a point m. For a nonnormal m the measure $\mu_{\mathfrak{g}}^m$ is not concentrated at any point in $\mathfrak{g}^*$.

5.1.26 Let us keep in mind that we have associated with any state $\omega \in \mathcal{S}(\mathfrak{A}^\Pi)$ a state (equally denoted) $\omega \in \mathcal{S}(\mathfrak{M}_G)$ which is the restriction to $\mathfrak{M}_G$ of the unique w^*-continuous extension to $(\mathfrak{A}^\Pi)^{**}$ of $\omega \in \mathcal{S}(\mathfrak{A}_\Pi)$. Such an $\omega \in \mathcal{S}(\mathfrak{M}_G)$ is necessarily normal: $\omega \in \mathcal{S}_*(\mathfrak{M}_G)$, and the corresponding measure $\mu_{\mathfrak{g}}^\omega := \mu_\omega \circ F_{\mathfrak{g}}^{-1}$ is purely atomic, 5.1.23. This reflects that fact that the described procedure maps into $\mathcal{S}(\mathfrak{N}_G)$ only such states on $(\mathfrak{A}^\Pi)^{**}$ which are describable by density matrices in $\mathcal{L}(P_G \mathcal{H}_\Pi)$.

Any state on $\mathfrak{M}_G$ can be, on the other hand, extended to some states on $(\mathfrak{A}^\Pi)^{**}$ (not normal—in general) and these determine their restrictions to $\mathfrak{A}^\Pi$ considered as a subalgebra of its bidual. In this way, we can obtain also those states on $P_G\mathfrak{A}^\Pi$ which are not expressible by density matrices. Hence to general finitely additive probability measures on $\mathcal{N}_*$ 'correspond', in some many-to-many way, arbitrary states on $P_G\mathfrak{A}^\Pi$. We intend now to change our ascription of states on $\mathfrak{M}_G$ to arbitrary states on $\mathfrak{A}^\Pi$ in such a way, that any state on $P_G\mathfrak{A}^\Pi$ will be mapped into $\mathcal{S}(\mathfrak{N}_G)$ (and not onto $m_\circ \in \mathcal{M}_*$ as before).

5.1.27 Quasilocal structure of $\mathfrak{A}^\Pi$: The algebra $\mathfrak{A}^\Pi$ has a natural *quasilocal structure* in the sense of 1.4.2. It is generated by *local algebras* $\mathfrak{A}_v := \mathfrak{A}^N$ ($N \in \Pi$), (1.4.1), where, in the notations of 5.1.3, $\mathfrak{A}^N$ is generated by $\pi_j(y)$ ($y \in \mathcal{L}(\mathcal{H})$, $j = 1, 2, \ldots N$) and is isomorphic to $\mathcal{L}(\mathcal{H}_N)$. $\mathcal{H}_N$ is here the N−fold tensor product of the Hilbert space $\mathcal{H}$, (5.1.10). Denote by $\mathfrak{A}_L$ the set of all finite linear combinations of finite products of arbitrary elements $y \in \mathfrak{A}^N$ for any finite N. The algebra $\mathfrak{A}_L := \cup_{\text{finite } N}\mathfrak{A}^N$ is called the 'local algebra' and its elements are '*local observables*'. The norm closure of $\mathfrak{A}_L$ is $\mathfrak{A}^\Pi$ = the *algebra of quasilocal observables* of our system.

A locally normal state $\omega \in \mathcal{S}(\mathfrak{A}^\Pi)$, i.e. a state the restriction of which to any local subalgebra $\mathfrak{A}^N$ is normal (cf. 1.4.3), can be calculated (with a use of natural isomorphisms) on all the elements $x \in \mathfrak{A}^N \subset \mathfrak{A}^\Pi$ with the help of density matrices ρ_ω^N on $\mathcal{H}_N$ ($N = 1, 2, \ldots$) via the usual formula

$$\omega(x) = Tr(\rho_\omega^N x), \quad x \in \mathcal{L}(\mathcal{H}_N), \tag{5.1.108}$$

where we have identified $\mathfrak{A}^N$ with $\mathcal{L}(\mathcal{H}_N)$. Let $\mathcal{S}_L(\mathfrak{A}^\Pi) =: \mathcal{S}_L$ denotes the set of all locally normal states on $\mathfrak{A}^\Pi$. The states expressible (globally) by a density matrix in the defining representation of $\mathfrak{A}^\Pi$ in $\mathcal{H}_\Pi$ are locally normal. $\mathfrak{A}^\Pi$ is simple, [53, 2.6.20].

5.1.28 Example. We shall illustrate here the fact that a strongly continuous one parameter group of unitaries $\exp(itP)$ acting on a Hilbert space $\mathcal{H}$ need not be continuous in certain other representations of $\mathcal{L}(\mathcal{H})$.

Let $\mathfrak{A} := \mathcal{L}(\mathcal{H})$ be the considered W^*-algebra, $\mathcal{H} := L^2(\mathbb{R})$, and Q (resp. P) be the selfadjoint operator on $\mathcal{H}$ defined on $\varphi \in C_0^1(\mathbb{R})$ by $(Q\varphi)(\lambda) := \lambda\varphi(\lambda)$ (resp. $(P\varphi)(\lambda) := -i\frac{d}{d\lambda}\varphi(\lambda)$), $\lambda \in \mathbb{R}$. Let $\mathfrak{M}$ be the maximal commutative W^*-algebra in $\mathcal{L}(\mathcal{H})$ generated by $\exp(itQ)$, $t \in \mathbb{R}$. Let χ_λ be the pure state on $\mathfrak{M}$ determined by

$$\chi_\lambda(\exp(itQ)) := \exp(it\lambda), \quad t \in \mathbb{R}. \tag{5.1.109}$$

Let ω_λ be an extension of χ_λ onto the whole W^*-algebra $\mathfrak{A}$. We claim that the function

$$t \mapsto \omega_\lambda(\exp(itP)), \quad t \in \mathbb{R}, \tag{5.1.110}$$

is discontinuous, hence the group $\pi_\lambda(\exp(itP))$ of unitaries in the cyclic representation π_λ of $\mathfrak{A}$ corresponding to the state $\omega_\lambda \in \mathcal{S}(\mathfrak{A})$ is not strongly continuous. Since

χ_λ is pure, it is a character on $\mathfrak{M}$. Consequently for any projector $q \in \mathfrak{M}$ it is

$$\chi_\lambda(q) = [\chi_\lambda(q)]^2, \ i.e. \ \chi_\lambda(q) \in \{0, 1\}, \ q^* = q^2 = q \in \mathfrak{M}. \qquad (5.1.111)$$

We obtain from the *Schwarz inequality*, for any $x \in \mathfrak{A}$,

$$\omega_\lambda(x) = \omega_\lambda(qx) = \omega_\lambda(xq), \ \text{for all} \ \{q = q^* = q^2 : q \in \mathfrak{M}, \ \chi_\lambda(q) = 1\}. \qquad (5.1.112)$$

Any element $z \in \mathfrak{M}$ can be expressed as a norm limit of finite linear combinations of projectors $q \in \mathfrak{M}$. Since the product $z \mapsto xz$ is norm-continuous and the state ω_λ is also continuous in the norm of $\mathfrak{A}$, we obtain from (5.1.112):

$$\omega_\lambda(xz - zx) \equiv \omega_\lambda([x, z]) = 0, \quad \forall x \in \mathfrak{A}, \ \text{and} \ \forall z \in \mathfrak{M}. \qquad (5.1.113)$$

Due to CCR we have

$$[\exp(itP), \exp(i\tau Q)] = (e^{it\tau} - 1) \exp(i\tau Q) \exp(itP), \qquad (5.1.114)$$

and after the substitution to (5.1.113):

$$0 = (e^{it\tau} - 1) \omega_\lambda(\exp(i\tau Q) \exp(itP)). \qquad (5.1.115)$$

The relation (5.1.115) is valid for all real t and τ. From an application of (5.1.113) to

$$\omega_\lambda \left(\exp(-i\tau Q)[\exp(itP), \exp(i\tau Q)]\right) = \omega_\lambda \left(\exp(-i\tau Q) \exp(itP) \exp(i\tau Q)\right)$$
$$- \omega_\lambda(\exp(itP))$$

we obtain the invariance of ω_λ with respect to the group σ^* of affine isometries of $\mathcal{S}(\mathfrak{A})$,

$$\sigma_\tau^*\omega(x) := \omega(\exp(-i\tau Q)x \exp(i\tau Q)), \ \tau \in \mathbb{R}, \ x \in \mathfrak{A}. \qquad (5.1.116)$$

This leads, together with the formula (4.1.9), to

$$\omega_\lambda(\exp(itP)) = e^{it\tau} \omega_\lambda(\exp(itP)) \quad \text{for all} \ t, \tau \in \mathbb{R}, \qquad (5.1.117)$$

what implies the discontinuity of (5.1.110).

The obtained formulas show also uniqueness of the extension ω_λ of χ_λ to the CCR-subalgebra of $\mathfrak{A}$ defined as the norm closed algebra generated by $\exp(i\tau Q)$ and $\exp(itP)$ $(t, \tau \in \mathbb{R})$.

5.1.29 We shall now change the definition 5.1.12 of the macroscopic algebra of the system $(\mathfrak{A}^\Pi, \sigma_G)$ in such a way that a larger subset of states from $\mathcal{S}(\mathfrak{A}^\Pi)$ will be mapped onto probability measures on $\mathfrak{g}^*$ than it was before, according to the ascription from 5.1.14. In the notations from 5.1.3 and 5.1.4, let

$$X_{\xi N} := \frac{1}{N} \sum_{j=1}^{N} \pi_j(X_\xi), \quad N = 1, 2, \ldots, \; \xi \in \mathfrak{g}. \tag{5.1.118}$$

Then the elements $\exp(it X_{\xi N}) \in \mathfrak{A}^N$ $(t \in \mathbb{R})$ are represented in the defining representation of $\mathfrak{A}^\Pi$ in $\mathcal{H}_\Pi$ by strongly continuous groups converging with $N \to \infty$ in the strong operator topology on the G-invariant subspace $P_G \mathcal{H}_\Pi$ of $\mathcal{H}_\Pi$ to strongly continuous central subgroups of $P_G \mathfrak{B}^\#$,

$$s\text{-}\lim_N \exp(it X_{\xi N}) P_G = \exp(it X_{\xi\Pi}) P_G, \tag{5.1.119}$$

see 5.1.7, 5.1.8 and 5.1.11. The algebra $\mathfrak{M}_G$ of macroscopic observables was built from spectral projectors $E_{\xi\Pi}^\#$ of $X_{\xi\Pi}$'s mapped into the center $\mathfrak{Z}$ of the bidual $(\mathfrak{A}^\Pi)^{**}$. We want to generalize this construction. We shall identify the bidual $(\mathfrak{A}^\Pi)^{**}$ with the weak closure of the *universal representation* of $\mathfrak{A}^\Pi$ (cf. [274, Def.1.16.5], [235, 3.7.6]). **Let p_G be the l.u.b. of all such projectors** $p \in \mathfrak{Z}$, for which the limits in $\sigma((\mathfrak{A}^\Pi)^{**}, (\mathfrak{A}^\Pi)^*)$-topology:

$$\exp(it X_{\xi\Pi}) p_G := \sigma\text{-}\lim_N \exp(it X_{\xi N}) p_G, \quad \forall \xi \in \mathfrak{g}, \tag{5.1.120}$$

exist (with $p_G \hookrightarrow p$). The symbol $X_{\xi\Pi}$ denotes here a selfadjoint operator acting on the subspace $p_G \mathcal{H}_u$ of the space $\mathcal{H}_u$ – the space of universal representation of $\mathfrak{A}^\Pi$. Here it is assumed, of course, that the groups $t \mapsto \exp(it X_{\xi N}) p_G$ are strongly continuous for all $N \in \Pi$. It is clear from the definitions of $X_{\xi N}$ and σ_G, 5.1.5, that

$$\sigma_G(p_G) = p_G. \tag{5.1.121}$$

The convergence in (5.1.120) means the convergence $X_{\xi N} \to X_{\xi\Pi}$ of selfadjoint operators on $p_G \mathcal{H}_u$ in the strong-resolvent sense, [262]. From

$$[\exp(it X_{\xi N}), \; y] = \left(\exp\left(\frac{it}{N} \sum_{j=1}^{K} \pi_j(X_\xi) \right) y \exp\left(-\frac{it}{N} \sum_{k=1}^{K} \pi_k(X_\xi) \right) - y \right) e^{it X_{\xi N}}, \tag{5.1.122}$$

which is valid for all $y \in \mathfrak{A}^K$ $(K \in \Pi)$ and $\xi \in \mathfrak{g}$, $t \in \mathbb{R}$, as well as from the assumed continuity of $p_G \exp(it X_{\xi N})$ we conclude that the limit $p_G \exp(it X_{\xi\Pi}) \in (\mathfrak{A}^\Pi)^{**}$ belongs to the center $\mathfrak{Z}$ of $(\mathfrak{A}^\Pi)^{**}$. Let now the Π-**macroscopic algebra of G - definiteness** of $(\mathfrak{A}^\Pi, \sigma_G)$ be **defined as the von Neumann subalgebra $\mathfrak{N}_G^\Pi$ of the center** $\mathfrak{Z}$ generated by all the spectral projectors $E_{\xi\Pi}(B)$ (Borel $B \subset \mathbb{R}$ and $\xi \in \mathfrak{g}$) of operators $X_{\xi\Pi}$ in $p_G \mathcal{H}_u$ (we hope that no confusion arises from the keeping an old notation for new objects!). The **algebra $\mathfrak{M}_G^\Pi$ is obtained from $\mathfrak{N}_G^\Pi$ by adjoining to** it the identity I of 3; it will be called the ΠG-**macroscopic algebra** of $(\mathfrak{A}^\Pi, \sigma_G)$. The relation between $\mathfrak{M}_G^\Pi$ and the previously introduced $\mathfrak{M}_G$, 5.1.12, is clear without any proof:

5.1.30 Lemma. $\mathfrak{N}_G = s_G \mathfrak{N}_G^\Pi = s_G \mathfrak{M}_G^\Pi = s_G \mathfrak{M}_G$, where the projector $s_G \in 3$ was introduced in 5.1.11.

5.1.31 We shall use concepts and notations connected with the usage of $\mathfrak{M}_G^\Pi$ in analogy to those connected with $\mathfrak{M}_G$, as they were introduced above. Let e.g., $B \subset \mathfrak{g}^*$ be a Borel set (with respect to the usual topology of a finite dimensional vector space), and ξ_j $(j = 1, 2, \ldots n)$ form a basis of $\mathfrak{g}$. Let

$$\xi_j B := \{\lambda \in \mathbb{R} : \ \lambda = F(\xi_j), \ F \in B\} \tag{5.1.123}$$

be the projection of B onto the j-th coordinate axis of the dual frame. If B has the form

$$B = \{F \in \mathfrak{g}^* : F(\xi_j) \in \xi_j B, \ \forall j \in \{1, 2, \ldots n := \dim G\}\}, \tag{5.1.124}$$

then we set

$$E_{\mathfrak{g}}^\Pi(B) := E_{\xi_1 \Pi}(\xi_1 B) E_{\xi_2 \Pi}(\xi_2 B) \ldots E_{\xi_n \Pi}(\xi_n B). \tag{5.1.125}$$

The W^*-algebra $\mathfrak{M}_G^\Pi$ is **generated by the projectors** $E_{\mathfrak{g}}^\Pi(B)$ from (5.1.125) and (5.1.124), and by the unit $I \in 3$.

The **algebra** $\mathfrak{M}_G^\Pi$, **contrary to** $\mathfrak{M}_G$, **cannot be built from projectors** (5.1.125) corresponding to one point sets $B := \{F\}$ $(F \in \mathfrak{g}^*)$ only. This can be seen as follows: Choose a probability Borel measure μ on $s_G \mathfrak{g}^*$, in the old notation from 5.1.16, such that any point (Dirac) measure is singular with respect to it: $\mu(\{F\}) = 0$ for all $F \in \mathfrak{g}^*$. Choose a product vector $\Psi(F) \in E_{\mathfrak{g}}^\#(F) P_G \mathcal{H}_\Pi$, one and only one for each such $F \in \mathfrak{g}^*$, for which $E_{\mathfrak{g}}^\#(F) \neq 0$. Denote by $\omega^F := \omega^{\Psi(F)}$ the corresponding state on $\mathfrak{A}^\Pi$. Assume, that all the functions

$$F \mapsto \omega^F(x), \ x \in \mathfrak{A}^\Pi, \tag{5.1.126}$$

are μ−measurable. This last assumption is trivially fulfilled, if μ is concentrated on an $Ad^*(G)$ orbit $G \cdot F \subset \mathfrak{g}^*$ and $\omega^{g \cdot F} := \sigma_g^* \omega^F$. Define then the state $\omega_\mu \in \mathcal{S}(\mathfrak{A}^\Pi)$ by

$$\omega_\mu(x) := \int_{\mathfrak{g}^*} \omega^F(x) \, \mu(dF). \tag{5.1.127}$$

In this way, we can construct **states** ω_μ **the central supports** $s_\mu \in 3$ **of which are contained in** p_G, $s_\mu p_G = s_\mu$, but $s_\mu s_G = 0$, as well as $s_\mu E_{\mathfrak{g}}^\Pi(F) = 0$ for all $F \in \mathfrak{g}^*$, in $\mathfrak{M}_G^\Pi$.

The last considerations show to us that $\mathfrak{M}_G$ and $\mathfrak{M}_G^\Pi$ are different from one another. The W^*-subalgebra of $\mathfrak{M}_G^\Pi$ generated by all $E_\mathfrak{g}^\Pi(F) := E_\mathfrak{g}^\Pi(\{F\})$ ($F \in \mathfrak{g}^*$) is naturally isomorphic to $\mathfrak{M}_G$. Hence $\mathfrak{M}_G^\Pi$ is larger than $\mathfrak{M}_G$ which can be injected into $\mathfrak{M}_G^\Pi$ via the last mentioned isomorphism.

5.1.32 Let us now introduce the mapping p_M:

$$p_M : \mathcal{S}(\mathfrak{A}^\Pi) \to \mathcal{S}_*(\mathfrak{M}_G^\Pi), \quad \omega \mapsto p_M \omega, \tag{5.1.128}$$

where $p_M \omega$ is the restriction to $\mathfrak{M}_G^\Pi$ of the canonical extension of the state $\omega \in \mathcal{S}(\mathfrak{A}^\Pi)$ to the normal state on $(\mathfrak{A}^\Pi)^{**}$. Any state $\omega \in \mathcal{S}(\mathfrak{A}^\Pi)$ (resp. $\omega \in \mathcal{S}(\mathfrak{M}_G^\Pi)$) can be uniquely decomposed as

$$\omega = \omega(p_G)\, p_G \omega + \omega(I - p_G)\, \omega_\circ, \tag{5.1.129}$$

where the symbols $p_G \omega(x)$ and $\omega_\circ(x)$ are given by

$$p_G \omega(x) := \frac{1}{\omega(p_G)} \omega(x\, p_G), \quad \omega_\circ(x) := \frac{1}{\omega(I - p_G)} \omega(x(I - p_G)).$$

Hence for $\omega(I - p_G) \neq 0$ it is $p_M \omega_\circ = m_\circ :=$ the **pure state in** $\mathcal{S}(\mathfrak{M}_G^\Pi)$ **sup** − **ported** by the minimal projector $I - p_G \in \mathfrak{M}_G^\Pi$.

Let

$$\mathcal{S}_\mathfrak{g}^\Pi := \{\omega \in \mathcal{S}(\mathfrak{A}^\Pi) : \omega(p_G) = 1\}. \tag{5.1.130}$$

In other words: $\mathcal{S}_\mathfrak{g}^\Pi = p_G \mathcal{S}(\mathfrak{A}^\Pi)$. For $\omega \in \mathcal{S}_\mathfrak{g}^\Pi$ one has $p_M \omega \in \mathcal{S}_*(\mathfrak{M}_G^\Pi)$. Conversely, each state in $\mathcal{S}_*(\mathfrak{M}_G^\Pi)$ is of the form $p_M \omega$ for some states $\omega \in \mathcal{S}_\mathfrak{g}^\Pi$.

5.1.33 Lemma. *The projector-valued additive function of intervals in $\mathfrak{g}^*$ introduced in (5.1.125) can be extended to a unique **projector-valued measure** $E_\mathfrak{g}^\Pi : B \mapsto E_\mathfrak{g}^\Pi(B)$ defined on all Borel sets B in $\mathfrak{g}^*$.*

Proof. The mapping

$$w : \mathfrak{g} \to \mathcal{L}(p_G \mathcal{H}_u), \quad \xi \mapsto w(\xi) := \exp(iX_{\xi\Pi})p_G, \tag{5.1.131}$$

see (5.1.120), is strongly continuous unitary representation of the abelian group $\mathfrak{g}$ (group multiplication is here the vector addition) in the subspace $p_G \mathcal{H}_u$ of the Hilbert space $\mathcal{H}_u$ of the universal representation of $\mathfrak{A}^\Pi$. This can be seen with a help of linearity of the mapping

$$\xi \mapsto X_{\xi\Pi}, \quad \xi \in \mathfrak{g}, \tag{5.1.132}$$

According to the *SNAG-theorem* ([266, Chap. X], [262, Theorem VIII.12], [120, Chap. IV]), there is unique projection measure $E_\mathfrak{g}^\Pi$ on the dual group $\hat{\mathfrak{g}}$ of $\mathfrak{g}$ representing this unitary representation in the standard fashion. The linear space $\mathfrak{g}^*$ can be identified with the group $\hat{\mathfrak{g}}$ of characters by the bijection associating with any $F \in \mathfrak{g}^*$ the character $\xi \mapsto \exp(i F(\xi))$ on $\mathfrak{g}$. It is clear that the restriction of $E_{\mathfrak{g}}^{\Pi}$ on intervals in $\mathfrak{g}^*$ coincides with (5.1.125). $\qquad\square$

5.1.34 Lemma. *All the nonzero projectors of the form* $E_{\mathfrak{g}}^{\Pi}(F) := E_{\mathfrak{g}}(\{F\})$, $F \in \mathfrak{g}^*$, *are minimal projectors in* $\mathfrak{N}_G^{\Pi}$ *and all minimal projectors in* $\mathfrak{N}_G^{\Pi}$ *are of this form.*

Proof. Let $q \in \mathfrak{N}_G^{\Pi}$ be a minimal projector. Since $q \in \mathfrak{Z}$, there is a state $\omega \in \mathcal{S}(\mathfrak{A}^{\Pi})$, the central projector of which is $s_\omega \leq q$. Choose such an ω. Then $\omega(x) = \omega(x s_\omega) = \omega(xq)$ for all $x \in \mathfrak{A}_{\Pi}$, and due to continuity properties of products in $(\mathfrak{A}^{\Pi})^{**}$ as well as of the normal extension $\omega \in \mathcal{S}_*((\mathfrak{A}^{\Pi})^{**})$, the same is true for all $x \in (\mathfrak{A}^{\Pi})^{**}$. The minimality of q in $\mathfrak{N}_G^{\Pi}$ implies that one of the following possibilities (i) or (ii) is valid

$$\text{(i)} \quad q E_{\mathfrak{g}}^{\Pi}(B) = q, \qquad \text{(ii)} \quad q E_{\mathfrak{g}}^{\Pi}(B) = 0 \qquad\qquad (5.1.133)$$

for any Borel $B \subset \mathfrak{g}^*$. Let us define a probability Borel measure $\mu_{\mathfrak{g}}^{\omega}$ on $\mathfrak{g}^*$ corresponding to the $\omega \in \mathcal{S}_{\mathfrak{g}}^{\Pi}$:

$$\mu_{\mathfrak{g}}^{\omega}(B) := p_M \omega(E_{\mathfrak{g}}^{\Pi}(B)), \text{ for all Borel } B \subset \mathfrak{g}^*. \qquad\qquad (5.1.134)$$

We see from (5.1.133) that for the chosen ω the values of $\mu_{\mathfrak{g}}^{\omega}$ lie in the two point set $\{0, 1\} \subset \mathbb{Z}_+$. Each of the projection measures $E_{\xi\Pi}$ ($\xi \in \mathfrak{g}$) and $E_{\mathfrak{g}}^{\Pi}$ are σ-additive, hence $\mathfrak{N}_G^{\Pi}$ is generated by those $E_{\mathfrak{g}}^{\Pi}(B)$ which correspond to *bounded* Borel subsets B of $\mathfrak{g}^*$. Hence $\mu_{\mathfrak{g}}^{\omega}$ is concentrated on a compact subset of $\mathfrak{g}^*$: $\mu_{\mathfrak{g}}^{\omega}(B_\circ) = 1$ for some compact $B_\circ$. The σ-additivity of $\mu_{\mathfrak{g}}^{\omega}$ implies then that $\mu_{\mathfrak{g}}^{\omega}$ is concentrated on a one-point set $F_\omega \in B_\circ$:

$$\mu_{\mathfrak{g}}^{\omega}(\{F_\omega\}) = p_M \, \omega(E_{\mathfrak{g}}^{\Pi}(F_\omega)) = 1. \qquad\qquad (5.1.135)$$

This implies $s_\omega \leq E_{\mathfrak{g}}^{\Pi}(F_\omega)$, and, due to (5.1.133) and due to our choice of s_ω:

$$q \leq E_{\mathfrak{g}}^{\Pi}(F_\omega). \qquad\qquad (5.1.136)$$

According to the definition of $\mathfrak{N}_G^{\Pi}$ in 5.1.29, q can be approximated, in $\sigma(\mathfrak{N}_G^{\Pi}, \mathfrak{N}_{G*}^{\Pi})$ topology, by a net $j \mapsto E_{\mathfrak{g}}^{\Pi}(B_j)$, where $F_\omega \in B_j$ for all j, due to (5.1.136). Coming to the Gel'fand representation $C(\mathcal{N}^{\Pi})$ of $\mathfrak{N}_G^{\Pi}$ and considering that **clopen sets in the spectrum space** $\mathcal{N}^{\Pi}$ **form a basis of topology**, e.g. 5.1.14 and [274], we see that the sets in $\mathcal{N}^{\Pi}$ corresponding to the projectors $E_{\mathfrak{g}}^{\Pi}(B_j)$ have in their intersection exactly one point $m_q \in \mathcal{N}^{\Pi}$ corresponding to the minimal projector q. All of $E_{\mathfrak{g}}^{\Pi}(B_j)$ contain, however, also $E_{\mathfrak{g}}^{\Pi}(F_\omega)$. This proves that $q = E_{\mathfrak{g}}^{\Pi}(F_\omega)$. $\square$

5.1.35 Lemma. *If* $\omega \in \mathcal{S}(\mathfrak{A}^{\Pi})$ *is pure or factor state, then also* $p_M \omega \in \mathcal{S}_*(\mathfrak{N}_G^{\Pi})$ *is pure.*

Proof. Verbally the same proof as that of 5.1.17, with $\mathfrak{M}_G \hookrightarrow \mathfrak{M}_G^\Pi$. $\qquad\square$

5.1.36 Definitions. *The* **generalized G-macroscopic phase space** *is the topological space* $\mathcal{M}^G := \mathfrak{g}^* \cup \{m_\circ\}$ *consisting of the finite (n-)dimensional topological vector space* $\mathfrak{g}^*$ *with the canonical symplectic forms defined on each orbit of the* $Ad^*(G)$*-action and of an isolated point* $m_\circ$. *A* **state on** $\mathcal{M}^G$ *is any probability* σ*-additive Borel measure* μ *on* $\mathcal{M}^G$. *We shall associate with any* $\omega \in \mathcal{S}(\mathfrak{A}^\Pi)$ *the* **G-macroscopic state** *on* $\mathcal{M}^G$ *determined by the measure*

$$\mu_\mathfrak{g}^\omega(B) := \omega(E_\mathfrak{g}^\Pi(B\backslash\{m_\circ\})) + \omega(I - p_G)\delta_{m_\circ}(B), \text{ for Borel } B \subset \mathcal{M}^G, \quad (5.1.137)$$

where on the right hand side ω *means the normal extension of the state on* $\mathfrak{A}^\Pi$ *to a normal state on* $(\mathfrak{A}^\Pi)^{**}$ *and* δ_m $(m \in \mathcal{M}^G)$ *means the Dirac measure concentrated on* $\{m\}$. *It is clear that every normal state* $\omega \in \mathcal{S}(\mathfrak{M}_G^\Pi)$ *can be transformed also into a state on* $\mathcal{M}^G$ *by the formula (5.1.137) and that its image* $\mu_\mathfrak{g}^\omega$ *uniquely determines* $\omega \in \mathcal{S}_*(\mathfrak{M}_G^\Pi)$. *It is also clear that the state on* $\mathcal{M}^G$ *corresponding to* $p_M\omega$, *5.1.32, in this way, coincides with* $\mu_\mathfrak{g}^\omega$. *The association* $\omega \mapsto \mu_\mathfrak{g}^\omega$ *is* **G-equivariant**, *i.e.*

$$\mu_\mathfrak{g}^{g\cdot\omega} = \sigma_g^*\mu_\mathfrak{g}^\omega, \text{ with } g \cdot \omega := \sigma_g^*\omega, \ g \in G, \quad (5.1.138)$$

and with

$$\sigma_g^*\mu(B) := \mu(Ad^*(g^{-1})(B\backslash\{m_\circ\})) + \mu(\{m_\circ\} \cap B), \quad (5.1.139)$$

for all $g \in G$ *and all Borel* $B \subset \mathcal{M}^G$. *We shall use also* $g \cdot \mu := \sigma_g^*\mu$. *This follows from the transformation properties of* $X_{\xi\Pi}$*'s and from*

$$\sigma_g E_\mathfrak{g}^\Pi(B) = E_\mathfrak{g}^\Pi(Ad^*(g)B), \quad (5.1.140)$$

compare 5.1.15.

Let us redefine some symbols introduced in 5.1.18. Let, (5.1.130),

$$\mathcal{S}_\mathfrak{g}^d := \{\omega \in \mathcal{S}_\mathfrak{g}^\Pi : \xi \in L^1(\mathcal{M}^G, \mu_\mathfrak{g}^\omega), \ \forall\xi \in \mathfrak{g}\}, \quad (5.1.141)$$

where ξ *is considered as the linear function* $F \mapsto \xi(F) := F(\xi)$ *on* $\mathfrak{g}^*$ $(\ni F)$. *Similarly, we shall define now* **G-macroscopically pure states** *to be elements* $\omega \in \mathcal{E}_\mathfrak{g} \subset \mathcal{S}(\mathfrak{A}^\Pi)$, *where*

$$\mathcal{E}_\mathfrak{g} := \{\omega \in \mathcal{S}_\mathfrak{g}^d : \mu_\mathfrak{g}^\omega(\xi^2) = [\mu_\mathfrak{g}^\omega(\xi)]^2, \ \forall\xi \in \mathfrak{g}\}. \quad (5.1.142)$$

Using (5.1.91) and (5.1.92), we can see that the Proposition 5.1.19 can be replaced by: $\omega \in \mathcal{E}_\mathfrak{g} \Leftrightarrow \mu_\mathfrak{g}^\omega = \delta_F$ *for some* $F \in \mathfrak{g}^*$.

5.1.37 Definitions. *A* **Poisson manifold** $\mathcal{M}$ *is a differentiable* C^∞*-manifold endowed with a bilinear mapping* $(f; g) \mapsto \{f, g\}$ *of couples of infinitely differentiable real functions* $f, g \in C^\infty(\mathcal{M}, \mathbb{R})$ *into* $C^\infty(\mathcal{M}, \mathbb{R})$, *the* **Poisson bracket**, *satisfying properties 1.3.5 (i)+(ii)+(iii)+(iv), i.e. the nondegeneracy 1.3.5(v) is* not *required. Due to 1.3.5(iv), the Poisson bracket* $\{f, g\}$ *depends on* df *and* dg *only, and can be uniquely expressed as the value of a two-contravariant* **tensor field** λ *on these one forms:*

$$\lambda(df, dg) := \{f, g\}. \tag{5.1.143}$$

To any $f \in C^\infty(\mathcal{M}, \mathbb{R})$ *corresponds then a unique vector field* σ_f *on* $\mathcal{M}$ *satisfying:*

$$dg(\sigma_f) := \lambda(df, dg), \text{ for all } g \in C^\infty(\mathcal{M}, \mathbb{R}). \tag{5.1.144}$$

σ_f *is the* **Hamiltonian vector field** *on* $\mathcal{M}$ *with the* Hamiltonian function f.

With $\mathcal{M} := \mathfrak{g}^*$, *the cotangent space* $T_F^*\mathfrak{g}^*$ *can be naturally identified, for any* $F \in \mathfrak{g}^*$, *with the Lie algebra* $\mathfrak{g}$ *of* G. *Then, with this identification,* $d_F f \in \mathfrak{g}$ *for any* f *and* F. *Then the Poisson bracket*

$$\{f, g\}(F) := -F([d_F f, d_F g]), \tag{5.1.145}$$

where on the right hand side is the value of $F \in \mathfrak{g}$ *on the Lie algebra commutator in* $\mathfrak{g}$, *defines a natural* Poisson *structure on* $\mathfrak{g}^*$. *In this way, also* $\mathcal{M}^G$ *is, naturally, a Poisson manifold. Hamiltonian vector fields* σ_f *are tangent to orbits of* $Ad^*(G)$-*action of* G *on* $\mathfrak{g}^*$ *at any point* $F \in \mathfrak{g}^*$, *compare [212]. The restriction of the Poisson structure (5.1.145) to any* $Ad^*(G)$-*orbit is the canonical symplectic structure on it.*

5.1.38 Theorem. *Let the system* $(\mathfrak{A}^\Pi, \sigma_G)$ *be defined by (5.1.21) and (5.1.25). Let* $\mathfrak{M}_G^\Pi$ *be the commutative* σ_G-*invariant* W^*-*subalgebra of* $\mathfrak{Z}$ (*:=* **the center of** $(\mathfrak{A}^\Pi)^{**}$) *defined in 5.1.29. Let* $p_M : \mathcal{S}(\mathfrak{A}^\Pi) \to \mathcal{S}_*(\mathfrak{M}_G^\Pi)$ *be the mapping (5.1.128). We shall write also*

$$p_M \omega := \mu_\mathfrak{g}^\omega, (5.1.137),$$

due to the existence of canonical embedding of $\mathcal{S}_*(\mathfrak{M}_G^\Pi)$ *into the space of probability* Radon *measures on* $\mathcal{M}^G$. *Then:*

(i) p_M *is affine,* $\sigma((\mathfrak{A}^\Pi)^*, (\mathfrak{A}^\Pi)^{**}) - \sigma((\mathfrak{M}_G^\Pi)^*, \mathfrak{M}_G^\Pi)$-*continuous surjection onto* $\mathcal{S}_*(\mathfrak{M}_G^\Pi) :=$ *the set of all normal states on* $\mathfrak{M}_G^\Pi$;

(ii) p_M *is* G-*equivariant, (5.1.138)*;

(iii) *Let* $\mathcal{S}_F := \{\omega \in \mathcal{S}(\mathfrak{A}^\Pi) : \mu_\mathfrak{g}^\omega = \delta_F\}$, *(here* $F \in \mathfrak{g}^*$, δ_F *is the Dirac measure concentrated at* F). *Then* $\mathcal{S}_F \subset \mathcal{E}_\mathfrak{g}$, *(5.1.142), and* $\mathcal{S}_F$ *is a weakly closed convex* **face**[4] *in* $\mathcal{S}(\mathfrak{A}^\Pi)$;

(iv) $\omega \in \mathcal{E}_\mathfrak{g}$ *implies* $\mu_\mathfrak{g}^\omega = \delta_F$ *for* $F = F_\omega \in \mathfrak{g}^*$, *and for any factor-state* $\omega \in \mathcal{S}(\mathfrak{A}^\Pi)$ *it is* $\mu_\mathfrak{g}^\omega = \delta_m$ *for some* $m \in \mathcal{M}^G$;

---

[4]A *face* S of a compact convex set K is defined to be a subset of K with the property that if $\omega = \sum_{i=1}^n \lambda_i \omega_i$ is a convex combination of elements $\omega_i \in K$ such that $\omega \in S$ then $\omega_i \in S$, $\forall i = 1, 2, \ldots n$.

(v) *Let μ_ω be the canonical measure on the spectrum space $\mathcal{M}^\Pi$ of $\mathfrak{M}_G^\Pi = C(\mathcal{M}^\Pi)$ corresponding to the state $p_M\omega \in \mathcal{S}_*(\mathfrak{M}_G^\Pi)$, $\omega \in \mathcal{S}(\mathfrak{A}^\Pi)$. Then there is a canonically defined μ_ω–measurable function $\hat{\omega}(m) =: \omega_m$ (spaces are taken with their w^*-topologies) such that the restriction $r_M\omega_m = m \in \mathcal{ES}(\mathfrak{M}_G^\Pi)$:*

$$r_M : (\mathfrak{A}^\Pi)^{***} \to (\mathfrak{M}_G^\Pi)^* \quad \text{is the natural restriction}$$

and

$$\omega(x) = \int_{\mathcal{M}^\Pi} \omega_m(x)\mu_\omega(dm) \ \text{for any } x \in (\mathfrak{A}^\Pi)^{**} \ [\supset \mathfrak{A}^\Pi]. \tag{5.1.146}$$

Proof. (i) is clear from the definition of p_M, compare also [53, 4.1.36]. (ii) is a rephrasing of (5.1.138). Since δ_F corresponds to a pure state on $\mathcal{M}^G$ and p_M is affine, $\mathcal{S}_F$ is a face. Closedness of $\mathcal{S}_F$ follows from the continuity of p_M, and convexity is clear. The rest of (iii) is contained in the concluding remark of 5.1.36 which implies also the first statement of (iv). A proof of the second statement of (iv) is an easy adaptation of that of 5.1.17 for the case of factor states. It remains to prove (v):

Let $\tilde{\omega} \in \mathcal{S}_*((\mathfrak{A}^\Pi)^{**})$ **be the unique normal extension of** $\omega \in \mathcal{S}(\mathfrak{A}^\Pi)$ and $(\pi_\omega, \mathcal{H}_\omega, \Omega_\omega)$ be the corresponding cyclic representation of $(\mathfrak{A}^\Pi)^{**}$. **Denote by** $\hat{\mu}_\omega$ the **orthogonal measure** (cf. [53, 4.1.20]) on $\mathcal{S}((\mathfrak{A}^\Pi)^{**})$ corresponding to the canonical decomposition of $\tilde{\omega}$ with respect to the subalgebra $\pi_\omega(\mathfrak{M}_G^\Pi)$ of the center of $\pi_\omega((\mathfrak{A}^\Pi)^{**})$, compare [53, 4.1.25]:

$$\tilde{\omega}(x) = \int \varphi(x)\,\hat{\mu}_\omega(d\varphi), \ \text{for all } x \in (\mathfrak{A}^\Pi)^{**}. \tag{5.1.147}$$

The **mapping**

$$y\,(\in \mathfrak{M}_G^\Pi) \mapsto \hat{y}\,(\in C(\mathcal{S}((\mathfrak{A}^\Pi)^{**}))), \ \hat{y}(\varphi) := \varphi(y),$$

restricted to the subalgebra $p_\omega\mathfrak{M}_G^\Pi$ (which is isomorphic to $\pi_\omega(\mathfrak{M}_G^\Pi)$ for the **uniquely determined projector** $p_\omega \in \mathfrak{M}_G^\Pi$) provides an isomorphism of the W^*-algebras $p_\omega\mathfrak{M}_G^\Pi$ and $L^\infty(\hat{\mu}_\omega)$, [118, Chap. I.9] and [53, 4.1.22]. Hence, for $y_j \in p_\omega\mathfrak{M}_G^\Pi$ ($j = 1, 2$) we have

$$\widehat{(y_1y_2)}(\varphi) = \hat{y}_1(\varphi)\hat{y}_2(\varphi), \ \text{for } \varphi \in \text{supp } \hat{\mu}_\omega. \tag{5.1.148}$$

Clearly, $\varphi(y) = 0$ for $y \in (I - p_\varphi)\mathfrak{M}_G^\Pi$ and $\varphi \in \text{supp } \hat{\mu}_\omega$. This fact together with (5.1.148) implies that the restriction $r_M\varphi$ is a pure state on $\mathfrak{M}_G^\Pi$ for $\varphi \in \text{supp } \hat{\mu}_\omega$, $r_M\varphi =: m_\varphi \in \mathcal{M}^\Pi$. The w^*-topology of the state space is Hausdorff and the clopen sets form a basis of the topology of $\mathcal{M}^\Pi$. This and the isomorphism of

$L^\infty(\hat\mu_\omega)$ with $p_\omega \mathfrak{M}^\Pi_G$ imply that the restriction of the mapping r_M onto supp $\hat\mu_\omega$ is a bijection onto

$$\text{supp } p_\omega := \{m \in \mathcal{M}^\Pi : m(p_\omega) = 1\}. \tag{5.1.149}$$

Denote $\omega_m := \varphi$ iff $r_M \varphi = m \in \mathcal{M}^\Pi$. Let μ_ω be the image of $\hat\mu_\omega$ under r_M:

$$\mu_\omega := \hat\mu_\omega \circ r_M^{-1} \quad \text{is a regular Borel measure on } \mathcal{M}^\Pi. \tag{5.1.150}$$

Since $p_M \omega = r_M \tilde\omega$, μ_ω is the measure specified in (v). The measurability of the function $\hat\omega : m \mapsto \omega_m$ defined on supp $p_\omega = $ supp μ_ω is clear, compare [53, 4.1.36]. The integral in (5.1.146) is then another form of (5.1.147). This concludes the proof. $\qquad\square$

5.1.39 Note. Let $r_\mathfrak{A} : \mathcal{S}((\mathfrak{A}^\Pi)^{**}) \to \mathcal{S}(\mathfrak{A}^\Pi)$ be **the restriction mapping. Let** $e_* : \mathcal{S}(\mathfrak{A}^\Pi) \to \mathcal{S}_*((\mathfrak{A}^\Pi)^{**})$ **be the normal extension,** $e_* \omega = \tilde\omega$. **Then** $p_M = r_M \circ e_*$. *For a general* $\varphi \in \mathcal{S}((\mathfrak{A}^\Pi)^{**})$, *it is*

$$r_M \varphi \neq (p_M \circ r_\mathfrak{A}) \varphi. \tag{5.1.151}$$

Since $\omega_m \in$ *supp* $\hat\mu_\omega$ *need not be normal, the inequality (5.1.151) holds also for* $\varphi = \omega_m$ *in general. The open question is, however, whether (under some conditions)* $(p_M \circ r_\mathfrak{A})\omega_m \in \mathcal{M}^\Pi = \mathcal{ES}(\mathfrak{M}^\Pi_G)$ *or, at least, when the canonical measure corresponding to* $(p_M \circ r_\mathfrak{A})\omega_m \in \mathcal{S}(C(\mathcal{M}^\Pi))$ *is concentrated on a set* $F_\mathfrak{g}^{-1}(F)$ *for some* $F \in \mathcal{M}^G$, **where**

$F_\mathfrak{g} : \mathcal{M}^\Pi \to \dot{\mathcal{M}}^G$ *is the natural mapping defined according to (5.1.87) $\wedge$ (5.1.93),*

and $\dot{\mathcal{M}}^G$ **is the one-point compactification of** $\mathcal{M}^G$. *Let us write down the* **definition of** $F_\mathfrak{g}$ **explicitly** *(see also proof of 5.1.19):*

(*) Let $E^\Pi_\mathfrak{g}$ *be the projection-valued measure defined on Borel subsets of* $\mathfrak{g}^*$ *with values in* $\mathfrak{Z}$, *as determined in 5.1.29 and in 5.1.33.* **Let** $\dot{\mathfrak{g}}^* := \mathfrak{g}^* \cup \{\infty\}$ *be the* **one-point compactification of** $\mathfrak{g}^*$ *and* $\dot{\mathcal{M}}^G := \dot{\mathfrak{g}}^* \cup \{m_\circ\}$, *where* $m_\circ$ **is an isolated point. Let** $\mathcal{M}^\Pi := \mathcal{ES}(\mathfrak{M}^\Pi_G)$ **be the spectrum space of the algebra** $\mathfrak{M}^\Pi_G = C(\mathcal{M}^\Pi)$ *generated by projectors* $E^\Pi_\mathfrak{g}(B)$ *(Borel* $B \subset \mathfrak{g}^*$), *i.e. by continuous functions* $m \mapsto m(E^\Pi_\mathfrak{g}(B))$, $m \in \mathcal{Z} := \mathcal{ES}(\mathfrak{Z})$. **Define the (continuous) mapping** $F_\mathfrak{g} : \mathcal{M}^\Pi \to \dot{\mathcal{M}}^G$ **by**

(i) $F_\mathfrak{g}(m) \in \mathfrak{g}^*$ *iff there is a* bounded *Borel* $B \subset \mathfrak{g}^*$ *such that*

$$\omega_m(E^\Pi_\mathfrak{g}(B)) \equiv m(E^\Pi_\mathfrak{g}(B)) = 1, \tag{5.1.152}$$

and, in this case, $F_\mathfrak{g}(m)(\xi) := m(X_{\xi\Pi} E^\Pi_\mathfrak{g}(B))$ *for all* $\xi \in \mathfrak{g}$. *Here* $X_{\xi\Pi}$ *are defined in 5.1.29. The character property of* m *ensures independency of* $F_\mathfrak{g}(m)$ *on* B *satisfying (5.1.152).*

(ii) $F_{\mathfrak{g}}(m) := m_o$ *iff* $m(I - p_G) = 1$, *i.e. iff* $m = m_o \in \mathcal{M}^\Pi$, *5.1.32.*
(iii) $F_{\mathfrak{g}}(m) := (\infty)$ *iff* $m(I - p_G) = 0$ *and (5.1.152) is false for all bounded* *Borel*
subsets $B \subset \mathfrak{g}^* : m(E_{\mathfrak{g}}^\Pi(B)) = 0$.

The same definition applied to all $m \in \mathcal{Z}$ *leads to the mapping*

$$F_{\mathfrak{g}} \circ r_M : \mathcal{Z} \to \dot{\mathcal{M}}^G, \quad \mathfrak{Z} := C(\mathcal{Z}), \tag{5.1.153}$$

which is continuous on the whole $\mathcal{Z}$.

The mapping $F_{\mathfrak{g}} \circ r_M$ *determines the projectors* $E_{\mathfrak{g}}^\Pi(B)$. *For bounded* B *we have*
$(\overline{B} := closure, \ B^\circ := interior)$[5]

$$m(E_{\mathfrak{g}}^\Pi(\overline{B})) = 1 \ \text{iff} \ m \in \overline{[(F_{\mathfrak{g}} \circ r_M)^{-1}(\overline{B})]^\circ} = [(F_{\mathfrak{g}} \circ r_M)^{-1}(\overline{B})]^\circ. \tag{5.1.154}$$

If we extend the $Ad^*(G)$ *to the whole* $\dot{\mathcal{M}}^G$ *by the requirement of* $Ad^*(G)$-
invariance of the points m_o *and* (∞) *we see, that* $F_{\mathfrak{g}}$ *is* G-*equivariant:*

$$F_{\mathfrak{g}}(\sigma_g^* m) = Ad^*(g) F_{\mathfrak{g}}(m), \ \text{for all} \ m \in \mathcal{M}^\Pi, \ g \in G. \tag{5.1.155}$$

5.1.40 The projection measure $E_{\mathfrak{g}}^\Pi$ on $\mathfrak{g}^*$ together with the Ad^*-action of G determine a **macroscopic limit** of the system $(\mathfrak{A}^\Pi, \sigma_G)$. This formulation together with
the mapping p_M of $\mathcal{S}(\mathfrak{A}^\Pi)$ into the classical macroscopic states of the system will
enable us to generalize the notion of the macroscopic limit to much more general
situations. We shall investigate also the dynamics of the system $(\mathfrak{A}^\Pi, \sigma_G)$ (resp. of
its generalizations) if the time evolution were not included in the action σ_G as the
action of a one parameter subgroup of G. The action σ_G of the 'kinematical group' G
allows us, as we shall show, to introduce rather wide class of 'mean-field-type' time
evolutions connected with noncompact groups G—at least for a large σ_G-invariant
subset of states in $\mathcal{S}(\mathfrak{A}^\Pi)$. Also automorphic time evolutions $\tau : t \mapsto \tau_t \in *$-Aut $\mathfrak{A}$
of a system $(\mathfrak{A}, \sigma_G, \tau_\mathbb{R})$ will be considered.

5.2 Generalized Macroscopic Limits

5.2.1 We have considered, in the preceding section, a macroscopic limit of the
system $(\mathfrak{A}^\Pi, \sigma_G)$. This system was of a rather special type: the algebra $\mathfrak{A}^\Pi$ was the
infinite tensor product of identical copies $\mathfrak{A}_j$ $(j \in \mathbb{Z}_+ =: \Pi)$ of a C^*-algebra $\mathfrak{A}_0$ and
the automorphism group σ_G left each of the copies $\mathfrak{A}_j$ invariant: $\sigma_g x \in \mathfrak{A}_j$ for each
$x \in \mathfrak{A}_j$, for all $g \in G$ and any $j \in \mathbb{Z}_+ \equiv \Pi$. We shall now generalize the procedure
of obtaining a macroscopic limit to much more general situations. We shall ignore
here possible quasilocal structures of the considered C^*-algebra $\mathfrak{A}$; the usage of the

---

[5]The relation (5.1.154) has been proved in the assumption that any projector $p \in \mathfrak{M}_G^\Pi$ is of the form
$p = E_{\mathfrak{g}}^\Pi(B)$ for some $B \subset \mathfrak{g}^*$, if $p(I - p_G) = 0$.

term 'macroscopic limit' can be here understood in an analogy with the preceding section.

The notion of the macroscopic limit introduced in this section is nonunique. A certain arbitrariness is contained, however, also in the corresponding notion of Sect. 5.1: The generators X_ξ^N of the restriction of σ_G to $\mathfrak{A}^N := \otimes_{j=1}^N \mathfrak{A}_j \subset \mathfrak{A}^\Pi$ are determined up to additive constants $a_N(\xi)$, $a_N \in \mathfrak{g}^*$, $\xi \in \mathfrak{g}$,[6] hence also the choice of $p_G \in \mathfrak{Z}$ was arbitrary in a certain sense. We shall avoid partly this kind of ambiguity in this section: we are dealing here just with the action of σ_G, and not with generators.

5.2.2 Let G be a connected Lie group, $\mathfrak{g}$ its Lie algebra, and $\mathfrak{g}^*$ the dual of $\mathfrak{g}$. Let $\mathfrak{A}$ be an arbitrary C^*-algebra, $\mathfrak{A}^{**}$ its double dual W^*-algebra, and $\mathfrak{Z}$ is the center of $\mathfrak{A}^{**}$. The algebra $\mathfrak{A}$ is naturally contained in $\mathfrak{A}^{**}$ as a $\sigma(\mathfrak{A}^{**}, \mathfrak{A}^*)$-dense C^*-subalgebra. Any state $\omega \in \mathcal{S}(\mathfrak{A}) :=$ the state space of $\mathfrak{A}$, has a **natural extension**

$$e_*\omega \in \mathcal{S}_*(\mathfrak{A}^{**}) := \text{ the normal states of } \mathfrak{A}^{**}.$$

If $\mathfrak{M}$ is a C^*-subalgebra of $\mathfrak{A}^{**}$, **then** $r_\mathfrak{M} : \mathcal{S}(\mathfrak{A}^{**}) \to \mathcal{S}(\mathfrak{M})$ is the **restriction mapping**; $r_\mathfrak{M}$ is $\sigma(\mathfrak{A}^{***}, \mathfrak{A}^{**}) - \sigma(\mathfrak{M}^*, \mathfrak{M})$ continuous and maps normal states onto normal states.

Let $\sigma : G \to {}^*\text{-Aut}\,\mathfrak{A}$, $g \mapsto \sigma_g$ **be a given action of** G; by the same symbol σ_G is denoted the canonical extension of $\sigma_G \subset {}^*\text{-Aut}\,\mathfrak{A}$ to the action on $\mathfrak{A}^{**}$—the double transpose of σ_G. This system **will be denoted by** $(\mathfrak{A}; \sigma_G)$. $\mathcal{Z}$ **will denote the spectrum space of** $\mathfrak{Z} = C(\mathcal{Z})$.

Let $\mathfrak{g}^*$ be endowed with the structure of a Poisson manifold, 5.1.37, given by a **tensor field** λ, usually $\lambda_F(\cdot, \cdot) := -F([\cdot, \cdot]) - \theta_F(\cdot, \cdot)$, i.e.

$$\{f, g\}(F) := -F([\mathrm{d}_F f, \mathrm{d}_F g]) - \theta_F(\mathrm{d}_F f, \mathrm{d}_F g), \ F \in \mathfrak{g}^*, \ \theta_F \equiv \theta, \qquad (5.2.1)$$

where $f, g \in C^\infty(\mathfrak{g}^*, \mathbb{R})$ and θ **is a two form** on $\mathfrak{g}$ satisfying

$$\theta(\xi_1, [\xi_2, \xi_3]) + \theta(\xi_2, [\xi_3, \xi_1]) + \theta(\xi_3, [\xi_1, \xi_2]) = 0, \qquad (5.2.2)$$

for all $\xi_j \in \mathfrak{g}$, $j = 1, 2, 3$. We assume that an action of G on $\mathfrak{g}^*$ is $\varphi : g \mapsto \varphi_g$, where φ_G is a 'maximal' group of Poisson morphisms, i.e. $\varphi_{gh} = \varphi_g \circ \varphi_h (g, h, \in G)$, $\varphi_e := \mathrm{id}_{\mathfrak{g}^*} (e :=$ the identity of G); each φ_h is a diffeomorphism of $\mathfrak{g}^*$ conserving the **Poisson structure**:

$$\varphi_h^*\{f, g\} = \{\varphi_h^* f, \varphi_h^* g\}, \quad f, g \in C^\infty(\mathfrak{g}^*, \mathbb{R}), h \in G, \qquad (5.2.3)$$

and $\varphi_G F (\forall F \in \mathfrak{g}^*)$ are the maximal integral submanifolds of λ, [212, Definition 3.1 and Thmeorem 3.4]. Usually, one takes

$$\varphi_h F := Ad^*(h)(F) + a_\theta(h), h \in G, F \in \mathfrak{g}^*, \qquad (5.2.4)$$

---

[6]a_N forms a zero-dimensional orbit of $Ad^*(G) : a_N([\xi, \eta]) \equiv 0$.

where a_θ is a unique differentiable mapping from G to $\mathfrak{g}^*$ with the properties,
[212]:

(i) $a_\theta(gh) = Ad^*(g)(a_\theta(h)) + a_\theta(g), \quad \forall g, h, \in G$,
(ii) $T_e a_\theta(\xi)(\eta) = \theta(\xi, \eta), \quad \forall \xi, \eta \in \mathfrak{g}$, where $T_e a_\theta : \mathfrak{g} \to \mathfrak{g}^*$ is the tangent map of
a_θ at $e \in G$.

The system $(\mathfrak{A}; \sigma_G)$ represents a quantal system **and $(\mathfrak{g}^*, \lambda; \varphi_G)$ is a (general-ized) classical system** which will play the role of a macroscopic limit of the system $(\mathfrak{A}; \sigma_G)$. Let us introduce candidates for this micro-macro connection:

5.2.3 Definitions. *Let $\mathcal{B}(\mathfrak{g}^*)$ be the set of all complex-valued uniformly bounded Borel functions on $\mathfrak{g}^*$ and let Σ_G be the Borel σ-algebra of subsets of $\mathfrak{g}^*$. Let the G-measure E (of the system $(\mathfrak{g}^*, \lambda; \varphi_G)$, resp. of $(\mathfrak{A}; \sigma_G)$) be any projection-valued measure on $\mathfrak{g}^*$ with values in 3, which is G-equivariant, i.e.*

$$E : \Sigma_G \to 3, \ B \mapsto E(B) = E(B)^* = E(B)^2 \in 3 \ (B \in \Sigma_G), \qquad (5.2.5a)$$

$$B_j \cap B_k = \emptyset \ (j \neq k, \ j, k \in \mathbb{Z}_+) \Rightarrow E(\cup_j B_j) = \sum_j E(B_j), \qquad (5.2.5b)$$

$$E(\varphi_g B) = \sigma_g E(B), \quad \text{for all } B \in \Sigma_G, \text{ and for all } g \in G. \qquad (5.2.5c)$$

Denote by $E(f) \in 3$ the integral of $f \in \mathcal{B}(\mathfrak{g}^*)$ over E.

Let $p_E := E(\mathfrak{g}^*)$, $I :=$ the unit of $\mathfrak{A}^{**}$.

Denote by $\mathfrak{N}(E)$ the W^*-subalgebra of 3 generated by $E(f)$, $f \in \mathcal{B}(\mathfrak{g}^*)$.

Let $\mathfrak{B}(E)$ denote the **Borel*-algebra** *[235, 4.5.5] in 3 generated by all the $E(f)$, $f \in \mathcal{B}(\mathfrak{g}^*)$; this means that $\mathfrak{B}(E)$ is the smallest C^*-subalgebra of 3 con-taining all the $E(B) (B \in \Sigma_G)$ and with each monotone (increasing or decreasing) sequence $x_j \in \mathfrak{B}(E)_s$ it is also s-$\lim x_j \in \mathfrak{B}(E)$. Clearly $\mathfrak{B}(E) \subset \mathfrak{N}(E)$.* **Here $\mathfrak{M}_s$ is the set of all selfadjoint elements of a** C^*-algebra $\mathfrak{M}$. The projector p_E is the common unit of $\mathfrak{B}(E)$ and $\mathfrak{N}(E)$. Any projector $q \in \mathfrak{B}(E)$ is of the form $q = E(B)$ for some $B \in \Sigma_G$, what need not be the case for $\mathfrak{N}(E)$. Projections in $\mathfrak{N}(E)$ separate various kinds of spectra of E (resp. of operators $E(f)$ etc.) what need not be the case of $\mathfrak{B}(E)$.
  Let supp $E \subset \mathfrak{g}^*$ be the minimal closed $B = \overline{B} \in \Sigma_G$ such that $E(B) = p_E$. On the other hand, supp $E(B) := \{m \in \mathcal{Z} = \mathcal{ES}(3) : m(E(B)) = 1\}$ is a clopen subset of $\mathcal{Z}$. **Let**

$$\dim(F) := \text{dimension of the orbit } \varphi_G F \subset \mathfrak{g}^*, \ \dim(F) = 2k \leq \dim \mathfrak{g}^*,$$

and $\dim(E) := \max\{\dim(F) : F \in \text{supp } E\}$. *The G-measure E is **trivial** iff $\dim(E) = 0$. The quantal system $(\mathfrak{A}; \sigma_G)$ has a **nontrivial macroscopic limit in***

the **classical system** $(\mathfrak{g}^*, \lambda; \varphi_G)$ *iff there is E such that* $\dim(E) \geq 2$. *If there is an E such that* $\dim(E) = n_G$, *and for any other G-measure E' it is* $\dim(E') \leq n_G$, *we say that the system* $(\mathfrak{A}; \sigma_G)$ *has G-**macroscopic limit of the dimension** n_G (in the classical system* $(\mathfrak{g}^*, \lambda; \varphi_G)$). **The number** $n_G =: 2k_G$ **is the G-macroscopic dimension** *of* $(\mathfrak{A}; \sigma_G)$ *and k_G is the G-**macroscopic number of degrees of freedom** *of the quantal system* $(\mathfrak{A}; \sigma_G)$.

5.2.4 We shall assume in the following that $n_G \geq 2$ and we shall consider only G-measures E with $\dim(E) = n_G$. The projectors p_E are, clearly, G-invariant:

$$\sigma_g(p_E) = p_E \quad \text{for all } g \in G \text{ and all } G\text{-measures } E. \tag{5.2.6}$$

Let $q \leq p_E$ be another G-invariant projector in $\mathfrak{Z}$. Then **we can define the restriction of E to q, the G-measure qE,** by

$$qE : \Sigma_G \to \mathfrak{Z}, \ B \mapsto qE(B); \ p_{qE} = qp_E. \tag{5.2.7}$$

If $p_E p_{E'} = 0$ for two G-measures E and E' then the mapping

$$E + E' : \ B \mapsto E(B) + E'(B) \quad (\forall B \in \Sigma_G) \tag{5.2.8}$$

is a G-measure with $\dim(E + E') = \max\{\dim(E), \dim(E')\}$, and $p_{E+E'} = p_E + p_{E'}$. **For any two G-measures E and E', there is a G-measure $E s E'$ given by**

$$E s E'(B) := E(B) + (I - p_E)E'(B) \quad \forall B \in \Sigma_G. \tag{5.2.9}$$

For the support projector $p_{EsE'}$ of the G-measure EsE' we have

$$p_{EsE'} = p_E + p_{E'} - p_E p_{E'} = p_{E'sE}, \tag{5.2.10}$$

although, in general, EsE' is different from $E'sE$. Now, one has $\dim(EsE') \geq \dim(E)$. Since $p_{EsE'} = p_E \vee p_{E'} := \text{l.u.b.}[p_E; p_{E'}]$, we can endow the set of classes $[E]$

$$[E] := \{E' : p_{E'} = p_E\} \tag{5.2.11}$$

with a partial ordering:

$$[E] \succ [E'] \quad \Leftrightarrow \quad p_E \geq p_{E'}. \tag{5.2.12}$$

This ordering makes the set $\{[E]\}$ of classes of G-measures a directed set.

The same ordering will be considered for any set of subclasses $[E]' \subset [E]$ determined by some further condition C, i.e. for classes

$$[E]' := \{E' : p_{E'} = p_E, \ C(E')\}. \tag{5.2.13}$$

Here $C(E')$ means "the G-measure E' satisfies the condition C", e.g. $C(E') :=$ $(\dim(E') = \dim(E_\circ))$, or $C(E) := (E(F) \neq 0 \Rightarrow \dim(F) \neq 0)$, etc. The classes (5.2.13) could also be denoted by $[E]$.

5.2.5 Lemma. *The function* $\dim : \mathfrak{g}^* \to \mathbb{R}$, $F \mapsto \dim(F)$ *is* lower semicontinuous. *Hence, the sets* $\{F \in \mathfrak{g}^* : \dim(F) \geq n\}$ *are open and the sets* $\{F \in \mathfrak{g}^* : \dim(F) \leq n\}$ *are closed in* $\mathfrak{g}^*$ *for any* $n \in \mathbb{Z}_+$. *Specifically, the set* $\{F \in \mathfrak{g}^* : \dim(F) = 0\}$ *is closed, and the set* $\{F \in \mathfrak{g}^* : \dim(F) = n\}$ *is Borel.*

Proof. It was assumed in 5.2.2 that the action φ_G is a 'maximal' Poisson action, i.e. the orbits of φ_G coincide with the maximal integral manifolds of the Poisson structure λ on $\mathfrak{g}^*$, [212]. The dimension $\dim(F)$ of $\varphi_G F$ is then given by the rank of the skew-symmetric 2-tensor λ_F (:= the value of λ in the point $F \in \mathfrak{g}^*$), i.e. by the rank of the mapping $\lambda_F : T_F \mathfrak{g}^* \to T_F^* \mathfrak{g}^*$, $v \mapsto \lambda_F(v, \cdot)$, **denoted by** $\mathrm{rank}(\lambda_F)$. Since λ depends smoothly on F, the function $F \mapsto \dim(F) = \mathrm{rank}(\lambda_F)$ is lower semicontinuous. The remaining assertions then follow. $\qquad\square$

5.2.6 Let $S_n := \{F \in \mathfrak{g}^* : \dim(F) \leq n - 1\}$, $1 \leq n \leq \dim G$. **For such a** G **-measure** E **with** $p_E \neq E(S_n)$ let $r_n E := (p_E - E(S_n))E$, see (5.2.7). Clearly, $\dim(E) = \dim(r_n E)$, if $r_n E \neq 0$. E is a **purely nontrivial** G-measure, if $0 \neq p_E$ and $r_1 E = E$. If $0 \neq E = r_n E$ and $r_{n+1} E = 0$, E is called a **purely n-dimensional** G-measure. For $n := \dim(E)$ the measure $r_n E$ is purely n-dimensional. The G-measures $E + E'$, $E s E'$ and $q E$ (with a G-invariant projector $q = q p_E \neq 0$) are purely n-dimensional together with E and E'. Let the ordering (5.2.12) be given for the set of classes $[E] := \{E' : p'_E = p_E \text{ and } E' = r_n E', r_{n+1} E' = 0\}$. In any linearly ordered subnet of such $[E]$'s there is a natural mapping

$$\pi_{EE'} : [E'] \to [E], \ E' \mapsto \pi_{EE'}(E') := p_E E' \in [E] \text{ for } p_{E'} \geq p_E. \quad (5.2.14)$$

The **mappings** $\pi_{EE'}$ **define a projective system** [73, Definition 20.1] on the linearly ordered subset J of classes $[E]$: $\pi_{EE''} = \pi_{EE'} \circ \pi_{E'E''}$ for $p_{E''} \geq p_{E'} \geq p_E$ and $\pi_{EE} = \mathrm{id}_{[E]}$. If $p \leq p_{E'}$, and $E := p E'$, then $E' = E s E'$. We want to show that J has an upper bound in the set of classes $[E]$ of purely n-dimensional G-measures E. This would imply, by the Zorn's lemma, the existence of *the* maximal element in the set (uniqueness of the maximal element follows from the directedness of the set).

5.2.7 Lemma. *Let L be a set of* **G-measures linearly ordered** *by* $E \leq E' \Leftrightarrow E' = E s E'$. *Then L has an upper bound.*

Proof. For any $0 \leq f \in \mathcal{B}(\mathfrak{g}^*)$ and $E' \geq E$ it is $E'(f) \geq E(f)$. Denote

$$E_L(f) := \mathrm{l.u.b.}\{E(f) : E \in L\} = s\text{-}\lim\{E(f) : E \in L\}. \quad (5.2.15)$$

The mapping E_L can be extended by linearity to $\mathcal{B}(\mathfrak{g}^*)$:

$$E_L : \mathcal{B}(\mathfrak{g}^*) \to 3, \ f \mapsto E_L(f); \quad (5.2.16)$$

it is bounded: $\|E_L(f)\| \leq \|f\| := \sup\{|f(F)| : F \in \mathfrak{g}^*\}$. Due to continuity of the product in the strong topology, the mapping E_L is a C^*-homomorphism of the commutative C^*-algebra $\mathcal{B}(\mathfrak{g}^*)$ into $\mathfrak{Z}$. This implies the σ-additivity of the set-function $E_L : B \mapsto E_L(B) := E_L(\chi_B)$, $B \in \Sigma_G$, hence, E_L is a projection measure. Since $\sigma_g \in {}^*\text{-Aut }\mathfrak{Z}$, and any automorphism of a W^*-algebra is σ-σ-continuous, E_L is a G-measure. Clearly $E_L \geq E$, $\forall E \in L$.   □

5.2.8 Proposition. *The directed set of classes of purely n-dimensional G-measures has a maximal element.*[7]

Proof. Let J be any linearly ordered subset of the directed set; cf. (5.2.12). We shall prove that it is possible to choose $E \in [E]$ in any $[E] \in J$ in such a way that $[E] \prec [E']$ iff $E' = EsE'$. Then the result will follow from the Lemma 5.2.7 and from the Zorn lemma. It is clear that the choice $E \in [E]$ of the desired kind can be made in any finite subset $K_\circ \subset J$, $[E] \in K_\circ$.

The desired choice (it will be called a 'consistent choice') can be made in the subset $K_E := \{[E'] \in J : [E'] \prec [E]\}$ of J by $E' := p_{E'}E$ for any $[E] \in J$, with any fixed $E \in [E]$. We have to prove existence of a consistent choice on the whole J. Let $J_\circ$ be a **well ordered cofinal subset** of J (the well ordering of $J_\circ$ is that one induced by the ordering of J–it is possible by the axiom of choice, and cofinality means that for any $[E] \in J$ there is an $[E_j] \in J_\circ : [E] \prec [E_j]$). Now we can choose $E_j \in [E_j]$ (for all $[E_j] \in J_\circ$) in a consistent way: For the successor $[E_{j+1}]$ of $[E_j]$ in $J_\circ$ we shall choose $E_{j+1} := E_j s E'_{j+1}$ with any $E'_{j+1} \in [E_{j+1}]$, if $[E_j]$ has been defined before. If $[E_j]$ is not a successor in $J_\circ$, put $E_j^\circ := \text{l.u.b.}\{E_k : [E_j] \succ [E_k] \in J_\circ$, all $E_k(\in [E_k])$ are mutually consistent$\}$, according to the Lemma 5.2.7, and choose $E_j := E_j^\circ s E'_j$ with any $E'_j \in [E_j]$. Then we can 'to fill gaps' by setting $E := p_E E_j$ for all $[E] \prec [E_j] \in J_\circ$. This provides a consistent choice $E \in [E]$ for all $[E] \in J$, if $J_\circ$ is considered as an initial segment of the set of all ordinals.   □

Note: The same proof applies to purely n-dimensional measures E of the form $E = qE$ for any fixed G–invariant projector $q \in \mathfrak{Z}$.

5.2.9 Let $[E]_G^\circ$ be the **maximal element of classes of purely n_G-dimensional G-measures** and let $p_G^\circ := p_E$ for $E \in [E]_G^\circ$. Let $[E]_G^k$ be **the maximal element of classes of purely $(n_G - 2k)$-dimensional G-measures** of the form $E = (I - \sum_{j=0}^{k-1} p_G^j)E$, and for $E \in [E]_G^k$ let $p_G^k := p_E$, $k = 1, 2, \ldots, \frac{n_G}{2}$. Define now the **class $[E]_G$ of maximal G-measures** by

$$[E]_G := \sum_{k=0}^{\frac{n_G}{2}}[E]_G^k, \text{ with } E \in [E]_G \text{ iff } E = \sum_{k=0}^{\frac{n_G}{2}} E_k, \ E_k \in [E]_G^k,$$

and the sum of mutually *orthogonal G-measures* is defined in (5.2.8). The choice of measures $E \in [E]_G$ for the realization of macroscopic limits corresponds to a

---

[7]The present author was informed about some important set-theoretical concepts connected with this Proposition by the late colleague Ivan Korec (1943–1998).

requirement of 'maximal sensitivity' of the corresponding macroscopic description of the system $(\mathfrak{A}; \sigma_G)$.

We shall not proceed further in an analysis of the set $[E]_G$ and we shall not try to specify some 'most convenient' element $E \in [E]_G$ as a representative of the macroscopic limit. Let us choose any fixed $E_{\mathfrak{g}} \in [E]_G$.

5.2.10 Definitions. *The projection-valued measure $E_{\mathfrak{g}} \in [E]_G$ on the Poisson manifold $(\mathfrak{g}^*, \lambda; \varphi_G)$, the G-action on which is 'maximal' (i.e. orbits $\varphi_G F$ are maximal symplectic immersed submanifolds of $\mathfrak{g}^*$ the Poisson bracket on which is given by λ, for any $F \in \mathfrak{g}^*$),* **with values in $\mathfrak{Z}$ (:= the center of $\mathfrak{A}^{**}$)** *is called the G-***macroscopic limit of the system** $(\mathfrak{A}; \sigma_G)$ *in the classical system* $(\mathfrak{g}^*, \lambda; \varphi_G)$*. The* **projector** $p_G := E_{\mathfrak{g}}(\mathfrak{g}^*)$ *is the* **support projector** *of the macroscopic limit. The dimension n_G will be called also* **the dimension of $E_{\mathfrak{g}}$.** *The Borel*- (resp. the W*-) algebra [235, 4.5.5] generated by $E_{\mathfrak{g}}$ (resp. by $E_{\mathfrak{g}}$ and $I \in \mathfrak{Z}$) will be called the B*- (resp. W*-)* **macroscopic algebra of G-definiteness** *(resp. the G-***macroscopic algebra**) of the system $(\mathfrak{A}; \sigma_G)$ and will be denoted (in the W*-cases) by $\mathfrak{N}_G$ (resp. by $\mathfrak{M}_G$).*

Denote by $p_M : \mathcal{S}(\mathfrak{A}) \to \mathcal{S}_*(\mathfrak{M}_G), \omega \mapsto p_M\omega := r_{\mathfrak{M}} \circ e_*(\omega)$, *where $r_{\mathfrak{M}}$ is the* **restriction** *of $\mathcal{S}(\mathfrak{A}^{**})$ to $\mathcal{S}(\mathfrak{M}_G)$ and e_* is the* **natural extension** *from $\mathcal{S}(\mathfrak{A})$ to $\mathcal{S}_*(\mathfrak{A}^{**})$. Let $\mu_{\mathfrak{g}}^{\omega}$ be the probability measure on $\mathcal{M}^G$ (:= $\mathfrak{g}^* \cup \{m_o\}$, m_o is an isolated point) given by*

$$\mu_{\mathfrak{g}}^{\omega}(B) := \omega(E_{\mathfrak{g}}(B\backslash\{m_o\})) + \omega(I - p_G)\delta_{m_o}(B), \ any \ Borel \ B \subset \mathcal{M}^G, \quad (5.2.17)$$

compare (5.1.137). Let us introduce the set

$$\mathcal{E}_{\mathfrak{g}} := \{\omega \in \mathcal{S}(\mathfrak{A}) : e_*\omega(p_G) = 1, \ \mu_{\mathfrak{g}}^{\omega}(\xi^2) = [\mu_{\mathfrak{g}}^{\omega}(\xi)]^2 < \infty, \ \forall \xi \in \mathfrak{g}\}, \quad (5.2.18)$$

compare (5.1.142), where $\xi \in \mathfrak{g}$ is considered as a linear function on $\mathfrak{g}^*$, since $\mathfrak{g} \subset \mathfrak{g}^{**}$.

We can introduce also unbounded operators $X_\xi := E_{\mathfrak{g}}(\xi)$ on the Hilbert space $\mathcal{H}_u$ of the universal representation of $\mathfrak{A}$. Then we have

5.2.11 Theorem. *The Theorem 5.1.38 as well as its proof are valid also after the omission of the index Π everywhere in its formulation and exchange of $Ad^*(G)$ by φ_G, with the interpretation of symbols according to 5.2.10.*

5.2.12 Note. *We could now, after the recognizing of the Theorem, to continue in the choices of $E_{\mathfrak{g}} \in [E]_G$ according to the following idea: Choose $E_{\mathfrak{g}}$ such that the sets $\mathcal{S}_F$ of states with sharp values of the macroscopic observables (cf. 5.1.38 (iii)) are in a certain sense 'maximal'. We shall not make this idea precise here. We believe, however, that continuing in this direction we could obtain $E_{\mathfrak{g}}$ 'essentially uniquely'—up to natural coordinate transformations in the $\mathfrak{g}^*$.*

5.2.13 A scheme of 'macroscopic quantization'.

Having once a classical limit in the form of the couple $\{(\mathfrak{g}^*, \lambda; \varphi_G), (\mathfrak{M}_G; \sigma_G)\}$, where $\sigma_G \subset {}^*\text{-}\operatorname{Aut}\mathfrak{M}_G$, we are interested in the question: Can the original algebra $\mathfrak{A}$ be reconstructed from this classical limit? Keeping in mind the model of Sect. 5.1 we propose the following scheme for obtaining the algebra $\mathfrak{A}$ of a system $(\mathfrak{A}; \sigma_G)$, the macroscopic limit of which is $(\mathfrak{M}_G; \sigma_G; E_\mathfrak{g})$ (here the measure $E_\mathfrak{g}$ symbolizes the connection with the classical system $(\mathfrak{g}^*, \lambda; \varphi_G)$), (let us denote by MQ the following scheme):

(MQ) *Find a faithful representation ρ of $\mathfrak{M}_G$ in a Hilbert space $\mathcal{H}_\rho$ (necessarily nonseparable) with the properties*:

(i) There is a simple C^*-subalgebra $\mathfrak{A}$ of $\mathcal{L}(\mathcal{H}_\rho)$ such that the center of its commutant $\mathfrak{A}'$ contains $\rho(\mathfrak{M}_G)$; σ_G extends to an automorphism group of $\mathfrak{A}$.

(ii) $\mathfrak{A}$ is expressible as the norm-closure of union of a net of von Neumann subalgebras $\mathfrak{A}_j$ ($j \in J :=$ a directed set): $j \prec k \Rightarrow \mathfrak{A}_j \subset \mathfrak{A}_k$.

(iii) Each $\mathfrak{A}_j$ is a σ_G-invariant subset of $\mathfrak{A}$ and the restriction of σ_G to any $\mathfrak{A}_j$ ($j \in J$) is unitarily implementable (i.e. it exists a strongly continuous unitary representation U^j of G in $\mathcal{H}_\rho$ such that $\sigma_g(x) = U^j(g)xU^j(g^{-1})$ for all $x \in \mathfrak{A}_j$, $g \in G$ and $j \in J$).

(iv) Each $\mathfrak{A}_k$ ($k \in J$) is generated by all $\mathfrak{A}_j$ with $j \prec k$ ($j \neq k$) as well as by the bounded Borel functions of the selfadjoint generators X_ξ^k ($\xi \in \mathfrak{g}$) of the one parameter groups $t \mapsto U^k(\exp(t\xi))$.

Hence the proposed 'quantization procedure' of the classical system $(\mathfrak{g}^*, \lambda; \varphi_G)$ consists in finding an *'imprimitivity system'* $(\mathfrak{M}_G, \sigma_G)$ (cf. [321]) determined by a choice of a G-measure $E_\mathfrak{g}$ (in some commutative C^*-algebra $\mathfrak{M}_G$, where $\sigma_G \subset {}^*\text{-}\operatorname{Aut}\mathfrak{M}_G$ is determined by $\sigma_g E_\mathfrak{g}(B) := E_\mathfrak{g}(\varphi_g B)$, $g \in G$, $B =$ Borel subsets in $\mathfrak{g}^*$), and afterwards applying the scheme (MQ) of *'macroscopic quantization'* to $(\mathfrak{M}_G; \sigma_G)$. We shall not investigate here conditions of existence and a 'degree of uniqueness' of this recipe. The scheme is nonempty, since it is fulfilled e.g. by the models considered in Sect. 5.1 if $U(G)$ is irreducible, 5.1.3.

The question of obtaining a microscopic quantum dynamics of this 'quantized macroscopic system' corresponding to its given classical time evolution is posed and solved in the next Chap. 6.

Chapter 6
Dynamics of Quantum Mechanical Macroscopic Systems

6.1 General Considerations

6.1.1 The formalism developed in Chap. 5 will be used in this chapter for a determination of a microscopic time evolution of an infinite quantum system from the macroscopic (classical) evolution. It is clear that such an unusual determination of *microscopic dynamics* is possible for a very special type of interactions only. We shall show that this is the case of a wide class of *quantum mean-field theories*,[1] at least in the time invariant subset $S_{\mathfrak{g}}^{\Pi}$ of the set $S(\mathfrak{A}^{\Pi})$ of all the microscopic states on the quasilocal algebra $\mathfrak{A}^{\Pi}$, cf. Sect. 5.1, esp. 5.1.32; cf. also '*classical states*' in [155]. The systems of the considered type are determined by the couple $(\mathfrak{A}; \sigma_G)$ consisting of a C^*-algebra $\mathfrak{A}$ $(:= \mathfrak{A}^{\Pi}$, e.g.; the upper indices Π will be usually omitted in this chapter) and of a representation $\sigma(G) := \sigma_G \subset {}^*\text{-}\mathrm{Aut}\,\mathfrak{A}$, cf. 5.2.2, as well as by a G-measure $E_{\mathfrak{g}}$, 5.2.3, and by a classical Hamiltonian function $Q \in C^{\infty}(\mathfrak{g}^*, \mathbb{R})$. A subclass of these systems consists of *thermodynamic limits* $N \to \infty$ of systems of the total number N of quantal (mutually equal) subsystems with dynamics described by *local Hamiltonians* Q^N. These local Hamiltonians are invariant with respect to any permutations of N subsystems and the k-body interaction constants (i.e. coefficients at products of k operators corresponding to k different subsystems) are proportional to N^{1-k}. We can construct such a sequence of the '*local time evolutions*' $\tau^N \subset {}^*\text{-}\mathrm{Aut}\,\mathfrak{A}$ in the following way:

Let us keep the notation of Sect. 5.1, and let a basis ξ_j $(j = 1, \ldots n)$ of $\mathfrak{g}$ be fixed, the dual basis being $\{f_j : j = 1, 2, \ldots n\} \subset \mathfrak{g}^*$. Let X_j $(j = 1, 2, \ldots n)$ be the selfadjoint generators of the one parameter unitary groups $t \mapsto U(\exp(t\xi_j))$ on $\mathcal{H}$, 5.1.3. Let Q be a polynomial in n variables and with a prescribed order of

---

[1] For some history, general meaning and technical construction of dynamics (given by full and correctly solved microscopic evolutions—without any approximations) of "Quantum mean-field theories" see also [40], and for some of its applications look in [41].

© Springer Nature Switzerland AG 2020, corrected publication 2020
P. Bóna, *Classical Systems in Quantum Mechanics*,
https://doi.org/10.1007/978-3-030-45070-0_6

multiplication of variables in such a way that the element $Q(\xi_1, \xi_2, \ldots \xi_n)$ of the Lie algebra envelope has the following property:

(SA) *Let* $\mathbf{Q} \equiv \sum_{k=0}^{q} \mathbf{Q}_k$, *where* $\mathbf{Q}_k$ *is a homogeneous polynomial of degree k. In any continuous unitary representation U of the group G on a separable Hilbert space* $\mathcal{H}$, *the operators* $Q_k(X_1, X_2, \ldots, X_n)$ *defined on analytic elements of U are essentially selfadjoint, for all* $k = 1, 2, \ldots, q$.

This *property (SA)* is fulfilled e.g. **if all the** $Q_k(\xi_1, \xi_2, \ldots, \xi_n)$ **are symmetric and elliptic**, cf. [13, Chap. 11]. Then we define the *local Hamiltonians* Q^N, with (5.1.14), denoting by N also an N-point subset of Π:

$$Q^N := N\, Q(X_{1N}, X_{2N}, \ldots, X_{nN}), \quad X_{jN} := \frac{1}{N} X_j^N, \quad N = 1, 2, \ldots; \quad (6.1.1)$$

cf. (5.1.118), i.e.

$$X_j^N := \sum_{k=1}^{|N|} \pi_k(X_j), \quad j = 1, 2, \ldots n, \ k \in N \subset \Pi,$$

which can be considered as (essentially) selfadjoint operators on $\mathcal{H}_\Pi$ $(= \mathcal{H}_N \otimes \mathcal{H}_{\Pi\setminus N} \equiv \Pi$-tuple tensor product). For any $x \in \mathfrak{A}$ $(:= \mathfrak{A}^\Pi) \subset \mathcal{L}(\mathcal{H}_\Pi)$ we set

$$\tau_t^N(x) := \exp(it Q^N)\, x\, \exp(-it Q^N), \quad t \in \mathbb{R}, \quad (6.1.2)$$

and these mappings τ_t^N clearly form a one parameter group of *-automorphisms of $\mathfrak{A}$ for each finite N. Systems of this type were introduced in [155] for the case of spin systems (i.e. dim $\mathcal{H}$ was finite). It was shown in [40, 155] that the sequence $\{\tau^N : N = 1, 2, \ldots\}$ determines an evolution τ^Q of the observables of the form $X_{\xi\Pi}$, cf. 5.1.7 and 5.1.9, which is expressed in our notation by the formula

$$\tau_t^Q(X_{\xi\Pi}) := w_0^* \text{-} \lim_{N \to \infty} \tau_t^N(X_{\xi N}) = \int f_\xi(\varphi_t^Q F)\, E_\mathfrak{g}(dF), \quad (6.1.3)$$

where w_0^*-topology on a von Neumann algebra containing $\mathfrak{A}$ and $X_{\xi\Pi}$ $(\xi \in \mathfrak{g})$ is determined by the set of the 'classical states'. The integral in (6.1.3) corresponds to the integral in [155, (2.29)], which, specified to our case, reads:

$$\lim_{N \to \infty} \omega(\tau_t^N(X_{\xi N})) = \int f_\xi(\varphi_t^Q F)\, \omega(E_\mathfrak{g}(dF)), \quad \omega \in \mathcal{S}_\mathfrak{g}. \quad (6.1.4)$$

We have used notation $f_\xi(F) := F(\xi)$ $(\xi \in \mathfrak{g}, F \in \mathfrak{g}^*)$, and φ^Q is the classical flow on $\mathfrak{g}^*$ corresponding to the Hamiltonian function $Q \in C^\infty(\mathfrak{g}^*, \mathbb{R})$, $Q(F) := Q(F_1, F_2, \ldots F_n)$, with $F_j := f_{\xi_j}(F) = F(\xi_j)$, $(F \in \mathfrak{g}^*)$; the introduction of the flow φ^Q will be discussed later in this section. The natural question is, however, whether the limits

$$\tau_t^Q(x) := (some \; topology) - \lim_{N \to \infty} \tau_t^N(x) \qquad (6.1.5)$$

exist for some $t > 0$ and for sufficiently many $x \in \mathfrak{A}$, so that τ_t^Q could be extended to a one parameter group (resp. semigroup) of mappings of $\mathfrak{A}$ (or of some of its completions) representing in a reasonable manner some time translations. We shall show that this is indeed the case, and not only for the spin systems. The resulting family of transformations τ^Q **does not consist**, however, (for general Q) **of automorphisms of the original** (i.e. that one used at the determination of the infinite system) **quasilocal** C^*-algebra $\mathfrak{A}$. The family of *-isomorphisms of $\mathfrak{A}$, τ^Q, can be extended to a one parameter group τ^Q of *-automorphism of a C^*-subalgebra of $\mathfrak{A}^{**}$ containing $\mathfrak{A}$ as a C^*-subalgebra. The resulting picture of the τ^Q-time evolution has the properties of the *quantum mean-field evolutions* according to the usual understanding, cf. also [40]. We shall write down explicit formulas for the evolution of an arbitrary element of the extended algebra of observables (including also an algebra of classical—intensive—observables) in terms of solutions of finite dimensional differential equations.

In the present section, we shall introduce some basic concepts used in the general construction of the automorphism group τ^Q. We shall sketch here also a scheme of the general construction. Details will be proved in the following sections of this chapter.

6.1.2 Let $(\mathfrak{g}^*, \lambda; \varphi_G)$ be a **Poisson manifold** with the **Poisson action** φ_G of the Lie group G the orbits of which coincide with the maximal integral submanifolds of the Poisson structure λ, cf. [212], and 5.1.37, and 5.2.2. **We shall assume, for simplicity, that** $\varphi_G := Ad^*(G)$ **and** $\lambda_F(\mathrm{d}f, \mathrm{d}g) := -F([\mathrm{d}_F f, \mathrm{d}_F g])$ **for all** $f, g \in C^\infty(\mathfrak{g}^*, \mathbb{R})$. Let $Q \in C^\infty(\mathfrak{g}^*, \mathbb{R})$ be such a fixed function on $\mathfrak{g}^*$, that the corresponding *Hamiltonian vector field* σ_Q on $\mathfrak{g}^*$, (5.1.144), is *complete*. This means that there is a one parameter group $t \mapsto \varphi_t^Q$ ($\varphi_{t+s}^Q = \varphi_t^Q \circ \varphi_s^Q$ for all $t, s \in \mathbb{R}$) of Poisson morphisms of $(\mathfrak{g}^*; \lambda)$ the derivative of which is σ_Q. Remember that σ_Q is complete for any Q in the case of compact groups G, in which case the $Ad^*(G)$-orbits are compact. The tangent spaces $T_F \mathfrak{g}^*$ ($F \in \mathfrak{g}^*$) will be identified with the linear manifold $\mathfrak{g}^*$ in the canonical way. Then we have also the canonical identification $T_F^* \mathfrak{g}^* = \mathfrak{g}$ of the cotangent spaces in any point $F \in \mathfrak{g}^*$ with the Lie algebra $\mathfrak{g}$ of G. Let $f_\xi \in C^\infty(\mathfrak{g}^*, \mathbb{R})$ **(for any** $\xi \in \mathfrak{g}$**) be the linear function**

$$f_\xi : F \mapsto f_\xi(F) := F(\xi) \, .$$

Any element ξ of the Lie algebra $\mathfrak{g}$ determines also a covector field on $\mathfrak{g}^*$:

$$\mathrm{d}f_\xi : F \mapsto \mathrm{d}_F f_\xi = \xi \in \mathfrak{g} = T_F^* \mathfrak{g}^* \, . \qquad (6.1.6)$$

The Hamiltonian (contravariant) vector field corresponding to the Hamiltonian function f_ξ coincides with the vector field σ_ξ determined by the flow

$$\varphi^\xi : (t; F) \mapsto \varphi_t^\xi F := Ad^*(\exp(t\xi))F \qquad (6.1.7)$$

on $\mathfrak{g}^*$. We have the relations:

$$\{h, f_\xi\}(F) = -F([dh, df_\xi]) = d_F f_\xi(\sigma_h) = -d_F h(\sigma_\xi), \quad h \in C^\infty(\mathfrak{g}^*, \mathbb{R}),$$
$$(6.1.8)$$

where σ_h is the Hamiltonian vector field corresponding to the Hamiltonian function h, cf. (5.1.144) and (5.1.145).

6.1.3 Let $g_Q : \mathbb{R} \times \mathfrak{g}^* \to G$, $(t; F) \mapsto g_Q(t, F)$ be a function determining the Hamiltonian flow φ_t^Q with the help of the action $\varphi_G := Ad^*(G)$ in the following sense:

$$Ad^*(g_Q(t, F))F = \varphi_t^Q(F) := \varphi_t^Q F, \text{ for all } t \in \mathbb{R}, \text{ and for all } F \in \mathfrak{g}^*. \quad (6.1.9)$$

Such functions g_Q exist due to φ^Q-invariance of the maximal integral submanifolds of φ_G (i.e. the orbits of $Ad^*(G)$) with respect to any Hamiltonian flow. Let us assume differentiability of g_Q **and set**

$$\beta_F^Q := \left.\frac{d}{dt}\right|_{t=0} g_Q(t, F), \text{ for all } F \in \mathfrak{g}^*. \qquad (6.1.10)$$

A necessary condition for fulfilment of (6.1.9) is the fulfilment of

$$F([\beta_F^Q, \eta]) = d_F Q(\sigma_\eta) \ (= -\Omega_F(\sigma_Q, \sigma_\eta) = -d_F f_\eta(\sigma_Q)), \quad \eta \in \mathfrak{g}, \ F \in \mathfrak{g}^*,$$
$$(6.1.11)$$

(cf. (6.1.8)), where Ω is the standard *Kirillov-Kostant symplectic form* on $\mathfrak{g}^*$, since the following relation is valid:

$$\left.\frac{d}{dt}\right|_{t=0} Ad^*(g_Q(t, F))F(\eta) = -F([\beta_F^Q, \eta]), \quad \eta \in \mathfrak{g}, F \in \mathfrak{g}^*. \qquad (6.1.12)$$

If we require, in addition to (6.1.11), fulfilment of the following '**cocycle identities**':

$$g_Q(s, \varphi_t^Q F)g_Q(t, F) = g_Q(t + s, F), \ g_Q(0, F) \equiv e, \qquad (6.1.13)$$

for all $t, s \in \mathbb{R}$ and all $F \in \mathfrak{g}^*$ (with $e :=$ the identity of G), then the condition (6.1.11) will be also sufficient for the validity of (6.1.9). **Let** $\beta^\circ : F \mapsto \beta_F^\circ \in \mathfrak{g}$ **be any differentiable function on** $\mathfrak{g}^*$ **satisfying**

$$F([\beta_F^\circ, \eta]) = 0, \quad \text{for all } F \in \mathfrak{g}^*, \ \eta \in \mathfrak{g}. \qquad (6.1.14)$$

Elements $\beta_F^\circ \in \mathfrak{g}$ determine one parameter subgroups of the stability groups of $F \in \mathfrak{g}^*$ for the coadjoint action $Ad^*(G)$, cf. Lemma 3.2.4. If a given β_F^Q satisfies (6.1.11), then also the substitution of

$$\beta_F^{'Q} := \beta_F^Q + \beta_F^\circ \tag{6.1.15}$$

in place of β_F^Q in (6.1.11) will give a valid equality. Let β_F^Q be an infinitely differentiable function of $F \in \mathfrak{g}^*$ with values in $\mathfrak{g}$ satisfying (6.1.11). The equation (6.1.13) with the condition (6.1.10) can be rewritten in the form of a differential equation on the group manifold G:

$$\frac{d}{dt} g_Q(t, F) = T_e(R_{g_Q(t,F)})\beta_{F_t}^Q, \quad \forall t \in \mathbb{R}, \ F \in \mathfrak{g}^*, \tag{6.1.16}$$

where $F_t := \varphi_t^Q(F)$, and R_G is the right action of the group G on itself: $R_g(h) := hg$ ($g, h \in G$); T_e is the tangent mapping restricted to the tangent space $T_e G = \mathfrak{g}$ of the group G at the identity $e \in G$, $T_e(f) : T_e G \to T_{f(e)}G$,

$$\xi \mapsto T_e(f)\xi := f_*\xi := \left.\frac{d}{dt}\right|_{t=0} f(\exp(t\xi))$$

for any differentiable function $f : G \to G$. According to the general theory of ordinary differential equations, there is a unique solution of (6.1.16) with the initial condition $g_Q(0, F) = e$. The solution g_Q depends, however, on the choice of the covector field β^Q which is, according to (6.1.15), nonunique in the general case.

The **cocycle** g_Q is, as we shall see later, the basic dynamical object determining fully the microscopic time evolutions in the mean-field theories of the considered type. Various choices of β^Q corresponding to the various possible choices of β° according to (6.1.15) will lead to the same classical evolution φ^Q of the subalgebra of classical (intensive) quantities of the extended algebra of quantal observables of the infinite system. The time evolutions of local (microscopic) observables corresponding to various choices of β° in (6.1.15) are, however, mutually different. We shall see that the **thermodynamic limits described in 6.1.1 correspond to the choice**

$$\beta_F^Q := d_F Q, \quad F \in \mathfrak{g}^*. \tag{6.1.17}$$

If we write $Q(F)$ in the terms of coordinate functions $F_j := F(\xi_j)$ as in 6.1.1, then we have

$$d_F Q = \sum_{j=1}^n \frac{\partial Q(F)}{\partial F_j} \xi_j \in \mathfrak{g}. \tag{6.1.18}$$

Let the structure constants of $\mathfrak{g}$ in the basis $\{\xi_j\}$ are $c_{kl}^j \in \mathbb{R}$, i.e.

$$[\xi_k, \xi_l] = c_{kl}^j \xi_j. \tag{6.1.19}$$

Then we have for the Poisson bracket of two classical Hamiltonians Q_1 and Q_2 the expression (called also the *Berezin bracket*):

$$\{Q_1, Q_2\}(F) := -F([d_F Q_1, d_F Q_2]) = -c^j_{km} \frac{\partial Q_1}{\partial F_k} \frac{\partial Q_2}{\partial F_m} F_j. \qquad (6.1.20)$$

6.1.4 Let us describe here, in a heuristic manner, the basic idea leading to the definition of the time evolutions τ^Q mentioned in 6.1.1 which will be described in Sect. 6.3 in details. It will be also shown in Sect. 6.3 that the evolutions obtained from the thermodynamic limits in the 'polynomial cases' (mentioned in 6.1.1 and investigated in Sect. 6.2) are special cases of the general definition of τ^Q based on the following general ideas.

The *cocycle* g_Q reproduces an arbitrary classical Hamiltonian evolution on the Poisson manifold $(\mathfrak{g}^*, \lambda; Ad^*(G))$ (since Q is an arbitrary Hamiltonian function) via the given (fixed!) action $Ad^*(G)$, cf. (6.1.9). We have given an action $\sigma(G) \in {}^*\text{-Aut}\,\mathfrak{A}$ and also the corresponding dual action $\sigma^*(G)$ on the set $\mathcal{S}(\mathfrak{A})$ of states on $\mathfrak{A}$, cf. (5.1.44). We have also a canonical decomposition of an arbitrary state $\omega \in \mathcal{S}(\mathfrak{A})$ into the states ω_m corresponding to classical phase space points $m \in \mathcal{M}$, namely (5.1.146), resp. the corresponding statement in 5.2.11. For $\omega \in \mathcal{S}_\mathfrak{g} := p_G \mathcal{S}(\mathfrak{A})$, the states ω_m lying in the support of the corresponding measure $\hat{\mu}_\omega$ on $\mathcal{S}(\mathfrak{A}^{**})$ can be indexed by $F_m \in \mathfrak{g}^*$, where the classical measure on $\mathfrak{g}^*$ corresponding to the state $\omega_m \in \mathcal{E}_\mathfrak{g}$ is concentrated on the one point set $\{F_m\}$, cf. 5.1.36 and 5.1.39. Hence we can use the family of mappings

$$t \mapsto \sigma^*(g_Q(t, F_m)), \ t \in \mathbb{R}, \ m \in \mathcal{M}, \qquad (6.1.21)$$

for a definition of time translations of the states ω_m. Such a definition makes sense since the projection measure $E_\mathfrak{g}$ (:= the G-macroscopic limit of the system $(\mathfrak{A}; \sigma(G))$ in $(\mathfrak{g}^*, \lambda; Ad^*(G))$, 5.2.10) is G-equivariant, (5.2.5c), what implies that the classical point-measure corresponding to $\sigma^*(g_Q(t, F_m))\omega_m \in \mathcal{E}_\mathfrak{g}$ is concentrated on $Ad^*(g_Q(t, F_m))F_m = \varphi^Q_t(F_m) \in \mathfrak{g}^*$; hence the cocycle identity (6.1.13) can be used to prove the group property of mappings (6.1.21). A heuristic definition of the time evolution τ^Q is then given with the help of the decomposition (5.1.146) by the formula:

$$\omega(\tau^Q_t(x)) := \int_\mathcal{M} \sigma^*(g_Q(t, F_m))\omega_m(x)\, \mu_\omega(dm), \quad \forall t \in \mathbb{R}, \ \omega \in \mathcal{S}_\mathfrak{g}. \qquad (6.1.22)$$

We shall see in Sect. 6.3 that this intuitive construction leads to a rigorously defined group τ^Q of *-automorphisms of a C^*-subalgebra of the W^*-algebra $p_G \mathfrak{A}^{**}$ containing the algebra $\mathfrak{A}$ as well as an algebra $\mathfrak{N}^c$ of classical observables in a natural manner. The algebra $\mathfrak{N}^c$ is then τ^Q-invariant: $\tau^Q_\mathbb{R}(\mathfrak{N}^c) = \mathfrak{N}^c$, contrary to the algebra $\mathfrak{A}$ (in a general case).

6.1.5 Remark. The general definition of mean-field time evolutions τ^Q based on the formula (6.1.22) depends on a topology determined by the subset $\mathcal{S}_\mathfrak{g} := p_G \mathcal{S}(\mathfrak{A})$

of states on $\mathfrak{A}$ (and their canonical normal extensions to $\mathfrak{A}^{**}$), so called 'classical states'. The reason why we cannot use the set of all states $\mathcal{S}(\mathfrak{A})$ for the definition of τ^Q can be seen from the thermodynamic limits of polynomial interactions described in Sect. 6.2: In the representations of $\mathfrak{A}$ containing the GNS-representations of states $\{\omega : \omega(p_G) \neq 1\}$ as their subrepresentations the thermodynamic limits of the local evolutions τ^N do not exist for a general Q. This fact can be seen from the definition of the projector p_G in (5.1.120) as well as from considerations in Sect. 6.2. Although the resulting (algebraic) concept of τ^Q can be used in certain cases to a definition of time evolution of all states on $\mathfrak{A}$, such a definition scarcely can be considered as a physically correct consequence of the given interaction Q. This interaction does not lead to any reasonable (from the point of view of physics) time evolution of states ω of the infinite system, the central supports s_ω of which are orthogonal to $p_G : s_\omega p_G = 0$, i.e. $\omega(p_G) = 0$. Since the set $\mathcal{S}_{\mathfrak{g}} = p_G \mathcal{S}(\mathfrak{A})$ is τ^Q-invariant (as will be clear later), the time evolution of states $\omega \in (I - p_G)\mathcal{S}(\mathfrak{A})$, where I is the identity of $\mathfrak{A}^{**}$, can be determined arbitrarily with a help of some group $\tau_{\mathbb{R}} \subset {}^*\text{-Aut}\,(I - p_G)\mathfrak{A}^{**}$. The group $\tau_{\mathbb{R}}$ has nothing to do, in a general case, with the evolution τ^Q. For special choices of the function Q, however, the evolution τ^Q can be defined on a larger subalgebra of $\mathfrak{A}^{**}$ than $p_G\mathfrak{A}^{**}$, hence also an evolution of a set of states larger than $\mathcal{S}_{\mathfrak{g}}$ can be defined in a natural way, cf. also [40, Sect. II.C]. This can be seen on the following (seemingly trivial) example.

6.1.6 Example. *An important class of 'mean-field' evolutions is obtained by choosing $Q := f_\eta \in \mathfrak{g}^{**}$, $f_\eta(F) := F(\eta), \eta \in \mathfrak{g}$. We have in this case*

$$g_Q(t, F) = g_\eta(t, F) := \exp(t\eta), \quad \forall F \in \mathfrak{g}^*, \ t \in \mathbb{R}. \tag{6.1.23}$$

The corresponding time evolution is (due to the independence of g_Q on $F \in \mathfrak{g}^*$):

$$\tau_t^Q = \tau_t^\eta := \sigma(\exp(-t\eta)) \in {}^*\text{-Aut}\,\mathfrak{A}, \ t \in \mathbb{R}. \tag{6.1.24}$$

This time evolution is 'representation independent' (contrary to the general case of an arbitrary Q) and the definition of the evolution of an arbitrary state $\omega \in \mathcal{S}(\mathfrak{A})$ is straightforward. Equally straightforward is the canonical extension of τ^η to the (equally denoted) group $\tau^\eta \in {}^*\text{-Aut}\,\mathfrak{A}^{**}$. This evolution (for unbounded X_η, especially that one obtained by the extension to $\mathfrak{A}^{**}$) is highly discontinuous, however, and some appropriate continuity properties can be found in a restriction to a properly chosen subset of states of $\mathcal{S}(\mathfrak{A})$ (this 'properly chosen set of states' will be possibly larger than $p_G\mathcal{S}(\mathfrak{A})$).

The group G in the cases of this example is a 'dynamical group' of the system $(\mathfrak{A}, \sigma(G))$ containing the time-evolution one parameter group as the subgroup $\{\exp(-t\eta) : t \in \mathbb{R}\} \subset G$.

6.2 Spin Systems with Polynomial Local Hamiltonians Q^N

6.2.1 Let us consider the system described in Sect. 5.1: The C^*-algebra of quasilocal observables $\mathfrak{A}$ is the C^*-inductive limit of the sequence of the von Neumann algebras $\mathfrak{A}^N := \mathcal{L}(\mathcal{H}_N)$, $\mathfrak{A} := \mathfrak{A}^\Pi$. A compact Lie group G acts on $\mathfrak{A}$ by the subgroup $\sigma(G) := \sigma_G$ of *-automorphisms of $\mathfrak{A}$ introduced in 5.1.5. It is assumed in this section that the generators X_ξ ($\xi \in \mathfrak{g}$) of the representation $U(G)$ in $\mathcal{H}$ introduced in 5.1.3 are bounded operators. We shall use the notation of the subsections from 5.1.2 to 5.1.28; we shall write Π for the set of all positive integers. It will be convenient for definiteness and for some technical reasons to work in the subrepresentation $s_G \pi_u$ of the universal representation π_u of the algebra $\mathfrak{A}$ in the Hilbert space $\mathcal{H}_u$. The bidual $\mathfrak{A}^{**}$ is canonically identified with the bicommutant $\pi_u(\mathfrak{A})''$ of $\pi_u(\mathfrak{A})$ in $\mathcal{L}(\mathcal{H}_u)$, and $s_G \in \mathfrak{Z} := \pi_u(\mathfrak{A})' \cap \pi_u(\mathfrak{A})'' \subset \mathcal{L}(\mathcal{H}_u)$ is defined in 5.1.11. The following considerations could be extended to the larger representation $p_G \pi_u$, where $p_G \in \mathfrak{Z}$ is introduced in 5.1.29. Hence we shall work in the framework of the von Neumann algebra $s_G \mathfrak{A}^{**}$ which is isomorphic with the subalgebra $P_G \mathfrak{B}^\#$ of $\mathcal{L}(\mathcal{H}_\Pi)$ via the mapping ρ_G, cf. 5.1.11. The quasilocal algebra $\mathfrak{A}$ will be identified with its representation $s_G \pi_u(\mathfrak{A})$ in the Hilbert space $s_G \mathcal{H}_u$ or, equivalently, with the corresponding C^*-subalgebra of the abstract W^*-algebra $s_G \mathfrak{A}^{**}$. Remember that $\mathfrak{A}$ is simple, hence any of its nonzero representations as a C^*-algebra is faithful.

Let us introduce notation for various elements and subsets of $s_G \mathfrak{A}^{**}$:

6.2.2 Notation. *Let us denote:*

(i) $E_\mathfrak{g}$ **denotes the projection measure** (G-measure, 5.2.3) on the linear space $\mathfrak{g}^*$ generated by $E_\mathfrak{g}(F)$ ($F \in \mathfrak{g}^*$) from 5.1.13; in the notation of 5.1.16 $E_\mathfrak{g}(B) = c(B)$ for any subset $B \equiv \boldsymbol{B}$ of $\mathfrak{g}^*$.

(ii) $E_\mathfrak{g}(f) := \int f(F) \, E_\mathfrak{g}(dF)$ for any complex valued function $f \in L^1(\mathfrak{g}^*, \mu_\mathfrak{g}^\omega)$ for all $\omega \in \mathcal{S}_*(s_G \mathfrak{A}^{**}) :=$ the normal states on $s_G \mathfrak{A}^{**}$, i.e. the integral $E_\mathfrak{g}(f)$ is assumed to converge in the w^*-sense.

(iii) $\mathfrak{B}_0^N := \mathfrak{A}^N \cup \{X_{\xi K} : \xi \in \mathfrak{g}, \ K \in \Pi\} \subset s_G \mathfrak{A}^{**}$, if the generators $X_\xi \in \mathcal{L}(\mathcal{H})$ are bounded, 5.1.3.

(iv) **Let** $\mathfrak{N}^c$ **be the** C^***-subalgebra of** $s_G \mathfrak{A}^{**}$ **generated** by all the elements $E_\mathfrak{g}(f)$ with uniformly bounded continuous $f \in C_b(\mathfrak{g}^*, \mathbb{C})$.

(v) $\mathfrak{C}^N :=$ the C^***- algebra generated by** $\mathfrak{A}^N$ and $\mathfrak{N}^c$; $\mathfrak{C}^N$ is isomorphic to the C^*-tensor product $\mathfrak{A}^N \otimes \mathfrak{N}^c$, the isomorphism being: $x \otimes z \mapsto xz \in \mathfrak{C}^N$ ($x \in \mathfrak{A}^N$, $z \in \mathfrak{N}^c$), cf [274, 1.22] and [306, IV.4.7].

(vi) $\mathfrak{C}$ **will denote the** C^***- algebra generated by** $\{\mathfrak{C}^N : N \in \Pi\}$; $\mathfrak{C}$ is isomorphic to the tensor product $\mathfrak{A} \otimes \mathfrak{N}^c$, cf. 6.2.13.

6.2.3 Notation. Let $\{\xi_j : j = 1, 2, \ldots n\}$ be a fixed basis of $\mathfrak{g}$. Let

$$X_j^N := |N| \, X_{jN} := \sum_{k=1}^{|N|} \pi_k(X(\xi_j)), \quad X(\xi) := X_\xi, \qquad (6.2.1)$$

for any $N \in \Pi$, be the selfadjoint element of $\mathfrak{A}$ introduced in 5.1.3 as a selfadjoint operator on $\mathcal{H}_\Pi$ and identified now with $s_G \pi_u(X_j^N)$. Let us denote

$$\mathbf{b} := \max\{1 + \|X(\xi_j)\| : j = 1, 2, \ldots n := \dim G\}. \tag{6.2.2}$$

We shall use the Einstein summation rule for the summation over repeated vector indices in $\mathfrak{g}$ and $\mathfrak{g}^*$. Let c_{jk}^m be the structure constants of $\mathfrak{g}$ in the given basis:

$$[\xi_j, \xi_k] = c_{jk}^m \xi_m. \tag{6.2.3}$$

Then we have from (5.1.1):

$$[X_j^K, X_k^K] = i\, c_{jk}^m X_m^K, \quad \text{for all } K \in \Pi. \tag{6.2.4}$$

Let Q be a polynomial specified in 6.1.1, hence satisfying the property 6.1.1(SA). Let Q be written in the form of linear combination of p **monomials of the maximal degree** q with the upper bound $M \geq 1$ of the absolute values of the coefficients. Let Q^K be given by (6.1.1) for all $K \in \Pi$. Let us introduce the notation:

$$\mathbf{c} := \max\{|c_{jk}^m| : j, k, m = 1, 2, \ldots n\}; \tag{6.2.5}$$

$$a_N := \max(n\mathbf{c}; 2|N|\mathbf{b}), \quad N \in \Pi; \tag{6.2.6}$$

$$\mathbf{b}(x) := \max(\mathbf{b}; \|x\|), \quad x \in \mathfrak{A}. \tag{6.2.7}$$

We shall use the standard notation for the multiple commutators:

$$[y, x]^{(m+1)} := [y, [y, x]^{(m)}], \ [y, x]^{(0)} := x, \ [y, x] := yx - xy, \tag{6.2.8}$$

for any $x, y \in \mathfrak{A}^{**}$. [We shall use also $|J| :=$ the number of elements of the set J.]

6.2.4 Lemma. *The following estimate is valid for any $x \in \mathfrak{B}_0^N$ and for all positive integers $N, K(\geq N), m$:*

$$\|[Q^K, x]^{(m)}\| < \frac{\mathbf{b}(x)}{q}(m-1)!\,(Mpq^2\mathbf{b}^{q-1}a_N)^m. \tag{6.2.9}$$

Proof. Each multiple commutator in (6.2.9) can be written in the form of a finite linear combination of monomials $P^{(m)}$ in the variables X_{jK} and y_r, where $y_r \in \mathfrak{B}_0^N$ is of one of the forms of the multiple commutators occurring in the two following formulas:

$$\|[X_{j_1}^N, [X_{j_2}^N, \ldots [X_{j_r}^N, x]\ldots]]\| \leq (2\mathbf{b}N)^r \|x\|, \ x \in \mathfrak{A}^N; \tag{6.2.10}$$

$$\|[X_{j_1}^K, [X_{j_2}^K, \ldots [X_{j_r}^K, X_{kL}]\ldots]]\| \leq (n\mathbf{c})^r \mathbf{b}, \ L \in \Pi. \tag{6.2.11}$$

These estimates of $\| y_r \|$ are easy consequences of the definitions as well as of the relations (6.2.4). Let $r \in \mathbb{Z}_+$ be called the degree of any of the variables denoted by y_r. Then the sum $\sum_j r_j$ of degrees of all the variables y_{r_j} occurring in any of the monomials $P^{(m)}$ is less or equal to m. The maximal degree of any of the monomials $P^{(m)}$ is $m(q-1)+1$, hence we have the estimate:

$$\| P^{(m)} \| \leq \mathbf{b}(x)(a_N \mathbf{b}^{q-1})^m, \tag{6.2.12}$$

where we have used the fact that a variable y_r of the form given in (6.2.10) occurs in any of the monomials $P^{(m)}$ at most in the first power (what implies the first power of $\mathbf{b}(x)$ in (6.2.12)), as well as the inequalities $a_N > \mathbf{b} > 1$ were used in the derivation of (6.2.12).

The maximal value of coefficients at the monomials $P^{(m)}$ is $< M^m$. The maximal number of monomials $P^{(m)}$ occurring in the expression of $[Q^K, x]^{(m)}$ can be calculated recursively, using the derivation property of the commutators. One has the identity

$$[x_{j_1} \ldots x_{j_q}, y_{k_1} \ldots y_{k_s}] = \sum_{i=1}^{q} \sum_{j=1}^{s} x_{j_1} \ldots x_{j_{i-1}} y_{k_1} \ldots y_{k_{j-1}}$$
$$[x_{j_i}, y_{k_j}] y_{k_{j+1}} \ldots y_{k_s} x_{j_{i+1}} \ldots x_{j_q}, \tag{6.2.13}$$

in which the commutator of two monomials of degrees q and s is expressed as a sum of qs monomials of degree $q + s - 1$ (some of the monomials could be equal to zero). If $[Q^K, x]^{(m)}$ is a sum of n_m monomials $const.P^{(m)}$ of the maximal degree $s_m := m(q-1)+1$, then $[Q^K, x]^{(m+1)}$ is a sum of n_{m+1} monomials, where

$$n_{m+1} \leq n_m p q s_m \leq n_m m p q^2. \tag{6.2.14}$$

Since $n_1 \leq pq$, we obtain the estimate:

$$n_m \leq \frac{(m-1)!}{q}(pq^2)^m. \tag{6.2.15}$$

After the multiplication of the right hand side of (6.2.12) by the right hand side of (6.2.15) and by the upper bound M^m of the coefficients at $P^{(m)}$, we obtain the estimate (6.2.9). $\square$

6.2.5 Lemma. *Let us define*

$$\kappa_N := (Mpq^2 \mathbf{b}^q a_N)^{-1}, \quad \text{for all } N \in \Pi. \tag{6.2.16}$$

Let $|t| \leq \kappa_N$, $x \in \mathfrak{B}_0^N$ for a given $N \in \Pi$. Then:

(i) The sums

$$\tau_t^K(x) := e^{itQ^K} x\, e^{-itQ^K} = \sum_{m=0}^{\infty} \frac{(it)^m}{m!} [Q^K, x]^{(m)}, \quad K \in \Pi, \qquad (6.2.17)$$

are convergent in the norm-topology of $\mathfrak{A}$, and this convergence is uniform on $\{K : K \in \Pi\} \times \{t : |t| \leq \kappa_N\} \times \{x : x \in \mathfrak{B}_0^N, \|x\| \leq a\}$ for any $a \in \mathbb{R}_+$.

(ii) The following limits exist in $s_G \mathfrak{A}^{**}$:

$$\tau_t^Q(x) := s^*\text{-}\lim_{|K|\to\infty} \tau_t^K(x), \qquad (6.2.18)$$

where the **convergence is understood in the** $s^*(s_G\mathfrak{A}^{**}, s_G\mathfrak{A}^*)$-**topology** *generated by the* **seminorms** $\hat{p}_\omega$ *and* $\hat{p}_\omega^*$ *for all* $\omega \in \mathcal{S}_*(s_G\mathfrak{A}^{**})$:

$$\hat{p}_\omega : x \mapsto \hat{p}_\omega(x) := \sqrt{\omega(x^*x)}, \quad \hat{p}_\omega^* : x \mapsto \hat{p}_\omega^*(x) := \sqrt{\omega(xx^*)}. \qquad (6.2.19)$$

Proof. The estimates (6.2.9) are independent of $K \in \Pi$ and the corresponding majorizing power series for (6.2.17) is uniformly convergent on the product of the disc $\{t : |t| \leq \kappa_N, t \in \mathbb{C}\}$ and the ball $\{x : x \in \mathfrak{B}_0^N, \|x\| \leq a\}$ for any nonnegative a. This proves (i). The definition of s_G in 5.1.11 implies the existence of the limits

$$X_{\xi\Pi} := s^*\text{-}\lim_{|K|\to\infty} X_{\xi K} = E_{\mathfrak{g}}(f_\xi) \in s_G\mathfrak{A}^{**}, \quad \xi \in \mathfrak{g}, \qquad (6.2.20)$$

what implies, in turn, together with the uniform boundedness in $K \in \Pi$ of the multiple commutators in (6.2.17), the existence of the limits

$$s^*\text{-}\lim_{|K|\to\infty} [Q^K, x]^{(m)} \in s_G\mathfrak{A}^{**}. \qquad (6.2.21)$$

The statement (i) together with these facts imply (ii).   □

6.2.6 Lemma. *Let $\mathfrak{B}^N$ be the C^*-subalgebra of $\mathfrak{A}$ generated by $\mathfrak{B}_0^N$. Each of the mappings $\tau_t^Q : \mathfrak{B}_0^N \to s_G\mathfrak{A}^{**}$ ($|t| \leq \kappa_N$) can be extended to a unique $*$-homomorphism of the C^*-algebra $\mathfrak{B}^N$ into $s_G\mathfrak{A}^{**}$.*

Proof. The mappings τ_t^K are inner automorphisms of $\mathfrak{A}$, and their canonical extensions to $\mathfrak{A}^{**}$ leave the center $\mathfrak{Z}$ elementwise invariant. Hence, we can consider τ_t^K as (inner) automorphisms of $s_G\mathfrak{A}^{**}$:

$$\tau_t^K \in {}^*\text{-Aut}\, s_G\mathfrak{A}^{**}, \quad \text{for all } t \in \mathbb{R}, \ K \subset \Pi. \qquad (6.2.22)$$

The properties of the s^*-limit imply that τ_t^Q ($|t| \leq \kappa_N$, $t \in \mathbb{R}$) are $*$-homomorphisms of the symmetric set $\mathfrak{B}_0^N$ into $s_G\mathfrak{A}^{**}$, as well as they are

*-homomorphisms of the minimal *-algebra in $\mathfrak{A}$ containing $\mathfrak{B}_0^N$ into $s_G\mathfrak{A}^{**}$. The obvious norm-boundedness of these homomorphisms gives by continuity the wanted (equally denoted) extensions τ_t^Q.   □

Note: The values $\tau_t^Q(x)$ can be calculated according to the formula (6.2.18) for all $x \in \mathfrak{B}^N$. This is a consequence of the norm-continuity of C^*-homomorphisms, and it is easily verified by an elementary calculation.

6.2.7 Lemma. *Let* $|t| \leq \kappa_1$, $\xi \in \mathfrak{g}$, $E_\mathfrak{g}(f_\xi) = X_{\xi\Pi} \in s_G\mathfrak{A}^{**}$, *cf. 6.2.2 (ii). Then the limits*

$$\tau_t^Q(E_\mathfrak{g}(f_\xi)) := s^*\text{-}\lim_{L\to\infty} \tau_t^Q(X_{\xi L}) \tag{6.2.23}$$

exist.

Proof. One has

$$\tau_t^Q(X_{\xi L}) = \sum_{m=0}^{\infty} \frac{(it)^m}{m!} s^*\text{-}\lim_{K\to\infty} [Q^K, X_{\xi L}]^{(m)}, \tag{6.2.24}$$

and the bounds (6.2.9) give the estimates independent of K and L. After the substitution of $x := X_{\xi L}$ into the sum in (6.2.17), this sum is norm-convergent uniformly in $(K; L) \in \Pi \times \Pi$. Hence we have

$$\tau_t^Q(E_\mathfrak{g}(f_\xi)) = \sum_{m=0}^{\infty} s^*\text{-}\lim_L s^*\text{-}\lim_K [Q^K, X_{\xi L}]^{(m)} \frac{(it)^m}{m!}, \tag{6.2.25}$$

cf. also (6.2.20) and (6.2.21), and the limit (6.2.23) exists (cf. also [40, Proposition 3.5]).   □

6.2.8 It will be shown next that the elements $E_\mathfrak{g}(f_\xi)$ $(\xi \in \mathfrak{g})$ of the algebra $\mathfrak{A}^{**}$ **generate the abelian** C^*-**algebra** $\mathfrak{N}^c$ **of (bounded continuous) classical observables,** cf. 6.2.2(iv), given on the support of $E_\mathfrak{g}$ in $\mathfrak{g}^*$. We shall show after this that the transformations τ_t^Q in (6.2.23) leave this C^*-algebra invariant, and that their unique extension for all $t \in \mathbb{R}$ reproduces the classical flow φ^Q, 6.1.2, restricted to the support supp $E_\mathfrak{g}$, 5.2.3. These results will lead to a natural definition of the **unique extension of** $\tau_t^Q : \mathfrak{A}^N \to s_G\mathfrak{A}^{**}$ for all $t \in \mathbb{R}$, such that these mappings together with the mappings (6.2.23) leave the tensor product $\mathfrak{C}^N = \mathfrak{A}^N \otimes \mathfrak{N}^c$, 6.2.2(v), invariant, and have a unique extension to a (equally denoted) one parameter group of *-automorphisms of this composite quantal ($\mathfrak{A}^N$) and classical ($\mathfrak{N}^c$) system.

Let $\varphi : \mathfrak{g}^* \to \mathfrak{g}^*$ be a Poisson automorphism, (5.2.3), leaving all the Ad^*-orbits invariant. Then, using the bicontinuity of φ and the G-equivariance of the G-measure $E_\mathfrak{g}$, one can prove that the $s_G\mathfrak{A}^{**}$-**valued function** $\hat{\varphi}E_\mathfrak{g}$ of Borel subsets $B \subset \mathfrak{g}^*$,

$$\hat{\varphi}E_\mathfrak{g} : B \mapsto \hat{\varphi}E_\mathfrak{g}(B) := E_\mathfrak{g}(\varphi^{-1}B), \tag{6.2.26}$$

is again a projection-valued mesure with the same support:

$$\text{supp } \hat{\varphi} E_{\mathfrak{g}} = \text{supp } E_{\mathfrak{g}}. \tag{6.2.27}$$

6.2.9 Proposition. *Let $E_{\mathfrak{g}}$ and φ be as above. Then the mapping*

$$E_{\mathfrak{g}} : f \mapsto E_{\mathfrak{g}}(f) := \int f(F) \, E_{\mathfrak{g}}(dF), \quad f \in C(\text{supp } E_{\mathfrak{g}}), \tag{6.2.28}$$

introduced in 6.2.2 (ii) is a C^*-isomorphism of the commutative C^*-algebra of continuous complex valued functions $C(\text{supp } E_{\mathfrak{g}})$ on the compact subset $\text{supp } E_{\mathfrak{g}}$ of $\mathfrak{g}^*$ (X_ξ's are now bounded!) onto $\mathfrak{N}^c$.

The C^*-algebra $\mathfrak{N}^c$ is generated by the finite set $E_{\mathfrak{g}}(f_{\xi_j})$, $j = 1, 2, \dots n$ of its elements (ξ_j's form a basis of $\mathfrak{g}$). The mapping

$$\varphi^* : f \mapsto \varphi^* f, \text{ with } \varphi^* f(F) := f(\varphi F), \tag{6.2.29}$$

restricted to $f \in C(\text{supp } E_{\mathfrak{g}})$ is a *-automorphism of $C(\text{supp } E_{\mathfrak{g}})$. One has

$$\hat{\varphi} : E_{\mathfrak{g}}(f) \mapsto \hat{\varphi}(E_{\mathfrak{g}}(f)) := \hat{\varphi} E_{\mathfrak{g}}(f) = E_{\mathfrak{g}}(\varphi^* f), \quad f \in C(\mathfrak{g}^*), \tag{6.2.30}$$

and the mapping $\hat{\varphi}$ in (6.2.30) is a *-automorphism of $\mathfrak{N}^c$.

Proof. Since $\text{supp } E_{\mathfrak{g}}$ is compact (due to the compactness of spectra of all the X_ξ's), the function set $C(\text{supp } E_{\mathfrak{g}})$ is a C^*-algebra **generated by polynomials in the variables** $F_j := F(\xi_j) = f_{\xi_j}(F)$ according to the classical Weierstrass theorem. The *-morphism property of $E_{\mathfrak{g}}$ in (6.2.28) is a consequence of the standard functional calculus of normal operators determined by a projection measure. One can show that if $f(F_0) \neq 0$ for some $F_0 \in \text{supp } E_{\mathfrak{g}}$ and a continuous f, then $E_{\mathfrak{g}}(f) \neq 0$, and this implies that the mapping $E_{\mathfrak{g}}$ in (6.2.28) is the C^*-isomorphism of $C(\text{supp } E_{\mathfrak{g}})$ onto $\mathfrak{N}^c$.

The mapping φ^* is a norm preserving *-morphism of $C(\text{supp } E_{\mathfrak{g}})$ into itself, hence, it is a *-automorphism.

The automorphism property of $\hat{\varphi}$ in (6.2.30) is then a consequence of the relation (6.2.27), since both the mappings $E_{\mathfrak{g}}^{-1} : \mathfrak{N}^c \to C(\text{supp } E_{\mathfrak{g}})$ and $\hat{\varphi} E_{\mathfrak{g}} : C(\text{supp } E_{\mathfrak{g}}) \to \mathfrak{N}^c$ are *-isomorphisms, and we have:

$$\hat{\varphi}(E_{\mathfrak{g}}(f)) = \hat{\varphi} E_{\mathfrak{g}} \circ E_{\mathfrak{g}}^{-1}(E_{\mathfrak{g}}(f)), \quad f \in C(\text{supp } E_{\mathfrak{g}}). \tag{6.2.31}$$

The equality in (6.2.30) can be obtained from (6.2.26) and the integral representation (6.2.28). This concludes the proof. □

6.2.10 Proposition. *The mappings τ_t^Q introduced in (6.2.23) leave the algebra $\mathfrak{N}^c$ invariant. The family τ^Q has a unique extension to a strongly continuous one parameter group of *-automorphisms of $\mathfrak{N}^c$. This group satisfies the equality*

$$\tau_t^Q(E_{\mathfrak{g}}(f)) = E_{\mathfrak{g}}(\varphi_t^{Q*} f), \quad f \in C(\text{supp } E_{\mathfrak{g}}), \tag{6.2.32}$$

where φ^Q is the classical flow corresponding to the Hamiltonian function Q, 6.1.2.

Proof. The classical flow φ^Q forms a group of Ad^*-orbits-preserving Poisson automorphisms of $\mathfrak{g}^*$. According to Proposition 6.2.9, the right side of (6.2.32) defines a one parameter group of *-automorphisms of $\mathfrak{N}^c$. The strong continuity of this group (i.e. the continuity in the norm of all the functions $t \mapsto E_{\mathfrak{g}}(\varphi_t^{Q*} f)$) follows from the differentiability (hence continuity) of

$$\varphi^Q : (F; t) \mapsto \varphi_t^Q(F), \tag{6.2.33}$$

what is uniformly continuous on compacts in $\mathfrak{g}^* \times \mathbb{R}$ ($\mathfrak{g}^*$ is endowed by the linear space topology), as well as from the norm-continuity of the isomorphism $E_{\mathfrak{g}}$. Hence, it suffices to prove the validity of the equation (6.2.32) for small t.

Let us calculate the limits in (6.2.25). We intend to prove

$$s\text{-}\lim_L s\text{-}\lim_K i^m [Q^K, X_{\xi L}]^{(m)} = E_{\mathfrak{g}}(\{Q, f_\xi\}^{(m)}), \quad \xi \in \mathfrak{g}, \ m \in \mathbb{Z}_+. \tag{6.2.34}$$

Here $\{Q, f\}^{(0)} := f, \{Q, f\}^{(m+1)} := \{Q, \{Q, f\}^{(m)}\}$, and $\{Q, f\}$ is the classical Poisson bracket on the Poisson manifold $\mathfrak{g}^*$. The limits in (6.2.34) do exist, cf. 6.2.5. The local Hamiltonians Q^K are polynomials of the form (6.1.1) and the commutators as well as the Poisson brackets are bilinear, antisymmetric, satisfying the Jacobi identity and the derivation property: $[a, bc] = [a, b]c + b[a, c]$.

We have also

$$s\text{-}\lim_L s\text{-}\lim_K i[X_\xi^K, X_{\eta L}] = s\text{-}\lim_L X_{[\eta, \xi] L} = E_{\mathfrak{g}}(\{f_\xi, f_\eta\}), \quad \xi, \eta \in \mathfrak{g}. \tag{6.2.35}$$

what can be seen from (5.1.5), (6.2.20) and (1.3.12). The morphism properties of $E_{\mathfrak{g}}$ then lead to the formula (6.2.34).

Inserting (6.2.34) into (6.2.25), we obtain

$$\tau_t^Q(E_{\mathfrak{g}}(f_\xi)) = \sum_{m=0}^{\infty} \frac{t^m}{m!} E_{\mathfrak{g}}(\{Q, f_\xi\}^{(m)}). \tag{6.2.36}$$

The estimates (6.2.9) and the isometry of the mapping $E_{\mathfrak{g}}$ from (6.2.28) give, with the help of (6.2.34), the norm-convergence (in the algebra $C(\text{supp } E_{\mathfrak{g}})$) of the sum defining the element $f_{\xi t} \in C(\text{supp } E_{\mathfrak{g}})$:

$$f_{\xi t}(F) := \sum_{m=0}^{\infty} \frac{t^m}{m!} \{Q, f_\xi\}^{(m)}(F), \quad F \in \text{supp } E_{\mathfrak{g}}, \ |t| \le \kappa_1. \tag{6.2.37}$$

The norm-continuity of the morphism $E_{\mathfrak{g}}$ then leads from (6.2.36) to

$$\tau_t^Q(E_{\mathfrak{g}}(f_\xi)) = E_{\mathfrak{g}}(f_{\xi t}), \ \xi \in \mathfrak{g}, \ |t| \le \kappa_1 := (M p q^2 b^q a_1)^{-1}. \tag{6.2.38}$$

The derivative of the function $t \mapsto f_{\xi t}$ is, according to (6.2.37):

$$\frac{d}{dt} f_{\xi t}(F) = \sum_{m=0}^{\infty} \frac{t^m}{m!} \{Q, \{Q, f_\xi\}^{(m)}\}(F), \tag{6.2.39}$$

the series in (6.2.39) being again absolutely and uniformly convergent in $F \in \text{supp } E_{\mathfrak{g}}$ and $|t| \le \kappa_1$, (6.2.9), i.e.

$$(t; F) \in \{u : u \in \mathbb{R}, \ |u| \le \kappa_1\} \times \text{supp } E_{\mathfrak{g}}. \tag{6.2.40}$$

The classical Hamilton equations written in the form of Poisson brackets for the case of the Hamiltonian function Q with the flow φ^Q have the form

$$\frac{d}{du} f(\varphi_u^Q F) = \{Q, f\}(\varphi_u^Q F), \quad F \in \mathfrak{g}^*, \ u \in \mathbb{R}. \tag{6.2.41}$$

Let us substitute $\varphi_u^Q F$ instead of F into the formula (6.2.39). From (6.2.41) we obtain

$$\frac{d}{dt} f_{\xi t}(\varphi_u^Q F) = \sum_{m=0}^{\infty} \frac{t^m}{m!} \frac{d}{du} \{Q, f_\xi\}^{(m)}(\varphi_u^Q F). \tag{6.2.42}$$

The uniform convergence in $u \in \mathbb{R}$ for any given $(t; F)$ from (6.2.40) and the known theorem on the differentiation of series of functions lead to the equality:

$$\frac{d}{dt} f_{\xi t}(\varphi_u^Q F) = \frac{d}{du} f_{\xi t}(\varphi_u^Q F) = \{Q, f_{\xi t}\}(\varphi_u^Q F), \tag{6.2.43}$$

where the second equality was obtained by an application of (6.2.41). Setting $u = 0$ in (6.2.43) and comparing with (6.2.41) we get:

$$f_{\xi t}(F) = f_{\xi 0}(\varphi_t^Q F) \equiv f_\xi(\varphi_t^Q F) = \varphi_t^{Q*} f_\xi(F), \tag{6.2.44}$$

since $f_{\xi 0} = f_\xi$ according to (6.2.37). Insertion of $f_{\xi t}$ from (6.2.44) into (6.2.38) gives (6.2.32) with $f := f_\xi$ ($\xi \in \mathfrak{g}$). The algebra $C(\text{supp } E_{\mathfrak{g}})$ is generated by f_ξ's, and φ_t^{Q*} is a *-isomorphism of $C(\text{supp } E_{\mathfrak{g}})$, (6.2.29). The norm-continuity of C^*-morphisms gives then the validity of (6.2.32) for the general $f \in C(\text{supp } E_{\mathfrak{g}})$. $\qquad\square$

6.2.11 Lemma. *The mappings τ_t^Q ($|t| \leq \kappa_N$) defined in 6.2.5 (ii) map the C^*-algebra $\mathfrak{A}^N$ into the C^*-algebra $\mathfrak{C}^N$ (which is generated in $s_G \mathfrak{A}^{**}$ by $\mathfrak{A}^N$ and $\mathfrak{N}^c$).*

Proof. We can write the definition of τ_t^Q ($|t| \leq \kappa_N$) on $\mathfrak{A}^N$, (6.2.17) and (6.2.18), in the form

$$\tau_t^Q(x) := \sum_{m=0}^{\infty} \frac{t^m}{m!} s^*\text{-}\lim_K [i\, Q^K, x]^{(m)}, \ x \in \mathfrak{A}^N. \tag{6.2.45}$$

Each multiple commutator in (6.2.45) can be expressed in the form of a polynomial in the variables $X_{\xi K}$ and some of the variables y_s of the form, cf. also (6.2.10) and (6.2.11):

$$y_s := [X_{j_1}^K, [X_{j_2}^K, \ldots [X_{j_s}^K, x]\ldots]] \in \mathfrak{A}^N, \ K \in \Pi, \tag{6.2.46}$$

with the coefficients independent of K. Due to (6.2.20) and the independence of any y_s of K, the strong limits in (6.2.45) are elements of $\mathfrak{C}^N$. The norm convergence of the sum in right hand side of (6.2.45) and the closeness of $\mathfrak{C}^N$ in the norm-topology give then the result. $\qquad \square$

6.2.12 Lemma. *For any $x \in \mathfrak{A}$ and any $z \in \mathfrak{N}^c$, the equality $xz = 0$ implies the validity of $\|x\| \cdot \|z\| = 0$.*

Proof. For $z \neq 0$, we have $z = E_{\mathfrak{g}}(f)$ with $|f(F_0)| \neq 0$ for some $f \in C(\text{supp } E_{\mathfrak{g}})$ and some $F_0 \in \text{supp } E_{\mathfrak{g}}$. Let, for the definiteness, be $f(F_0) > 0$. Then there is a subset $B_0 \subset \mathfrak{g}^*$ such that $E_{\mathfrak{g}}(B_0) \neq 0$ and $f(F) > \frac{1}{2} f(F_0)$ for all $F \in B_0$. Since $\mathfrak{N}^c$ is in the commutant of $\mathfrak{A}$ in $s_G \mathfrak{A}^{**}$, the product of the positive (i.e. nonnegative) operator $x^* x \in \mathfrak{A}$ with the positive operator $(E_{\mathfrak{g}}(f) - \frac{1}{2} f(F_0)) E_{\mathfrak{g}}(B_0) \in \mathfrak{N}^c$ is a nonnegative operator in $\mathfrak{C}$. Then $xz = 0$ implies

$$0 \leq x^* x \, (E_{\mathfrak{g}}(f) - \frac{1}{2} f(F_0)) E_{\mathfrak{g}}(B_0) = -\frac{1}{2} \, f(F_0) \, x^* x \, E_{\mathfrak{g}}(B_0). \tag{6.2.47}$$

Hence we have $x \, E_{\mathfrak{g}}(B_0) = 0$. The mapping: $x \mapsto x \, E_{\mathfrak{g}}(B_0)$ is a nonzero (nondegenerate) representation of the simple C^*-algebra $\mathfrak{A}$ in $\mathfrak{A}^{**}$, hence $x = 0$. $\qquad \square$

6.2.13 Lemma. *Let $\mathfrak{A}^N \otimes \mathfrak{N}^c$ and $\mathfrak{A} \otimes \mathfrak{N}^c$ be the C^*-**products** (uniquely defined, since $\mathfrak{N}^c$ is abelian, [274, 1.22.5.]), with the canonical inclusion $\mathfrak{A}^N \otimes \mathfrak{N}^c \subset \mathfrak{A} \otimes \mathfrak{N}^c$. Let λ_0^{-1} be the homomorphism of $\mathfrak{A} \otimes \mathfrak{N}^c$ into $\mathfrak{C}$, 6.2.2, determined by the association:*

$$\lambda_0^{-1} : \sum_j x_j \otimes z_j \mapsto \sum_j x_j z_j \in \mathfrak{C}, \ x_j \in \mathfrak{A}, \ z_j \in \mathfrak{N}^c. \tag{6.2.48}$$

*Then λ_0^{-1} can be extended to a unique *-isomorphism $\lambda_0^{-1} =: (\lambda_0)^{-1}$ of the C^*-algebra $\mathfrak{A} \otimes \mathfrak{N}^c$ onto $\mathfrak{C}$, the restrictions of which to the subalgebras $\mathfrak{A}^N \otimes \mathfrak{N}^c$ ($N \in \Pi$) are *-isomorphisms onto $\mathfrak{C}^N$ ($N \in \Pi$), cf. 6.2.2.*

Proof. The existence of an isomorphism onto $\mathfrak{C}$ extending λ_0^{-1} is a direct consequence of [306, Exercise IV.2], due to our Lemma 6.2.12. The uniqueness is the trivial consequence of the norm-continuity of C^*-homomorphisms, since the finite sums in (6.2.48) form dense sets in the corresponding C^*-algebras. The same considerations are applicable to the restrictions to $\mathfrak{A}^N \otimes \mathfrak{N}^c$, hence we have the assertions of the Lemma. $\qquad\qquad\square$

6.2.14 Lemma. *Let* τ_K $(K \in \Pi)$, *resp.* τ_c, *be a* *-**homomorphism of** $\mathfrak{A}^K$, *resp.* **of** $\mathfrak{N}^c$, *into* $\mathfrak{C}^K$. *Assume that* $\tau_c(\mathfrak{N}^c) \subset \mathfrak{N}^c$. *Then there is a unique* *-*homomorphism* $\tau : \mathfrak{C}^K \to \mathfrak{C}^K$ *such that:*

$$\tau(xz) = \tau_K(x)\tau_c(z), \quad \text{for all } x \in \mathfrak{A}^K, \ z \in \mathfrak{N}^c. \tag{6.2.49}$$

Proof. Let $\lambda_0 : xz \mapsto x \otimes z$ be the isomorphism of $\mathfrak{C}^K$ onto $\mathfrak{A}^K \otimes \mathfrak{N}^c$ determined in 6.2.13. According to [306, IV.4.7.], there is a unique homomorphism τ_0 of $\mathfrak{A}^K \otimes \mathfrak{N}^c$ into $\mathfrak{C}^K$ such that

$$\tau_0(x \otimes z) = \tau_K(x)\tau_c(z), \quad x \in \mathfrak{A}^K, \ z \in \mathfrak{N}^c. \tag{6.2.50}$$

Since the C^*-norm on $\mathfrak{A} \otimes \mathfrak{N}^c$ is a *cross* norm (see [306, IV.]), the *-property of τ_0 follows from the norm continuity and from the *-property of τ_K and τ_c. We shall define τ as the composition

$$\tau := \tau_0 \circ \lambda_0. \tag{6.2.51}$$

The uniqueness of τ is then a consequence of linearity and continuity in the norm-topology. $\qquad\qquad\square$

6.2.15 Proposition. *There is a unique family* $\tau^Q := \{\tau_t^Q \, ; \, |t| \leq \kappa_N, \ t \in \mathbb{R}\}$ *of* C^*-*morphisms of* $\mathfrak{C}^N$ *into itself such that their restriction to* $\mathfrak{A}^N \subset \mathfrak{C}^N$ *is given by* (6.2.45), *and their restriction to* $\mathfrak{N}^c \subset \mathfrak{C}^N$ *is given by* (6.2.32). **This family** τ^Q **has a unique extension to an (equally denoted) one parameter group of** *-**automorphisms of** $\mathfrak{C}^N$, *for any* $N \in \Pi$.

Proof. After the identification of τ_K (resp. τ_c) from 6.2.14 with τ_t^Q from (6.2.45) (resp. with τ_t^Q from (6.2.32)) for any real $t : |t| \leq r_K$ $(K \in \Pi)$, the wanted morphism $\tau_t^Q : \mathfrak{C}^K \to \mathfrak{C}^K$ is obtained by its identification with τ from (6.2.49). It suffices to prove the group property of these morphisms τ_t^Q of $\mathfrak{C}^N$ into itself (with $N \in \Pi$) for small $t \in \mathbb{R}$. Since the restrictions of τ^Q to $\mathfrak{N}^c$ form an automorphism group of $\mathfrak{N}^c$, and the algebra $\mathfrak{N}^c$ is in the center of $\mathfrak{C}^N$, it suffices to prove

$$\tau_{t_1+t_2}^Q(x) = \tau_{t_1}^Q(\tau_{t_2}^Q(x)) \text{ for all } x \in \mathfrak{A}^N, \tag{6.2.52}$$

and for all sufficiently small nonzero t_j (e.g., for all $t_j : \max(|t_1|, |t_2|) < \frac{1}{2}\kappa_N$). For such t_j's, we have according to 6.2.5(ii) and (6.2.45):

$$\tau_{t_1}^Q(\tau_{t_2}^Q(x)) = \tau_{t_1}^Q(s\text{-}\lim_{K\to\infty}\tau_{t_2}^K(x)) = \sum_{m=0}^{\infty}\frac{(it_2)^m}{m!}\tau_{t_1}^Q(s\text{-}\lim_{K\to\infty}[Q^K,x]^{(m)}), \quad (6.2.53)$$

where the norm continuity of $\tau_{t_1}^Q$ and the norm-convergence of the series were used. (We write here $s\text{-}\lim$ instead of $s^*\text{-}\lim$, where the $s(s_G\mathfrak{A}^{**}, s_G\mathfrak{A}^*)$-topology is generated by the seminorms $\hat{p}_\omega$ from (6.2.19). This notation is used for brevity only; the existence and equality of both the limits $s\text{-}\lim$ and $s^*\text{-}\lim$ is clear from the proof of Lemma 6.2.5.) Considering the structure of the multiple commutators in (6.2.53) according to the discussion in the proof of 6.2.11, by the morphism property of $\tau_{t_1}^Q$ on $\mathfrak{C}^N$ as well as the definition (6.2.23) with (6.2.32) we obtain:

$$\tau_{t_1}^Q(s\text{-}\lim_K[Q^K,x]^{(m)}) = s\text{-}\lim_K\tau_{t_1}^Q([Q^K,x]^{(m)}) \qquad (6.2.54)$$

$$= s\text{-}\lim_K[\tau_{t_1}^Q(Q^K),\tau_{t_1}^Q(x)]^{(m)}. \qquad (6.2.55)$$

Since any $*$-morphism τ^Q is a *contraction*, the bounds from 6.2.4 are valid also for the multiple commutators in (6.2.55). From the norm-convergence of the sums we obtain consequently:

$$\tau_{t_1}^Q(\tau_{t_2}^Q(x)) = s\text{-}\lim_K\sum_{m=0}^{\infty}\frac{(it_2)^m}{m!}[\tau_{t_1}^Q(Q^K),\tau_{t_1}^Q(x)]^{(m)}$$

$$= s\text{-}\lim_K\tau_{t_1}^Q(\tau_{t_2}^K(x)). \qquad (6.2.56)$$

One has also

$$\tau_{t_2}^K(x)\in\mathfrak{B}^N \text{ for all } x\in\mathfrak{A}^N, \text{ and for all } K\in\Pi. \qquad (6.2.57)$$

Then, according to the 6.2.6 and the formula (6.2.18), one obtains:

$$\tau_{t_1}^Q(\tau_{t_2}^K(x)) = s\text{-}\lim_L\tau_{t_1}^L(\tau_{t_2}^K(x))$$

$$= s\text{-}\lim_L\sum_{k,m=0}^{\infty}\frac{(it_1)^k}{k!}\frac{(it_2)^m}{m!}[Q^L,[Q^K,x]^{(m)}]^{(k)}. \qquad (6.2.58)$$

The norms of the multiple commutators in (6.2.58) for $L\geq K\geq N$ are bounded from above according to the estimate (cf. also 6.2.3)

$$\|[Q^L,[Q^K,x]^{(m)}]^{(k)}\| < \frac{\mathbf{b}(x)}{q}(m+k-1)!\,(Mpq^2b^{q-1}a_N)^{m+k}, \qquad (6.2.59)$$

what can be obtained by the considerations analogous to those used in the proof of 6.2.4. Hence the sum in (6.2.58) converges in norm, uniformly in $(K;L)\in\Pi\times\Pi$

with $L \geq K \geq 1$. Then the continuity of the product of elements of a W^*-algebra in the s-topology leads to:

$$s\text{-}\lim_{K \to \infty} \tau_{t_1}^Q(\tau_{t_2}^K(x)) = s\text{-}\lim_K \sum_{k,m=0}^{\infty} \frac{(it_1)^k}{k!} \frac{(it_2)^m}{m!} [Q^K, [Q^K, x]^{(m)}]^{(k)}$$

$$= s\text{-}\lim_K \sum_{p=0}^{\infty} \frac{(t_1 + t_2)^p}{p!} [iQ^K, x]^{(p)}$$

$$= s\text{-}\lim_K \tau_{t_1+t_2}^K(x) = \tau_{t_1+t_2}^Q(x). \qquad (6.2.60)$$

The relations (6.2.56) and (6.2.60) give the desired group property (6.2.52) for all sufficiently small nonzero t_1, t_2, hence $\tau_t^Q \in *\text{-}\mathrm{Aut}\ \mathfrak{C}^N$ (due to the consequent invertibility of τ_t^Q on $\mathfrak{C}^N$), and τ^Q is a one-parameter group of automorphisms of $\mathfrak{C}^N$ (for any given $N \in \Pi$). □

6.2.16 Note. *We have worked in this section in the framework of the subalgebra* $s_G \mathfrak{A}^{**}$ *of the von Neumann algebra* $\mathfrak{A}^{**}$. *The* **only properties of the projector** $s_G \in \mathfrak{Z}$ *we have used in the previous considerations was the existence of the limits* $X_{\xi\Pi} := s^*\text{-}\lim_N s_G X_{\xi N}$ *for all* $\xi \in \mathfrak{g}$ *(here the elements* $X_{\xi N} \in \mathfrak{A}$ *are identified with* $\pi_u(X_{\xi N})$, *cf. 6.2.1) as well as the* $\sigma(G)$-*invariance:* $\sigma(g)(s_G) = s_G$ *for all* $g \in G$. **Any projector** $s_\pi \in \mathfrak{Z}$ **with these two properties**, *i.e.* s_π *such that:*

(i) the limits $s^*\text{-}\lim_N X_{\xi N} s_\pi$ *exist in* $s^*(\mathfrak{A}^{**}, \mathfrak{A}^*)$-*topology for all* $\xi \in \mathfrak{g}$,

(ii) s_π *is* $\sigma(G)$-*invariant:* $\sigma(g)(s_\pi) = s_\pi$ *for all* $g \in G$,

could be used instead of s_G *in the considerations of this section. Such projectors form a lattice in* $\mathfrak{Z}$ *with the maximal element* p_G *defined in 5.1.29. The G -measure corresponding to* p_G *was introduced in 5.1.33 and denoted by* $E_{\mathfrak{g}}^\Pi$. *Then the G -measure used up to now in this section was* $E_{\mathfrak{g}} = s_G E_{\mathfrak{g}}^\Pi$, *and the G-measure* E_G^π *corresponding to* **another projector** $s_\pi \in \mathfrak{Z}$ *satisfying (i) and (ii)* **equals to** $s_\pi E_{\mathfrak{g}}^\Pi$. *The algebra* $\mathfrak{N}_\pi^c := E_{\mathfrak{g}}^\pi(C_b(\mathfrak{g}^*, \mathbb{C}))$ *corresponding to the projector* s_π, *hence also the quasilocal algebra* $\mathfrak{C}_\pi := \mathfrak{A} \otimes \mathfrak{N}_\pi^c$, *depend nontrivially on the choice of* s_π. *If, however,* $s_G \leq s_\pi \leq p_G$, *then* $\mathfrak{N}_\pi^c$ *is isomorphic to* $\mathfrak{N}^c$. *This is an immediate consequence of the Proposition 6.2.9 as well as of the following 6.2.17.*

6.2.17 Lemma. *Let the projector* $s_\pi \in \mathfrak{Z}$ *(:= the center of* $\mathfrak{A}^{**}$*) satisfy 6.2.16 (i)+(ii). Let* $s_G \leq s_\pi \leq p_G$, *and let* $E_{\mathfrak{g}}^\pi := s_\pi E_{\mathfrak{g}}^\Pi$. *Then* $\mathrm{supp}\ E_{\mathfrak{g}}^\pi = \mathrm{supp}\ E_{\mathfrak{g}}^\Pi (= \mathrm{supp}\ E_{\mathfrak{g}},$ *consequently).*

Proof. Let $sp(X_\xi) \subset \mathbb{R}$ ($\xi \in \mathfrak{g}$) be spectrum of the bounded selfadjoint operator $X_\xi \in \mathcal{L}(\mathcal{H})$. **Let** conv$(B)$ **be the convex hull** of the subset B of a linear space. We have $X_{\xi N} \in \mathfrak{A}$ ($N \in \Pi$), hence the **spectrum** $sp(\pi(X_{\xi N}))$ **does not depend of the representation** π of $\mathfrak{A}$ ($\mathfrak{A}$ is simple). From the construction of $X_{\xi \Pi}$ in 5.1.7 and 5.1.8 we obtain successively:

$$sp(X_{\xi N}) \subset \text{conv}(sp(X_\xi)), \quad \xi \in \mathfrak{g}, \ N \in \Pi, \tag{6.2.61}$$

what can be seen from [262, Theorem VIII. 33]; from the spectral resolution of X_ξ with a help of 5.1.8 one has

$$\{\lambda \in \mathbb{C} : \ \lambda = (\varphi, X_\xi \varphi), \ \|\varphi\| = 1, \ \varphi \in \mathcal{H}\} = \text{conv}(sp(X_\xi)) \subset sp(X_{\xi \Pi}); \tag{6.2.62}$$

hence by [262, Theorem VIII.24]:

$$sp(X_{\xi \Pi}) = \text{conv}(sp(X_\xi)). \tag{6.2.63}$$

The equality (6.2.63) is independent of such representations π of $\mathfrak{A}$ in which (6.2.62) is valid, i.e. for

$$X_{\xi \pi} := s^*\text{-}\lim_N s_\pi X_{\xi N} \in \mathfrak{A}^{**} \tag{6.2.64}$$

we have the implication:

$$\text{conv}(sp(X_\xi)) \subset sp(X_{\xi \pi}) \Rightarrow sp(X_{\xi \pi}) = \text{conv}(sp(X_\xi)). \tag{6.2.65}$$

We have $X_{\xi \Pi} := X_{\xi \pi}$ for $s_\pi := s_G$ and the spectrum of $X_{\xi \pi}$ cannot decrease with increasing s_π. This proves the conclusion of (6.2.65) for all $s_\pi \geq s_G$, $\xi \in \mathfrak{g}$. Hence the spectra of $X_{\xi \pi}$ are independent of s_π for $s_G \leq s_\pi \leq p_G$. The construction of the projection measure $E_{\mathfrak{g}}^\pi$ according to (5.1.125) and 5.1.33 shows that $F \in \text{supp } E_{\mathfrak{g}}^\pi$ implies $F(\xi) \in sp(X_{\xi \pi})$:

$$X_{\xi \pi} = \int F(\xi) \, E_{\mathfrak{g}}^\pi(dF) = E_{\mathfrak{g}}^\pi(f_\xi). \tag{6.2.66}$$

This formula shows also that $\lambda \in sp(X_{\xi \pi})$ implies the existence of such an $F \in \text{supp } E_{\mathfrak{g}}^\pi$ that $F(\xi) = \lambda$. We shall show in the next Lemma that supp $E_{\mathfrak{g}}$ is a convex subset of $\mathfrak{g}^*$. Let

$$B_{\mathfrak{g}} := \{F \in \mathfrak{g}^* : \ F(\xi) \in \text{conv}(sp(X_\xi)), \ \forall \xi \in \mathfrak{g}\}. \tag{6.2.67}$$

The set $B_{\mathfrak{g}}$ is convex and closed in $\mathfrak{g}^*$. We have

$$\text{supp } E_{\mathfrak{g}}^\pi \subset B_{\mathfrak{g}} \text{ for any } s_\pi \geq s_G \ (s_\pi \leq p_G). \tag{6.2.68}$$

Let $B = \bar{B} = \mathrm{conv}(B) \subset B_\mathfrak{g}$ be such that for any $\xi \in \mathfrak{g}$ the following implication is valid:

$$\lambda \in \mathrm{conv}(sp(X_\xi)) \Rightarrow \exists\, F \in B : F(\xi) = \lambda. \qquad (6.2.69)$$

The set $B := \mathrm{supp}\, E_\mathfrak{g}$, and also $B := B_\mathfrak{g}$ has the property (6.2.69).

Let $F_0 \in \mathfrak{g}^*$ does not belong to $B :$ $F_0 \notin B$. Then, according to [157, Lemma (B.26)], there is an element of $\mathfrak{g}^{**} = \mathfrak{g}$, $\xi_0 \in \mathfrak{g}$, such that

$$\inf\{F(\xi_0) :\ F \in B\} > F_0(\xi_0). \qquad (6.2.70)$$

But from (6.2.69) and from $B \subset B_\mathfrak{g}$ we see that $\{F(\xi) : F \in B\} = \mathrm{conv}(sp(X_\xi))$ for all $\xi \in \mathfrak{g}$, hence $F_0(\xi_0) \notin \mathrm{conv}(sp(X_{\xi_0}))$, and this implies that $F_0 \notin B_\mathfrak{g}$. We have proved that $B = B_\mathfrak{g}$, hence $\mathrm{supp}\, E_\mathfrak{g} = B_\mathfrak{g}$. But

$$s_G \leq s_\pi \Rightarrow E_\mathfrak{g} \leq E_\mathfrak{g}^\pi \Rightarrow \mathrm{supp}\, E_\mathfrak{g} \subset \mathrm{supp}\, E_\mathfrak{g}^\pi, \qquad (6.2.71)$$

what with the help of (6.2.68) gives now the desired result.   □

6.2.18 Lemma. $\mathrm{supp}\, E_\mathfrak{g}$ is convex.

Proof. The projection measure $E_\mathfrak{g}$ introduced in 6.2.2.(i) is built of its values $E_\mathfrak{g}(F)$, (5.1.56), calculated on one point sets $\{F\} \subset \mathfrak{g}^*$. The measure $E_\mathfrak{g}$ is isomorphically mapped onto the measure $E_\mathfrak{g}^\# := \rho_G \circ E_\mathfrak{g}$ acting in the Hilbert subspace $P_G \mathcal{H}_\Pi$ of the infinite (complete) tensor product space $\mathcal{H}_\Pi$, cf. 5.1.11. According to the definitions in 5.1.7, 5.1.9 and 5.1.11, $F \in \mathrm{supp}\, E_\mathfrak{g}$ means that there is a product-vector $\Psi \in \mathcal{H}_\Pi$:

$$\Psi := \bigotimes_{k \in \Pi} \varphi_k,\ \varphi_k \in \mathcal{H}_k := u_k \mathcal{H},\ \|\varphi_k\| = 1,\ \text{for all } k \in \Pi, \qquad (6.2.72)$$

such that the following relations are valid:

$$\lim_{N \to \infty} \frac{1}{N} \sum_{k=1}^{N} (\varphi_k,\, \pi_k(X_\xi)\varphi_k) = F(\xi),\ \text{for all } \xi \in \mathfrak{g}. \qquad (6.2.73)$$

Let $F^{(j)} \in \mathrm{supp}\, E_\mathfrak{g}$ $(j = 1, 2)$ be determined according to (6.2.73) by the product vectors $\Psi^{(j)} := \otimes_{k \in \Pi}\varphi_k^{(j)} \in \mathcal{D}_\Pi(\mathfrak{g})$. We shall construct a product vector $\Psi \in \mathcal{D}_\Pi(\mathfrak{g})$, for any rational number $c : 0 < c = \frac{r}{s} < 1$, such that the corresponding value of $F \in \mathfrak{g}$, cf. (6.2.72) and (6.2.73), is

$$F = cF^{(1)} + (1 - c)F^{(2)}. \qquad (6.2.74)$$

This will prove the convexity of $\mathrm{supp}\, E_\mathfrak{g}$, since $\mathrm{supp}\, E_\mathfrak{g}$ is a closed subset of $\mathfrak{g}^*$.

We shall construct the sequence $\{\varphi_k : k \in \Pi\}$ defining Ψ according to (6.2.72) from the sequence $\{\varphi_k^{(j)} : k \in \Pi, \ j = 1, 2\}$ for any two natural numbers $0 < r < s$ as follows:

$$\varphi_{ms+j} := \varphi_{mr+j}^{(1)}, \quad \text{for } j = 1, 2, \ldots r; \ m \in \mathbb{Z}_+; \tag{6.2.75}$$

$$:= \varphi_{m(s-r)+j-r}^{(2)}, \quad \text{for } j = r+1, r+2, \ldots s; \ m \in \mathbb{Z}_+.$$

(Here we have identified $\mathcal{H}_k$ with $\mathcal{H}$ ($k \in \Pi$). The formally correct rewriting of the formula (6.2.75) includes, e.g., $\varphi_{ms+j} := u_{ms+j} u_{mr+j}^{-1} \varphi_{mr+j}^{(1)}$.)

Let

$$\Psi_k^{(j)}(\xi) := (\varphi_k^{(j)}, \pi_k(X_\xi)\varphi_k^{(j)}), \ j = 1, 2; \ \Psi_k(\xi) := (\varphi_k, \pi_k(X_\xi)\varphi_k). \tag{6.2.76}$$

Inserting from (6.2.75) into the left hand side of (6.2.73) we obtain:

$$\frac{1}{ms+j} \sum_{k=1}^{ms+j} \Psi_k(\xi) = \frac{1}{ms+j} \sum_{k=1}^{j} \Psi_{ms+k}(\xi) \tag{6.2.77}$$

$$+ \frac{ms}{ms+j}\left(\frac{r}{s}\frac{1}{mr}\sum_{k=1}^{mr} \Psi_k^{(1)}(\xi) + \frac{s-r}{s}\frac{1}{m(s-r)}\sum_{k=1}^{m(s-r)} \Psi_k^{(2)}(\xi)\right).$$

Taking the limit $m \to \infty$ on both sides of (6.2.77) ($j \in \{1, 2, \ldots s\}$), we obtain (6.2.74). $\qquad\square$

6.2.19 Proposition. *Let $s_\pi \le p_G$ be a $\sigma(G)$-invariant projector in the center $\mathfrak{Z}$ of $\mathfrak{A}^{**}$. Let $E_{\mathfrak{g}}^\pi := s_\pi E_{\mathfrak{g}}^\Pi$ be the corresponding G-measure. Then $\mathfrak{N}_\pi^c :=$ $E_{\mathfrak{g}}^\pi(C_b(\mathfrak{g}^*, \mathbb{C})) \subset \mathfrak{N}^c$, cf. 6.2.2(iv). Specifically, $\mathfrak{N}_\pi^c = \mathfrak{N}^c$ for $s_\pi \ge s_G$. (Here we have identified $*$-isomorphic C^*-algebras.)*

Proof. If $s_{\pi j}$ ($j = 1, 2$) are two such projectors s_π with $s_{\pi 1} \le s_{\pi 2}$, then for the corresponding G-measures one has $\operatorname{supp} E_{\mathfrak{g}}^{\pi 1} \subset \operatorname{supp} E_{\mathfrak{g}}^{\pi 2} \subset \operatorname{supp} E_{\mathfrak{g}}$, cf. 6.2.17. The Proposition 6.2.9 and its proof is applicable to any G-measure in the case of bounded generators X_ξ ($\xi \in \mathfrak{g}$). Since $C(\operatorname{supp} E_{\mathfrak{g}}^{\pi 1}) \subset C(\operatorname{supp} E_{\mathfrak{g}}^{\pi 2}) \subset C(\operatorname{supp} E_{\mathfrak{g}})$, and $\mathfrak{N}_\pi^c = E_{\mathfrak{g}}^\pi(C(\operatorname{supp} E_{\mathfrak{g}}^\pi))$ is an isomorphic image of $C(\operatorname{supp} E_{\mathfrak{g}}^\pi)$, the result follows. $\qquad\square$

Note: With a help of this proposition one can show that s_G can be replaced by s_π, with $s_G \le s_\pi \le p_G$, everywhere in this Sect. 6.2.

6.2.20 Theorem. *Let $\mathfrak{A} := \mathfrak{A}^\Pi$ be the quasilocal algebra introduced in 5.1.4; $\sigma(G) \subset *\text{-Aut } \mathfrak{A}$ is generated by the continuous unitary representation $U(G)$ in $\mathcal{H}$ of a Lie group G with bounded generators $X_\xi = X_\xi^*$ ($\xi \in \mathfrak{g}$), cf. 5.1.5 and 5.1.3. Let $s_\pi \le p_G$ be a $\sigma(G)$-invariant central projector in $\mathfrak{A}^{**}$, where p_G is introduced in 5.1.29. Let $E_{\mathfrak{g}}^\pi := s_\pi E_{\mathfrak{g}}^\Pi$, where $E_{\mathfrak{g}}^\Pi$ is defined in 5.1.33. Let $\mathfrak{N}_\pi^c$ and $\mathfrak{C}_\pi$ be defined as in 6.2.16 and $\mathfrak{C}_\pi^N := \mathfrak{A}^N \otimes \mathfrak{N}_\pi^c$; the algebras $\mathfrak{N}_\pi^c$, $\mathfrak{C}_\pi^N$, and $\mathfrak{C}_\pi$ are considered as*

C^*-subalgebras *of* $s_\pi \mathfrak{A}^{**}$ *in the canonical way, cf.* 6.2.1, 6.2.2 *and* 6.2.13. *Let Q be a polynomial with the property (SA) of* 6.1.1. *Then one has:*

(i) The sequence $\{\tau^K : K \in \Pi\}$ *of the one parameter* $*$-*automorphism groups of* $\mathfrak{A}$ *generated by* Q^K *according to* (6.1.2) *determines a unique one parameter group* $\tau^Q \subset$ $*$-$\mathrm{Aut}\,\mathfrak{C}$ *(with* $\mathfrak{C} := \mathfrak{C}_\pi$ *for* $s_\pi := p_G$*) such that for any* $N \in \Pi$ *and for all* $|t| \leq \kappa_N$ *(cf.* 6.2.16)

$$\tau_t^Q(x) = s^*\text{-}\lim_{K \to \infty} \tau_t^K(x), \quad \forall x \in \mathfrak{A}^N := p_G \pi_u(\mathfrak{A}^N). \tag{6.2.78}$$

The $s^*(p_G \mathfrak{A}^{**}, p_G \mathfrak{A}^*)$-**topology** *is determined by the seminorms from* (6.2.19) *with* $\omega \in \mathcal{S}_*(p_G \mathfrak{A}^{**})$.

(ii) The C^*-*subalgebras* $\mathfrak{N}_\pi^c$, $\mathfrak{C}_\pi$ *and* $\mathfrak{C}_\pi^N$ *(*$N \subset \Pi$, $s_\pi \leq p_G$*) of* $\mathfrak{C}$ *are invariant with respect to* τ^Q. **Let the restriction of** τ^Q **to** $\mathfrak{C}_\pi$ **be denoted by** τ^π. **(***Note: We have changed the notation here. It was denoted by* τ^Q *the group* τ^π *with* $s_\pi := s_G$ *in the preceding subsections.*)

(iii) The restriction of τ^π *to* $\mathfrak{N}_\pi^c$ *reproduces the classical flow* φ^Q *corresponding to the Hamiltonian function Q on the Poisson manifold* $\mathfrak{g}^*$ *in the sense that*

$$\tau_t^\pi(E_\mathfrak{g}^\pi(f)) = E_\mathfrak{g}^\pi(\varphi_t^{Q*}f), \quad f \in C(\mathfrak{g}^*). \tag{6.2.79}$$

(iv) The group τ^Q *is a strongly continuous subgroup of* $*$-$\mathrm{Aut}\,\mathfrak{C}$, *i.e. the functions*

$$t \mapsto \tau_t^Q(y) \tag{6.2.80}$$

are norm-continuous for all $y \in \mathfrak{C}$: *The triple* $\{\mathfrak{C}, \mathbb{R}, \tau^Q\}$ *is a* C^*-*dynamical system,* [53, 2.7.1].

(v) τ^π *(for any* s_π *specified above) is a* $\sigma(\mathfrak{C}_\pi, s_\pi \mathfrak{A}^*)$-*continuous group of automorphis of* $\mathfrak{C}_\pi$, *i.e. the functions*

$$t \mapsto \omega(\tau_t^\pi(y)) \tag{6.2.81}$$

are continuous for all states $\omega \in s_\pi \mathfrak{A}^*$ *(*$:= \{f \in \mathfrak{A}^* : f(s_\pi x) = f(x), \forall x \in \mathfrak{A}^{**}\}$*) and for all* $y \in \mathfrak{C}_\pi$, *and for all such* ω *one has:*

$$\omega \circ \tau_t^\pi \in s_\pi \mathfrak{A}^*, \quad \forall t \in \mathbb{R}. \tag{6.2.82}$$

(vi) The infinitesimal generator of τ^π *is the* **derivation** δ_π *on* $\mathfrak{C}_\pi$ *such that*

$$\delta_\pi(y) = i \sum_{j=1}^{n} E_{\mathfrak{g}}^\pi(\partial_j Q)\,[X_j^N, y], \quad \text{for all } y \in \mathfrak{A}^N, \tag{6.2.83a}$$

$$\delta_\pi(E_{\mathfrak{g}}^\pi(f)) = E_{\mathfrak{g}}^\pi(\{Q, f\}) \text{ for } f \in C^1(\mathfrak{g}^*), \tag{6.2.83b}$$

where the square bracket in (6.2.83a) *is the commutator, and* $\mathfrak{A}^N$ *is considered there as* $s_\pi \pi_u(\mathfrak{A}^N)$ *(* $\mathfrak{A}$ *is simple!), and the partial derivatives* $\partial_j Q$ *denote the differentiation of Q with respect to the components* $F_j := F(\xi_j)$ *of* $F \in \mathfrak{g}^*$ *in the dual basis to the basis* $\{\xi_j : j = 1, 2, \ldots n\}$ *of* $\mathfrak{g}$, $X_j := X_{\xi_j}$. *The compound bracket in* right hand side *in* (6.2.83b) *denotes the classical Poisson bracket on* $\mathfrak{g}^*$. *The operator* δ_π *determined by* (6.2.83) *determines the group* $\tau^\pi \in *\text{-}\mathrm{Aut}\,\mathfrak{C}_\pi$ *uniquely:*

$$\tau_t^\pi(y) = \sum_{m=0}^{\infty} \frac{t^m}{m!}\,\delta_\pi^m(y), \quad \text{for all } y \in \mathfrak{B}^\#,\ |t| \leq \kappa_N,\ N \in \Pi. \tag{6.2.84}$$

Proof. We shall use here the fact mentioned in the Note in Proposition 6.2.19 that in the assertions of this section we can replace s_G by p_G; we shall refer to the assertions and their proofs in Sect. 6.2 as if they were reformulated with this replacement.

(i) The restrictions of τ^Q to the subalgebras $\mathfrak{C}^N$ given in Proposition 6.2.15 determine a unique group $\tau^Q \subset *\text{-}\mathrm{Aut}\,\mathfrak{C}$, since each of the mappings

$$\tau_t^Q : y \mapsto \tau_t^Q(y),\ y \in \mathfrak{C}^N,\ N \in \Pi,\ t \in \mathbb{R}, \tag{6.2.85}$$

is norm-continuous and $\{y :\ y \in \mathfrak{C}^N,\ N \in \Pi\}$ is norm-dense in $\mathfrak{C}$.

(ii) After the replacement of s_G by s_π (hence also $E_{\mathfrak{g}}$ by $E_{\mathfrak{g}}^\pi$) in Propositions 6.2.10 and 6.2.15 we obtain the invariance of $\mathfrak{N}_\pi^c$ and of $\mathfrak{C}_\pi$ due to $\sigma(G)$-equivariance of E_G^π. The τ^Q-invariance of $\mathfrak{C}_\pi^N$ is clear.

(iii) Immediately from Proposition 6.2.10, since $\tau_t^\pi(E_{\mathfrak{g}}^\pi(f)) = s_\pi \tau_t^Q(E_{\mathfrak{g}}^\Pi(f))$.

(iv) It suffices to prove the continuity in (6.2.80) for $t \to 0$. With $y := x \in \mathfrak{A}^N$ the continuity is given by the uniform convergence in (6.2.45), and this implies the continuity for all $x \in \mathfrak{A}$ (by an $\epsilon/3$-argument). For $y := E_{\mathfrak{g}}^\pi(f)$, $f \in C(\mathrm{supp}\,E_{\mathfrak{g}}^\pi)$, it suffices to prove

$$\lim_{t \to 0} \|\varphi_t^{Q*} f - f\| = 0, \tag{6.2.86}$$

since $f \mapsto E_{\mathfrak{g}}^\pi(f)$ is a C^*-morphism. The validity of (6.2.86) is a consequence of the joint continuity of the classical flow φ^Q,

$$\varphi^Q : (t; F) \mapsto \varphi_t^Q(F) \in \mathfrak{g}^*, \tag{6.2.87}$$

as well as of the compactness of supp $E_\mathfrak{g}^\pi$ and of the continuity of f.

(v) The continuity in (6.2.81) is a consequence of (iv). Let us consider τ_t^π ($t \in \mathbb{R}$) as a family of representations of $\mathfrak{A} := \pi_u(\mathfrak{A}) \subset \mathfrak{A}^{**}$ in the subalgebra $s_\pi \mathfrak{A}^{**}$ of $\mathfrak{A}^{**}$. The unique $\sigma(\mathfrak{A}^{**}, \mathfrak{A}^*) - \sigma(s_\pi \mathfrak{A}^{**}, s_\pi \mathfrak{A}^*)$-continuous extensions of these representations to $\mathfrak{A}^{**}$, [274, 1.21.13], **will be denoted by** $\overline{\tau}_t^Q$ (**resp.** $\overline{\tau}_t^\pi$ **for** $s_\pi < p_G$). From Proposition 6.2.10 and from its proof one can see

$$\overline{\tau}_t^\pi(s_\pi) = \overline{\tau}_t^\pi(E_\mathfrak{g}^\pi(\mathfrak{g}^*)) = s_\pi. \tag{6.2.88}$$

We have also $\overline{\tau}_t^\pi(\mathrm{id}_\mathfrak{A} - s_\pi) = 0$, and the restrictions of $\overline{\tau}_t^\pi$ ($t \in \mathbb{R}$) to the $\overline{\tau}_t^\pi$-invariant subalgebra $s_\pi \mathfrak{A}^{**}$ form a family of *-automorphisms which are automatically σ-continuous, [274, 4.1.23]. The definition τ^π with a help of strong limits, cf. (6.2.23), shows that the restriction of $\overline{\tau}_t^\pi$ to $\mathfrak{C}_\pi$ coincides with the above defined $\tau_t^\pi \in$ *-Aut $\mathfrak{C}_\pi$. This proves the normality of $\omega \circ \tau_t^\pi$ for any normal ω (i.e. $\omega \in s_\pi \mathfrak{A}^*$), hence (6.2.82).

(vi) The automorphism group τ^π of $\mathfrak{C}_\pi$ is determined uniquely by the determination of $\tau_t^\pi(x)$ for all $x \in \mathfrak{B}^N$ (cf. 6.2.6 and 6.2.2(iii) for notation), for $|t| \leq \kappa_N$, $N \in \Pi$; this is clear from 6.2.5 and from its consequences. The series in the formula

$$\frac{d}{dt}\tau_t^\pi(x) = i \sum_{m=0}^\infty \frac{(it)^m}{m!} s^*\text{-}\lim_K [Q^K, [Q^K, x]^{(m)}] \tag{6.2.89}$$

converges uniformly in the disc $|t| \leq \kappa_N$ ($x \in \mathfrak{B}^N$), hence the equality (6.2.89) is valid. Considerations similar to those used in the dealing with (6.2.58) lead to the equalities:

$$s\text{-}\lim_K [iQ^K, [iQ^K, x]^{(m)}] = s\text{-}\lim_K s\text{-}\lim_L [iQ^L, [iQ^K, x]^{(m)}]$$

$$= s\text{-}\lim_K \delta_\pi([iQ^K, x]^{(m)}), \tag{6.2.90}$$

where for all $N \in \Pi$:

$$\delta_\pi(x) := s\text{-}\lim_L i[Q^L, x], \text{ for all } x \in \mathfrak{B}^N := s_\pi \pi_u(\mathfrak{B}^N). \tag{6.2.91}$$

The derivation property of commutators and the polynomial form of Q together with (6.2.20) lead to the expression (6.2.83a) for δ_π in (6.2.91). Setting $t = 0$ in (6.2.89), we see that so defined $\delta_\pi(x)$ is the value of the generator δ_π of τ^π on $x \in \mathfrak{A}^N$ ($N \in \Pi$). By the differentiation of (6.2.32) with $f \in C^1(\text{supp } E_\mathfrak{g}^\pi)$ we obtain (6.2.83b), cf. (6.2.43) and notes in [53] above 3.2.29. From the continuity properties of τ^π and the corresponding closedness of δ_π, cf. [53, 3.1.6], we obtain by the repeated use of (6.2.90):

$$s\text{-}\lim_K \delta_\pi([i\,Q^K, x]^{(m)}) = \delta_\pi(s\text{-}\lim_K [i\,Q^K, x]^{(m)}) = \delta_\pi^{m+1}(x). \qquad (6.2.92)$$

Insertion from (6.2.91) and (6.2.92) into (6.2.45), cf. the note following (6.2.53), gives for $x \in \mathfrak{B}^N$, $|t| \le \kappa_N$ the norm-convergent series:

$$\tau_t^\pi(x) = \sum_{m=0}^{\infty} \frac{t^m}{m!}\, \delta_\pi^m(x). \qquad (6.2.93)$$

This proves that the operator δ_π from (6.2.83) determines τ^π. □

6.3 Time Evolution in Generalized Mean-Field Theories

6.3.1 We shall construct in this section a general class of time evolutions τ^Q of the infinite quantum systems $(\mathfrak{A}; \sigma(G))$ defined in Sect. 5.2. The time evolution τ^Q is determined in a canonical way by an arbitrary classical Hamiltonian function Q on the (generalized) homogeneous classical phase space $\mathfrak{g}^*$ as well as by the automorphism group $\sigma(G)$ of $\mathfrak{A}$. It will be shown later that the here presented construction leads to the same evolution what was denoted by τ^Q in Sect. 6.2 in the case of $\mathfrak{A} := \mathfrak{A}^\Pi$, $\sigma(G)$ being defined according to 5.1.5, and with Q being a polynomial in a basis of $\mathfrak{g}^*$ dual to any fixed basis $\{\xi_j,\ j = 1, 2, \ldots n\}$; the generators of the continuous representation $U(G)$ in the 'one-spin space' $\mathcal{H}$, 5.1.3, are supposed to be bounded in this special case.

We shall start with the general case, the specifications to the cases considered in Sect. 5.1, and the further specification to the cases of Sect. 6.2 will be made later on. Let us fix here some general assumptions valid throughout of this section.

Using the notation of Sect. 5.2, let $E_\mathfrak{g}$ be a fixed nontrivial G-measure associated with the system $(\mathfrak{A}; \sigma(G))$ such that, with $p_G := E_\mathfrak{g}(\mathfrak{g}^*)$, the following implication is valid:

$$\omega \in p_G \mathcal{S}(\mathfrak{A}) =: \mathcal{S}_\mathfrak{g} \Rightarrow g\ (\in G) \mapsto \omega(\sigma(g)(x))\ \text{ is continuous for all } x \in \mathfrak{A}.$$
$$(6.3.1)$$

It will be shown in 6.3.10 that this assumption is fulfilled by $\sigma(G)$ from 5.1.5 with p_G from 5.1.29. We shall assume that $\mathfrak{A}$ is a unital C^*-algebra which is simple (this last assumption is made only for brevity of our expression). The nontriviality of $E_\mathfrak{g}$ means a certain 'breaking of symmetries' occurring in the system, cf. our 5.2.3 for basic definitions, and for illustration of the phenomenon of "spontaneous symmetry breaking" see [106, 265], [53, Sect. 4.3.4], [41, IV.A], 6.5.5.

The time evolution τ^Q will be defined with a help of the group-valued function $g_Q(t, F)$ on $\mathbb{R} \times \mathfrak{g}^*$ defined in 6.1.3 with (6.1.17) (another possible choice of β_F^Q will not change the general construction of τ^Q, so that the nonuniqueness of β_F^Q leads to various possibilities for the definition of the time evolutions τ^Q). The notation introduced in 6.1.2 and 6.1.3 will be used here. Let us note that the equation (6.1.16) for g_Q can be written in any continuous unitary representation $U(G)$ in $\mathcal{H}$ in the form

$$i \frac{d}{dt} U(g_Q(t, F)) = X(\beta_{F_t}^Q) U(g_Q(t, F)), \quad F \in \mathfrak{g}^*, \quad t \in \mathbb{R}, \tag{6.3.2}$$

where $F_t := \varphi_t^Q(F)$; $X(\xi) := X_\xi$ $(\xi \in \mathfrak{g})$ are the selfadjoint generators of $U(G)$, and $\beta_F^Q \in \mathfrak{g}$ was introduced in 6.1.3. The equation (6.3.2) is of the form of (linear) quantum-mechanical evolution equation with the time-dependent Hamiltonian operator $X(\beta_F^Q)$. The equation (6.3.2) describes, in the setting of Sect. 5.1, the time evolution of any 'individual' quantum subsystem placed in any fixed site $k \in \Pi$ in the surrounding 'mean field' $\varphi_t^Q(F) \in \mathfrak{g}^*$ generated by the whole collection of the quantal subsystems (for all the sites $k \in \Pi$) interacting by an 'infinitely weak and of infinitely long-range' interaction with each other. The equation (6.3.2) will be useful in the analysis of thermodynamic properties of the considered systems.

6.3.2 Definitions.

(i) $C_b := C_b(\text{supp } E_\mathfrak{g}, \mathbb{C})$ **will denote** the set of all uniformly bounded complex-valued continuous functions on $\text{supp } E_\mathfrak{g} \subset \mathfrak{g}^*$, see 5.2.3 for the definition of $\text{supp } E_\mathfrak{g}$.

(ii) The s^*-topology on $\mathfrak{A}$ is determined by seminorms $\hat{p}_\omega$, $\hat{p}_\omega^*$ (cf. (6.2.19)) for all $\omega \in p_G \mathcal{S}(\mathfrak{A})$.

(iii) Let $\mathfrak{C}_{bs} := \mathfrak{C}_{bs}(\text{supp } E_\mathfrak{g}, \mathfrak{A})$ be the set of all $\mathfrak{A}$-valued, uniformly bounded s^*-continuous functions on $\text{supp } E_\mathfrak{g}$, i.e. $f \in \mathfrak{C}_{bs}$ means that the function

$$f : F(\in \text{supp } E_\mathfrak{g}) \mapsto f(F) (\in \mathfrak{A}) \tag{6.3.3}$$

is bounded in the sense

$$\|f\| := \sup\{\|f(F)\| : F \in \text{supp } E_\mathfrak{g}\} < \infty, \tag{6.3.4}$$

and all the functions

$$F \mapsto \omega\left((f(F) - f(F_0))^*(f(F) - f(F_0))\right), \quad \omega \in p_G \mathcal{S}(\mathfrak{A}), \quad F_0 \in \text{supp } E_\mathfrak{g}, \tag{6.3.5a}$$

$$F \mapsto \omega\left((f(F) - f(F_0))(f(F) - f(F_0))^*\right), \quad \omega \in p_G \mathcal{S}(\mathfrak{A}), \quad F_0 \in \text{supp } E_\mathfrak{g}, \tag{6.3.5b}$$

converge to zero for F converging to F_0 in the norm-topology of $\mathfrak{g}^*$.

(iv) For any $\sigma(G)$-invariant C^*-subalgebra $\mathfrak{A}^J$ of $\mathfrak{A}$:

$$\sigma(g)(x) := \sigma_g(x) \in \mathfrak{A}^J \text{ for all } x \in \mathfrak{A}^J, \ g \in G, \qquad (6.3.6)$$

let $\mathfrak{C}_{bs}^J := \mathfrak{C}_{bs}(\operatorname{supp} E_{\mathfrak{g}}, \mathfrak{A}^J)$ be defined equally as it was defined $\mathfrak{C}_{bs}$ in (iii) with the replacement of $\mathfrak{A}$ by $\mathfrak{A}^J$.

(v) Let $\mathfrak{C}_{bs}^G$ (resp. $\mathfrak{C}_{bs}^{GJ}$) be the C^*-subalgebra (cf. 6.3.4) of $\mathfrak{C}_{bs}$ (resp. of $\mathfrak{C}_{bs}^J$) generated by all the functions $f_0 \in \mathfrak{C}_{bs}$ of the form

$$f_0 : F \mapsto \sigma_{g_0(F)}(x) f(F), \ f \in C_b, \ g_0 \in C(\operatorname{supp} E_{\mathfrak{g}}, G), \qquad (6.3.7)$$

*with any $x \in \mathfrak{A}$ (resp. any $x \in \mathfrak{A}^J$). **The set** $C(\operatorname{supp} E_{\mathfrak{g}}, G)$ consists of all continuous G-valued functions on $\operatorname{supp} E_{\mathfrak{g}}$.*

*(vi) We shall use also $K := \operatorname{supp} E_{\mathfrak{g}}$, resp. $K \subset \mathfrak{g}^*$ **will denote** any $Ad^*(G)$-invariant closed subset of the generalized classical phase space in more general cases. **We shall identify** $C_b := C_b(K, \mathbb{C})$ with the subset $\mathfrak{C}_{bs}(K, \mathbb{C} \operatorname{id}_{\mathfrak{A}})$ of $\mathfrak{C}_{bs}$ in the canonical way: $f \in C_b$ is identified with the function*

$$f : F \mapsto \operatorname{id}_{\mathfrak{A}} f(F), \ F \in K, \ \operatorname{id}_{\mathfrak{A}} \text{ is the identity of } \mathfrak{A}. \qquad (6.3.8)$$

6.3.3 Proposition. *The set $\mathfrak{C}_{bs}$ is a $*$-algebra with respect to the natural (pointwise) algebraic operations determined by the corresponding operations in the range $\mathfrak{A}$ of the elements $f \in \mathfrak{C}_{bs}$:*

$$(f_1 + \lambda f_2)(F) := f_1(F) + \lambda f_2(F), \ (f_1 f_2)(F) := f_1(F) f_2(F),$$
$$f^*(F) := [f(F)]^*, \forall F \in K, \lambda \in \mathbb{C}, \ f_j \text{ and } f \in \mathfrak{C}_{bs}, \qquad (6.3.9)$$

and it is a normed algebra with the norm $\|f\|$ of $f \in \mathfrak{C}_{bs}$ given by (6.3.4). This normed $*$-algebra $\mathfrak{C}_{bs}$ is a C^*-algebra, and its subsets $\mathfrak{C}_{bs}^J$ and C_b endowed with the induced algebraic operations and the norm are C^*-subalgebras of $\mathfrak{C}_{bs}$.

Proof. The continuity properties of the product in $\mathfrak{A}$ with respect to the s^*-topology are given by Proposition 1.8.12 and Theorem 1.8.9 of [274]. Then the uniform boundedness of $f \in \mathfrak{C}_{bs}$ and the continuity of the $*$-operation in the s^*-topology gives the invariance of $\mathfrak{C}_{bs}$ with respect to the algebraic operations (6.3.9). The norm properties of the function given in (6.3.4) are easily verified, and the C^*-property of the norm:

$$\|f\|^2 = [\sup_F \|f(F)\|]^2 = \sup_F \|f(F)\|^2 = \sup_F \|f(F)^* f(F)\| = \|f^* f\|,$$
$$(6.3.10)$$

is valid too. We shall verify completness of $\mathfrak{C}_{bs}$ in this norm. For any Cauchy sequence $\{f_n; n \in \mathbb{Z}_+\}$ in $\mathfrak{C}_{bs}$, the sequence $\{f_n(F), n \in \mathbb{Z}_+\}$ is Cauchy in $\mathfrak{A}$ for any $F \in K$. The completness of $\mathfrak{A}$ gives the existence of pointwise limits

$$f(F) := \text{n-}\lim_k f_k(F) \in \mathfrak{A}, \quad F \in K. \tag{6.3.11}$$

By defining the norm of any function $f : K \to \mathfrak{A}$ ($\|f\|$ could be infinite in general) by (6.3.4), we have the norm-convergence of f_k to f from (6.3.11): If $\|f_n - f_m\| < \delta$ for all $n, m > n_\delta$, then $\|f_n - f\| < \delta$ for all $n > n_\delta$, for any positive δ, since $\lim_m \|f_n(F) - f_m(F)\| = \|f_n(F) - f(F)\|$ for all $F \in K$. Considering the cyclic representation $(\pi_\omega, \mathcal{H}_\omega, \Omega_\omega)$ corresponding to any $\omega \in p_G S(\mathfrak{A})$ as a subrepresentation of the universal representation π_u in $p_G \pi_u(\mathfrak{A})$, we have with the identification of $\mathfrak{A}$ with $p_G \pi_u(\mathfrak{A})$ (cf. (6.2.19)):

$$\hat{p}_\omega(f(F) - f(F_0)) = \|(f(F) - f(F_0))\Omega_\omega\|$$
$$\leq 2\|f_m - f\| + \|(f_m(F) - f_m(F_0))\Omega_\omega\|, \tag{6.3.12}$$

and the s-continuity of f_m's gives the s-continuity of f. A use of the norm-continuity of the *-operation gives us the s^*-continuity of f, i.e. $f \in \mathfrak{C}_{bs}$. The remaining assertions of the proposition follow now easily. $\qquad\qquad\square$

6.3.4 Lemma. *The functions f_0 from (6.3.7) belong to $\mathfrak{C}_{bs}$. Hence, $\mathfrak{C}_{bs}^G$ and $\mathfrak{C}_{bs}^{GJ}$ are C^*-subalgebras of $\mathfrak{C}_{bs}$.*

Proof. Since $f \in C_b$ can be considered as an element of $\mathfrak{C}_{bs}$, it suffices to prove $f_0 \in \mathfrak{C}_{bs}$ for f_0 given by (6.3.7) with $f :=$ constant function. This will be proved by proving the s^*-continuity of $\sigma(G)$. For any $x \in \mathfrak{A}$ and any $\omega \in p_G S(\mathfrak{A})$, we have

$$\hat{p}_\omega \left(\sigma_g(x) - \sigma_{g0}(x)\right)^2 = \omega \left((\sigma_g(x^*) - \sigma_{g0}(x^*))(\sigma_g(x) - \sigma_{g0}(x))\right) =$$
$$= \omega \left(\sigma_g(x^*x) - \sigma_{g0}(x^*x)\right) +$$
$$\omega \left((\sigma_{g0}(x^*) - \sigma_g(x^*)) \sigma_{g0}(x)\right) + \omega \left(\sigma_{g0}(x^*)(\sigma_{g0}(x) - \sigma_g(x))\right), \tag{6.3.13}$$

and the s-continuity follows from the assumption (6.3.1) by repeated use of the polarization identity (expressing nondiagonal matrix elements of bounded operators in a Hilbert space by a finite linear combination of the diagonal ones). The s^*-continuity is then obtained by the replacement of x by x^* in the above considerations. $\qquad\qquad\square$

6.3.5 The quasilocal C^*-algebra $\mathfrak{A}$ of quantum (microscopic) observables is naturally embedded into $\mathfrak{C}_{bs}$ as a C^*-subalgebra by the identification of any $x \in \mathfrak{A}$ with a constant function $f \in \mathfrak{C}_{bs}^G$:

$$f(F) := x = \mathbf{1}(F)\sigma_e(x), \quad F \in K, \tag{6.3.14}$$

where $\mathbf{1}(F) := 1$ for all $F \in \mathfrak{g}^*$. The classical (macroscopic) observables are embedded into $\mathfrak{C}_{bs}^G$ according to the formula (6.3.8), where the classical observables are represented by functions belonging to $C_b(K, \mathbb{C})$. We can (and we shall) consider $\mathfrak{C}_{bs}^G$, or $\mathfrak{C}_{bs}$, as the (extended) C^*-algebra of observables of the systems with 'mean-field' dynamics. It might be useful, however, to embed this new algebra of observables in a canonical way into the W^*-algebra $\mathfrak{A}^{**}$, since there is a canonical bijection between the set of all states $\omega \in \mathcal{S}(\mathfrak{A})$ and the set of all normal states $\omega \in \mathcal{S}_*(\mathfrak{A}^{**})$ on the double dual $\mathfrak{A}^{**}$ of $\mathfrak{A}$: any $\omega \in \mathcal{S}(\mathfrak{A})$ corresponds to its (equally denoted) canonical normal extension $\omega \in \mathcal{S}_*(\mathfrak{A}^{**})$. Hence, after obtaining an **embedding of $\mathfrak{C}_{bs}$ into** $\mathfrak{A}^{**}$ such that $\mathfrak{A} \subset \mathfrak{C}_{bs}$ is mapped onto $\pi_u(\mathfrak{A}) \subset \mathfrak{A}^{**}$ or onto its subrepresentation, we shall obtain a certain canonical extension of any state $\omega \in \mathcal{S}(\mathfrak{A})$ (or of any state $\omega \in p_G \mathcal{S}(\mathfrak{A})$, where $p_G \in \mathfrak{Z}$ is the projector onto the above mentioned subrepresentation of π_u) to a state on $\mathfrak{A}^{**}$ (resp. on $p_G \mathfrak{A}^{**}$), and this in turn gives to us a certain canonical extension of states on $\mathfrak{A}$ to states on $\mathfrak{C}_{bs}$. Such an embedding is given in the following proposition.

6.3.6　Proposition. *Let us consider the integral decomposition of any $\omega \in p_G \mathcal{S}_*(\mathfrak{A}^{**})$ [where $\mathfrak{A}$ is simple] given by the formula 5.1.146 according to Theorem 5.2.11, and let $F_{\mathfrak{g}} : M \to \dot{\mathfrak{g}}^*$ be given as in 5.1.39. There is a C^*-isomorphism of $\mathfrak{C}_{bs}$ into $p_G \mathfrak{A}^{**}$ formally written in the form*

$$E_{\mathfrak{g}} : f (\in \mathfrak{C}_{bs}) \mapsto E_{\mathfrak{g}}(f) := \int f(F)\, E_{\mathfrak{g}}(dF), \qquad (6.3.15)$$

where $E_{\mathfrak{g}}$ denotes the G-measure (as before) as well as the presently introduced isomorphism. The isomorphism $E_{\mathfrak{g}}$ is uniquely determined by the formula[2]

$$\omega(E_{\mathfrak{g}}(f)) := \int \omega_m(f(F_m))\, \mu_\omega(dm), \quad \forall \omega \in p_G \mathcal{S}_*(\mathfrak{A}^{**}), \qquad (6.3.16)$$

where the decomposition (5.1.146) was used, and (cf. 5.1.39)

$$F_{\mathfrak{g}} : m \mapsto F_m := F_{\mathfrak{g}}(m), \quad m \in \mathcal{N} \subset \mathcal{M}, \qquad (6.3.17)$$

*is defined on the **spectrum space** $\mathcal{N}$ of the (commutative) subalgebra $\mathfrak{N}(E_{\mathfrak{g}})$ of $p_G \mathfrak{A}^{**}$, cf. 5.2.3.*

*　　The mapping $E_{\mathfrak{g}}$ leaves $\mathfrak{A} := p_G \pi_u(\mathfrak{A})$ invariant and maps C_b onto a C^*-subalgebra $\mathfrak{N}^c$ of $\mathfrak{N}(E_{\mathfrak{g}}) =: \mathfrak{N}_G$, see also 5.2.10.*

Proof. Let $B \subset \mathcal{N}$ be any Borel set and χ_B is its characteristic function. The functions

$$m \mapsto \omega_m(x)\,\chi_B(m), \quad x \in \mathfrak{A}^{**}, \qquad (6.3.18)$$

---

[2]Note: For noncompact supp $E_{\mathfrak{g}}$, the integral is a limit of integrals over bounded subsets $B \subset \mathfrak{g}^*$: $\int \cdots := \lim_{B \uparrow \mathfrak{g}^*} \int \omega_m(E_{\mathfrak{g}}(B) f(F_m))\, \mu_\omega(dm)$.

are Borel functions on $\mathcal{N}$ for any $\omega \in p_G S_*(\mathfrak{A}^{**})$. Since the function $F_\mathfrak{g}$ in (6.3.17) is continuous, the measurability of the functions

$$m \mapsto \omega_m(f \circ F_\mathfrak{g}(m)), \quad f \in \mathfrak{C}_{bs}, \tag{6.3.19}$$

can be proved with a help of a sequence $F_\mathfrak{g}^{(n)}$ of functions from $\mathcal{N}$ into the one point compactification $\dot{\mathfrak{g}}^*$ of $\mathfrak{g}^*$ assuming each only a finite number of values and pointwise converging to $F_\mathfrak{g}$ in the natural topology of $\dot{\mathfrak{g}}^*$. Then the functions

$$m \mapsto \omega_m(f \circ F_\mathfrak{g}^{(n)}(m)), \quad f \in \mathfrak{C}_{bs}, \ \omega \in p_G S_*(\mathfrak{A}^{**}) \tag{6.3.20}$$

are finite sums of functions of the form (6.3.18), hence the functions (6.3.20) are measurable. The s^*-continuity of f implies then the pointwise convergence of the functions (6.3.20) to the function (6.3.19) for $n \to \infty$. According to a known theorem in measure theory, cf. e.g. [223, 6.10.VII.], the pointwise limit of uniformly bounded measurable functions is measurable, hence (6.3.19) are Borel functions. We have proved the existence of the integrals in (6.3.16) for any $f \in \mathfrak{C}_{bs}$. The function

$$E_\mathfrak{g}(f) : \ \omega (\in S(\mathfrak{A})) \mapsto \omega(E_\mathfrak{g}(f)) \in \mathbb{C}, \tag{6.3.21}$$

is affine: The extension mapping e_* of $S(\mathfrak{A})$ onto $S_*(\mathfrak{A}^{**})$ is affine, and the association of subcentral (hence orthogonal, hence regular Borel) measures to the states $\omega \in S_*(\mathfrak{A}^{**})$ defined by (cf. 5.1.147)

$$\hat{\mu} : \ \omega \mapsto \hat{\mu}_\omega \in \{\text{probability measures on } S(\mathfrak{A}^{**})\}, \tag{6.3.22}$$

where the measure $\hat{\mu}_\omega$ corresponds to the decomposition of $\omega \in S(\mathfrak{A}^{**})$ given by the commutative subalgebra $\pi_\omega(\mathfrak{M}_G)''$ in $\mathcal{L}(\mathcal{H}_\omega)$ (cf. [53, 4.1.25.], and for the definition of $\mathfrak{M}_G$ see 5.2.10), is also affine. The affinity of (6.3.22) can be proved on the basis of the fact that all the measures in (6.3.22) are obtained from the same algebra $\mathfrak{M}_G \subset 3$ by considering $\hat{\mu}_\omega$ and $\lambda_1 \hat{\mu}_{\omega_1} + \lambda_2 \hat{\mu}_{\omega_2}$ (with $\omega := \lambda_1 \omega_1 + \lambda_2 \omega_2$) as limits of the nets of measures which correspond to the net of finite dimensional subalgebras of $\mathfrak{M}_G$, compare Lemma 4.1.26 in [53]:

$$\omega(x) = \sum_j \omega(p_j x) = \lambda_1 \sum_j \omega_1(p_j x) + \lambda_2 \sum_j \omega_2(p_j x), \quad \sum_j p_j = \mathrm{id}_\mathfrak{A},$$
$$\tag{6.3.23}$$

for any finite set of mutually orthogonal projectors $p_j \in \mathfrak{M}_G$. Hence, (6.3.22) is an affine mapping:

$$\hat{\mu}_\omega = \lambda_1 \hat{\mu}_{\omega_1} + \lambda_2 \hat{\mu}_{\omega_2}, \ \text{for } \omega := \lambda_1 \omega_1 + \lambda_2 \omega_2. \tag{6.3.24}$$

The relation (6.3.24) has a unique extension to all $\omega_j \in \mathfrak{A}^*$ ($\lambda_j \in \mathbb{C}$). Writing for $\omega \in p_G S(\mathfrak{A})$:

$$\omega(E_{\mathfrak{g}}(f)) = \int \varphi\left(f(F_{\mathfrak{g}} \circ r_M(\varphi))\right) \hat{\mu}_\omega(\mathrm{d}\varphi), \qquad (6.3.25)$$

what is meaningful for $\varphi \in \operatorname{supp} \hat{\mu}_\omega$ (cf. the proof of 5.1.38), we obtain now affinity of (6.3.21) which can be uniquely extended to linearity on the whole $\mathfrak{A}^*(\ni \omega)$. The boundedness of the mapping (6.3.21) is a direct consequence of (6.3.25) as well as of the boundedness of the function f. This proves that $E_{\mathfrak{g}}(f) \in \mathfrak{A}^{**}$, where the linear extension of (6.3.21) is denoted by the same symbol. We shall consider $\mathfrak{A}^{**}$ as a W^*-algebra in the canonical way: $\mathfrak{A}^{**} := \pi_u(\mathfrak{A})'' \subset \mathcal{L}(\mathcal{H}_u)$. We shall prove the morphism property of $E_{\mathfrak{g}}$ in (6.3.15). The linearity of (6.3.15) is clear from (6.3.16) and from the linearity of each of ω_m. By a 'polarization procedure' one can prove

$$\omega(E_{\mathfrak{g}}(f)y) = \int \omega_m(f(F_m)y)\,\mu_\omega(\mathrm{d}m). \ y \in \mathfrak{A}^{**}, \ \omega \in p_G S_*(\mathfrak{A}^{**}). \qquad (6.3.26)$$

Since $\omega_m(yE_{\mathfrak{g}}(f)) = \omega_m(yf(F_m))$ for all $\omega \in p_G S_*(\mathfrak{A}^{**})$, $m \in \operatorname{supp} \mu_\omega$, $y \in \mathfrak{A}^{**}$ and $f \in \mathfrak{C}_{bs}$, we have also

$$\omega(E_{\mathfrak{g}}(f_1)E_{\mathfrak{g}}(f_2)) = \int \omega_m(f_1(F_m)E_{\mathfrak{g}}(f_2))\,\mu_\omega(\mathrm{d}m) \qquad (6.3.27)$$

$$= \int \omega_m(f_1(F_m)f_2(F_m))\,\mu_\omega(\mathrm{d}m) = \omega(E_{\mathfrak{g}}(f_1 f_2)),$$

which proves $E_{\mathfrak{g}}(f_1 f_2) = E_{\mathfrak{g}}(f_1)E_{\mathfrak{g}}(f_2)$ for all $f_j \in \mathfrak{C}_{bs}$ $(j = 1, 2)$. The *-property follows by the decomposition of $f \in \mathfrak{C}_{bs}$ into the real and imaginary parts in (6.3.16).

We shall show that the kernel of the morphism $E_{\mathfrak{g}} : \mathfrak{C}_{bs} \to p_G\mathfrak{A}^{**}$ is trivial. We shall use here the simplicity of the C^*-algebra $\mathfrak{A}$. Let $f > 0$ be a positive element of $\mathfrak{C}_{bs}$, $\|f\| > 0$. If $f(F_0) \neq 0$, $F_0 \in K$, then there is a state $\omega \in S(\mathfrak{A})$ with $\omega(f(F_0)) \neq 0$. The s-continuity of $f \in \mathfrak{C}_{bs}$ implies that the set

$$B := \{F \in K : \omega(f(F)) > \frac{1}{2}\omega(f(F_0))\} \subset \mathfrak{g}^* \qquad (6.3.28)$$

is open in $K := \operatorname{supp} E_{\mathfrak{g}}$. Hence $E_{\mathfrak{g}}(B) \neq 0$, and

$$\|f(F)\| > \frac{1}{2}|\omega(f(F_0))| > 0, \text{ for all } F \in B. \qquad (6.3.29)$$

Any state $\omega_0 \in S(\mathfrak{A})$ supported by $E_{\mathfrak{g}}(B) : \omega_0(x) = \omega_0(E_{\mathfrak{g}}(B)x)$ $(x \in \mathfrak{A})$, is decomposed according to (5.1.146) into the states ω_m with $F_m \in B$ for all $m \in \operatorname{supp} \mu_{\omega_0}$. Since $\mathfrak{A}$ is simple, there is an element $x_m \in \mathfrak{A}$ for any such ω_m that

$$\omega_m(x_m^* x_m) = 1, \text{ and } \omega_m(x_m^* f(F_m)x_m) \neq 0. \qquad (6.3.30)$$

The state $\varphi_m \in \mathcal{S}(\mathfrak{A})$, $\varphi_m(y) := \omega_m(x_m^* y x_m)$ is also supported by $E_{\mathfrak{g}}(B_m)$ with any open $B_m \subset K$ containing F_m. Hence the decomposition (5.1.146) of $\omega := \varphi_m$ is concentrated on the one point set $\{m\}$. This means that

$$\varphi_m(E_{\mathfrak{g}}(f)) := \varphi_m(f(F_m)) \neq 0, \qquad (6.3.31)$$

hence $E_{\mathfrak{g}}(f) \neq 0$ for any nonzero $f \in \mathfrak{C}_{bs}$. This proves the isometry of $E_{\mathfrak{g}}$, hence $E_{\mathfrak{g}}$ is a C^*-isomorphism of $\mathfrak{C}_{bs}$ into $E_{\mathfrak{g}}(\mathfrak{g}^*)\mathfrak{A}^{**} = p_G \mathfrak{A}^{**}$. The remaining assertions are clearly valid. $\qquad \square$

6.3.7 Lemma. *Let $f \in \mathfrak{C}_{bs}$, $\omega \in p_G \mathcal{S}(\mathfrak{A})$. Then the function*

$$(g; F) \mapsto \omega\left(\sigma_g^{-1}(f(F))\right) \in \mathbb{C}, \; (g; F) \in G \times K, \qquad (6.3.32)$$

is jointly continuous on the topological product $G \times \operatorname{supp} E_{\mathfrak{g}}$.

Proof. Let $f := f_0$, cf. (6.3.7). Then

$$\sigma_g^{-1}(f_0(F)) = \sigma(g^{-1} g_0(F))(x) \, f(F), \qquad (6.3.33)$$

and the joint continuity of the group operation

$$(g_1; g_2) \, (\in G \times G) \mapsto g_1^{-1} g_2 \in G \qquad (6.3.34)$$

gives the joint continuity in (6.3.32) with $f := f_0$. It can be verified directly, cf. e.g. (6.3.13), that the function in (6.3.33) is even s^*-continuous in the couple $(g; F) \in G \times K$. But the finite algebraic combinations as well as the uniform limits of s^*-continuous bounded functions are s^*-continuous. Since $\mathfrak{C}_{bs}^G$ is generated by functions of the form f_0, we have proved that the functions

$$(g; F) \mapsto \sigma_g^{-1}(f(F)) \in \mathfrak{A}, \text{ for all } f \in \mathfrak{C}_{bs}^G, \qquad (6.3.35)$$

are even s^*-continuous. $\qquad \square$

6.3.8 Proposition. *Let, with the notation of 6.1.3, be $f \in \mathfrak{C}_{bs}^G$, and for a fixed $Q \in C^\infty(\mathfrak{g}^*, \mathbb{R})$ and for any $t \in \mathbb{R}$, $F \in K$, let*

$$f_t(F) := \sigma(g_Q^{-1}(t, F))(f(\varphi_t^Q F)). \qquad (6.3.36)$$

Then $f_t \in \mathfrak{C}_{bs}^G$ and the mappings $f \mapsto f_t$ form a one-parameter group of $*$-automorphisms of $\mathfrak{C}_{bs}^G$: $f_{t+s} = (f_t)_s$, for all $t, s \in \mathbb{R}$.

Proof. From the continuity properties of g_Q and φ^Q (g_Q and φ^Q depend smoothly on t and F), and from the s^*-continuity of functions (6.3.35), we have $f_t \in \mathfrak{C}_{bs}$ for any $f \in \mathfrak{C}_{bs}$. The $*$-morphism properties of the mapping $f \mapsto f_t$ are fulfilled due to the morphism properties of the pull-back φ^* by any diffeomorphism φ of K,

$$\varphi^*: f \mapsto \varphi^* f, \ \varphi^* f(F) := f(\varphi F), \ F \in K \subset \mathfrak{g}^*, \tag{6.3.37}$$

as well as of $\sigma(g) \in *\text{-Aut } \mathfrak{A}$. The group property follows immediately from the group property of the flow φ^Q and from the cocycle property (6.1.13) of g_Q. The group property implies invertibility, hence isometry of the considered mappings. $\square$

6.3.9 We have just proved existence of a certain 'time evolution' in the C^*-algebra $\mathfrak{C}_{bs}$ containing $\mathfrak{A}$ and $\mathfrak{N}^c$. This evolution is determined by an arbitrary classical Hamiltonian function Q and by the representation $\sigma(G)$ of the group G of 'macroscopic symmetries' with the help of the formula (6.3.36). To have possibility to see connections with the 'mean-field evolutions' discussed in Sect. 6.2, we shall transfer this evolution into $\mathfrak{A}^{**}$ by a use of the isomorphism $E_\mathfrak{g}$ from (6.3.15). We shall see that the time evolutions defined by a limiting procedure in Sect. 6.2 can be defined directly by the formula (6.3.36) (transferred into $p_G \mathfrak{A}^{**}$). The same possibility of a definition of 'mean-field evolutions' arises in all the systems considered in Sect. 5.1. To make this possibility clear, let us prove the property (6.3.1) for those systems.

6.3.10 Lemma. *Let us consider the systems determined with a help of infinite tensor product considered in Sect. 5.1. Then the group $\sigma(G) \subset *\text{-Aut } \mathfrak{A}$ ($\mathfrak{A} := \mathfrak{A}^\Pi$) has the property (6.3.1): The functions $g \mapsto \sigma_g(x)$ on G are s^*-continuous for all $x \in \mathfrak{A}$, the s^*-continuity being determined by the seminorms $\hat{p}_\omega$ and $\hat{p}_\omega^*$ from (6.2.19) with $\omega \in p_G \mathcal{S}(\mathfrak{A})$, and p_G was defined in 5.1.29.*

Proof. The implication "(6.3.1) $\Rightarrow$ s^*-continuity" was proved in Lemma 6.3.4. Since the set of local elements $x \in \cup_{N \subset \Pi} \mathfrak{A}^N$ is norm-dense in $\mathfrak{A}$, it suffices to prove the continuity in (6.3.1) for x local. We have assumed in 5.1.29 the existence of the generators X_ξ^N ($\xi \in \mathfrak{g}, N \subset \Pi$) of all one parameter subgroups of the unitary group $V_N(G)$ acting in $\mathcal{H}_N$, cf. 4.3.8 and 5.1.2, as well as the existence of (equally denoted) generators for the unitary groups $p_G \pi_u(V_N(\exp(\xi t)))$ for all $\xi \in \mathfrak{g}$. For $\omega \in p_G \mathcal{S}(\mathfrak{A})$ and $x \in \mathfrak{A}^N$ we have

$$\omega(\sigma(\exp(\xi t))(x)) = (\Omega_\omega, \exp(-it X_\xi^N)\pi_u(x)\exp(it X_\xi^N)\Omega_\omega), \tag{6.3.38}$$

what continuously depends on t. We have to prove the strong-continuity of the group $U(g) := p_G \pi_u(V_N(g))$ from the strong continuity of all one parameter subgroups $U(\exp \xi t) =: \exp(-it X_\xi)$, ($\xi \in \mathfrak{g}$); we write here X_ξ instead of X_ξ^N. Let $\xi_j \in \mathfrak{g}$, $j = 1, 2, \ldots n$ be a fixed basis in $\mathfrak{g}$ and set $X_j := X_{\xi_j}$. Let us parametrize $g \in G$ in a neighbourhood of the unity $e \in G$ by $\mathbf{t} := (t_1, t_2, \ldots t_n) \in \mathbb{R}^n$ in the following way, cf. [152, Lemma II.2.4]:

$$g \equiv g(\mathbf{t}) := \exp(t_1 \xi_1)\exp(t_2 \xi_2) \ldots \exp(t_n \xi_n). \tag{6.3.39}$$

Now we can prove weak continuity of $U(g(\mathbf{t}))$ in $\mathbf{t} = 0 \in \mathbb{R}^n$ from the known strong continuity of $U_j(t) := U(\exp \xi_j t) = exp(-it X_j)$, for all $j = 1, 2, \ldots n$. Since U is a representation of G, we can write

$$U(g(\mathbf{t})) - I = \prod_{j=1}^{n} U_j(t_j) - I = \sum_{k=1}^{n} \left[\prod_{j=1}^{k-1} U_j(t_j) \right] (U_k(t_k) - I), \qquad (6.3.40)$$

where I is the unit operator in the Hilbert space of the representation and the product of zero number of factors equals to I. Since the unitary operators do not change the norm of vectors, we have for any unit vectors Ψ_1 and Ψ_2 in the Hilbert space:

$$|(\Psi_1, (U(g(\mathbf{t})) - I)\Psi_2)| \leq \sum_{k=1}^{n} \|(U_k(t_k) - I)\Psi_2)\|. \qquad (6.3.41)$$

This estimate gives weak, hence strong continuity of $U(g)$. $\qquad \square$

6.3.11 Definition. *Let $E_{\mathfrak{g}}$ be the *-isomorphism of $\mathfrak{C}_{bs}$ into $p_G \mathfrak{A}^{**}$ described in (6.3.15). Let $\tau_t^Q \in$ *- Aut $p_G \mathfrak{A}^{**}$ ($t \in \mathbb{R}$) denote the one-parameter group determined by*

$$\tau_t^Q(E_{\mathfrak{g}}(f)) := E_{\mathfrak{g}}(f_t), \ t \in \mathbb{R}, \ f \in \mathfrak{C}_{bs}^G, \qquad (6.3.42)$$

where $f_t \in \mathfrak{C}_{bs}^G$ was introduced in (6.3.36) . The uniqueness of the extension of (6.3.42) to the whole $P_G \mathfrak{A}^{**}$ is given by uniqueness of the normal extension of the representations $\tau_t^Q : \mathfrak{A} \to p_G \mathfrak{A}^{**}$ to the representations of $\mathfrak{A}^{**}$ in $p_G \mathfrak{A}^{**}$, [274, 1.21.13], and the automorphism property of these extensions is given by the τ^Q-invariance of p_G (hence, $\tau_t^Q(\mathrm{id}_{\mathfrak{A}^{**}} - p_G) = 0$ for all t and Q). The automorphism group τ^Q will be called the **mean-field time evolution** *of the system $(\mathfrak{A}; \sigma(G))$* **determined by the classical Hamiltonian function Q.**

6.3.12 Theorem. *Let $E_{\mathfrak{g}}$ be a nontrivial G-measure associated with the system $(\mathfrak{A}; \sigma(G))$, cf. 5.2.3, with $K := \mathrm{supp}\, E_{\mathfrak{g}} \subset \mathfrak{g}^*$ such that $\sigma(G) \subset$ *- Aut $\mathfrak{A}$ is $\sigma(\mathfrak{A}, p_G \mathfrak{A}^*)$-continuous ($p_G := E_{\mathfrak{g}}(K)$). Let $\tau^Q \subset$ *- Aut $E_{\mathfrak{g}}(\mathfrak{C}_{bs}^G)$ be the mean-field time evolution of $(\mathfrak{A}; \sigma(G))$ determined by any $Q \in C^\infty(\mathfrak{g}^*, \mathbb{R})$. Let $\mathfrak{A}^J$ be any $\sigma(G)$-invariant C^*-subalgebra of $\mathfrak{A}$. Then:*

(i) $\mathfrak{N}^c := E_{\mathfrak{g}}(C_b)$ and $\mathfrak{C}^J := E_{\mathfrak{g}}(\mathfrak{C}_{bs}^{GJ})$ are τ^Q-invariant C^*-subalgebras of the 'algebra of mean-field observables' $\mathfrak{C} := E_{\mathfrak{g}}(\mathfrak{C}_{bs}^G) \subset p_G \mathfrak{A}^{**}$.

(ii) τ^Q is a $\sigma(\mathfrak{C}, \mathcal{S}_{\mathfrak{g}})$-continuous group, i.e. for any $y \in \mathfrak{C}$ and for any $\omega \in p_G \mathcal{S}_*(\mathfrak{A}^{**}) =: \mathcal{S}_{\mathfrak{g}}$ the function $t \mapsto \omega(\tau_t^Q(y))$ is continuous and the states $\omega \circ \tau_t^Q : y \mapsto \omega(\tau_t^Q(y))$ belong to $\mathcal{S}_{\mathfrak{g}}$, $\omega \circ \tau_t^Q \in p_G \mathfrak{A}^*$.

(iii) Let $\{\xi_j : j = 1, \ldots n\}$ be a fixed basis of $\mathfrak{g}$ and $F_j := F(\xi_j)$ be the coordinates of $F \in \mathfrak{g}^*$ in the dual basis. Let $\delta_{\xi_j} : \mathfrak{A} \to \mathfrak{A}$ be **the derivations** *(defined on $\sigma(\mathfrak{A}, p_G \mathfrak{A}^*)$-dense domains in $\mathfrak{A}$) of the one parameter subgroups $\sigma(\exp t\xi_j)$ of $\sigma(G)$. Then the infinitesimal generator of the group τ^Q is the* **derivation** *δ_Q on $\mathfrak{C}$ expressed by:*

$$\delta_Q(E_{\mathfrak{g}}(f)) := \left. \frac{d}{dt} \right|_{t=0} \tau_t^Q(E_{\mathfrak{g}}(f)) = \tag{6.3.43}$$

$$= \sum_{j=1}^{n} \int \left(\partial_j f(F)\{Q, F_j\}(F) - \partial_j Q(F) \delta_{\xi_j}(f(F)) \right) E_{\mathfrak{g}}(dF),$$

where the derivation is taken in the $\sigma(\mathfrak{C}, S_{\mathfrak{g}})$-topology, the symbol $\partial_j f(F)$ means the derivative of a function on $\mathfrak{g}^*$ with respect to the variable F_j in the point $F \in \mathfrak{g}^*$, and the meaning of the integral is explained in Proposition 6.3.6. $\{Q, F_j\}$ is here the Poisson bracket on $\mathfrak{g}^*$, 6.1.2.

(iv) If the group $\sigma(G)$ is strongly continuous (i.e. $g \mapsto \sigma_g(x)$ is continuous in norm for each $x \in \mathfrak{A}$), and if K is compact, then the group τ_t^Q will be strongly continuous.

Proof. The group τ^Q is considered here as an automorphism group of the τ^Q-invariant subalgebra $E_{\mathfrak{g}}(\mathfrak{C}_{bs}^G) =: \mathfrak{C}$ of $p_G \mathfrak{A}^{**}$.

(i) The invariance of $\mathfrak{N}^c$ is given by the invariance of C_b with respect to the transformations (6.3.36), which is valid due to the invariance of scalars in $\mathfrak{A}$ with respect to $\sigma(G) : \sigma_g(\lambda \, \mathrm{id}_{\mathfrak{A}}) = \lambda \, \mathrm{id}_{\mathfrak{A}}, \lambda \in \mathbb{C}, \forall g \in G$. Similarly, the relation $\sigma(G)(\mathfrak{A}^J) = \mathfrak{A}^J$ gives the τ^Q-invariance of $\mathfrak{C}^J$.

(ii) The continuity of the functions $t \mapsto \omega(\tau_t^Q(y))$ ($\omega \in S_{\mathfrak{g}}$, $y \in \mathfrak{C}$) can be obtained from the definition of the evolution $f \mapsto f_t$ in $\mathfrak{C}_{bs}^G$ as well as from the definition (6.3.16) of $E_{\mathfrak{g}}(f)$ as follows:
    Due to the s^*-bicontinuity of the mappings (6.3.35) and due to the (bi-)continuity of the functions g_Q and φ^Q, the functions

$$\Psi(m) : t \mapsto \Psi_t(m) := \omega_m(f_t(F_m)), \quad m \in \operatorname{supp} \mu_\omega, \tag{6.3.44}$$

are continuous for any fixed $\omega \in S_{\mathfrak{g}}$ and $f \in \mathfrak{C}_{bs}^G$. We have proved in (6.3.19) the measurability of all the functions $\Psi_t : m \mapsto \Psi_t(m)$. Since $|\Psi_t(m)| \leq \|f\|$ ($t \in \mathbb{R}$, $m \in \operatorname{supp} \mu_\omega$) and μ_ω is finite, an application of the Lebesgue dominated convergence theorem gives

$$\lim_{t \to 0} \omega(\tau_t^Q(E_{\mathfrak{g}}(f))) = \lim_{t \to 0} \int \Psi_t(m) \, \mu_\omega(dm) = \int \Psi_0(m) \, \mu_\omega(dm) = \omega(E_{\mathfrak{g}}(f)). \tag{6.3.45}$$

This gives the desired continuity.
    Any τ_t^Q can be considered as a *-automorphism of the W^*-algebra $p_G \mathfrak{A}^{**}$, and each such automorphism is $\sigma(p_G \mathfrak{A}^{**}, S_{\mathfrak{g}}) - \sigma(p_G \mathfrak{A}^{**}, S_{\mathfrak{g}})$-continuous, cf. [274, 4.1.23]. This implies that the state $\omega \circ \tau_t^Q$ is a normal state on $p_G \mathfrak{A}^{**}$ together with ω, hence $\omega \in S_{\mathfrak{g}}$ implies that $\omega \circ \tau_t^Q \in S_{\mathfrak{g}}$.

(iii) We shall calculate the derivation δ_Q from (6.3.43) by calculating the derivatives of the functions $\Psi(m)$ in (6.3.44). For 'sufficiently nice' elements $E_{\mathfrak{g}}(f) \in D(\delta_Q)$ (:= the domain of δ_Q) we have:

$$\frac{d}{dt}\bigg|_{t=0} \omega(f_t(F)) = \frac{d}{dt}\bigg|_{t=0} \omega(f(\varphi_t^Q F)) + \frac{d}{dt}\bigg|_{t=0} \omega\left(\sigma(g_Q^{-1}(t, F))(f(F))\right).$$
(6.3.46)

For the calculation of the first term we shall use the classical evolution equation (6.2.41), where we shall consider $f(F)$ as a function of coordinates $F_j :=$ $F_j(0)$, $F_j(t) := F_j(\varphi_t^Q F) := \varphi_t^Q F(\xi_j)$:

$$\frac{d}{dt} f(\varphi_t^Q F) = \sum_{j=1}^{n} \partial_j f(\varphi_t^Q F) \frac{d}{dt} F_j(\varphi_t^Q F) = \sum_{j=1}^{n} \partial_j f(\varphi_t^Q F)\{Q, F_j\}(\varphi_t^Q F).$$
(6.3.47)

Insertion of $f(F) := \omega(f(F))$ into (6.3.47) and setting $t = 0$ we obtain

$$\frac{d}{dt}\bigg|_{t=0} \omega(f(\varphi_t^Q F)) = \sum_{j=1}^{n} \partial_j \omega(f(F))\{Q, F_j\}(F). \tag{6.3.48}$$

The second term in (6.3.46) can be calculated with a help of (6.1.10) + (6.1.17) + (6.1.18), and by considering that for any $\xi \in \mathfrak{g}$ we have defined

$$\frac{d}{dt}\bigg|_{t=0} \omega(\sigma(\exp t\xi)(x)) = \omega(\delta_\xi(x)), \quad x \in D(\delta_\xi) \subset \mathfrak{A}. \tag{6.3.49}$$

One obtains

$$\frac{d}{dt}\bigg|_{t=0} \omega(\sigma(g_Q(t, F))(x)) = \sum_{j=1}^{n} \partial_j Q(F)\omega(\delta_{\xi_j}(x)), \ x \in \bigcap_{j=1}^{n} D(\delta_{\xi_j}). \tag{6.3.50}$$

Combining (6.3.48) and (6.3.50), where we set $\omega := \omega_m$, $F := F_m$ and $x := f(F_m)$, we obtain for the 'sufficiently nice' $f \in \mathfrak{C}_{bs}^G$:

$$\frac{d}{dt}\bigg|_{t=0} \omega(\tau_t^Q E_{\mathfrak{g}}(f)) = \sum_{j=1}^{n} \int \omega_m \left(\partial_j f(F_m)\{Q, F_j\}(F_m) - \partial_j Q(F_m)\delta_{\xi_j}(f(F_m))\right) \mu_\omega(dm).$$
(6.3.51)

The change of the sign is caused by the replacement of g_Q by g_Q^{-1} in (6.3.50). The comparison of (6.3.43) with (6.3.51) gives the result.

(iv) We have to prove that the functions

$$t \mapsto \| f_t - f \| \text{ for all } f \in \mathcal{C}_{bs}^Q \tag{6.3.52}$$

are continuous at $t = 0$. Let us write

$$\| f_t(F) - f(F) \| = \| \sigma^{-1}(g_Q(t, F))(f(\varphi_t^Q F)) - f(F) \| \leq$$
$$\leq \| \sigma(g_Q^{-1}(t, F))(f(F) - f(F_0)) \| + \| \sigma(g_Q^{-1}(t, F))(f(F_0)) - f(F_0) \| + \| f(F_0) - f(F) \|$$
$$= 2 \| f(F_0) - f(F) \| + \| \sigma(g_Q^{-1}(t, F))(f(F_0)) - f(F_0) \|. \tag{6.3.53}$$

The strong continuity of $\sigma(G)$ and the joint continuity of g_Q lead to existence of an open interval $I(F_0, \varepsilon) \subset \mathbb{R}$ containing $t = 0$ as well as of an open neighbourhood of F_0, $\mathcal{U}(F_0, \varepsilon) \subset K$, corresponding to any $F_0 \in K$ and to any $\varepsilon > 0$, such that

$$\| \sigma(g_Q^{-1}(t, F))(f(F_0)) - f(F_0) \| < \frac{\varepsilon}{3}, \text{ for all } (t; F) \in I(F_0, \varepsilon) \times \mathcal{U}(F_0, \varepsilon). \tag{6.3.54}$$

The strong continuity of $\sigma(G)$ leads also to norm continuity of the functions f_0 in (6.3.7) which generate $\mathcal{C}_{bs}^G$, hence all $f \in \mathcal{C}_{bs}^G$ are continuous in norm in the present case. This shows that we can choose the neighbourhoods $\mathcal{U}(F_0, \varepsilon)$ in such a way that

$$\| f(F) - f(F_0) \| < \frac{\varepsilon}{3}, \text{ if } F \in \mathcal{U}(F_0, \varepsilon), \text{ for any } F_0 \in K. \tag{6.3.55}$$

Since K is compact, we can find a finite set $\{ F_p : p = 1, 2, \dots P \} \subset K$ such that the union of $\{ \mathcal{U}(F_p, \varepsilon) : p = 1, 2, \dots P \}$ covers K. Let $I(\varepsilon)$ be the intersection of the intervals $\{ I(F_p, \varepsilon) : p = 1, 2, \dots P \}$. Then

$$\| f_t(F) - f(F) \| < \varepsilon, \text{ for all } (t; F) \in I(\varepsilon) \times K. \tag{6.3.56}$$

Taking supremum in (6.3.56) we obtain the desired continuity in (6.3.52).   □

6.3.13 To compare the derivations δ_Q from (6.3.43) with δ_π from the formulas (6.2.83), it suffices to take $f \in \mathcal{C}_{bs}^{GJ}$ where $\mathfrak{A}^J := \mathfrak{A}^N$ is a $\sigma(G)$-invariant 'local algebra'. For such an f we have

$$\sigma(\exp t\xi)(f(F)) = \exp(-it X_\xi^N) f(F) \exp(it X_\xi^N), \ t \in \mathbb{R}, \ F \in \mathfrak{g}^*, \tag{6.3.57}$$

for any $\xi \in \mathfrak{g}$; here we made the usual identifications, cf. notation in 6.3.10. Then we have

$$\delta_\xi(f(F)) = -i [X_\xi^N, f(F)], \tag{6.3.58}$$

where the commutator is taken between operators in the Hilbert space $p_G \mathcal{H}_u$. We can sea easily now that the derivations δ_π and δ_Q are expressed by identical formulas. This proves the identity of the time evolutions determined in Sect. 6.2 with the evolutions from the present section in the case of the **UHF-algebra** $\mathfrak{A} := \mathfrak{A}^\Pi$ (cf. [53, 2.6.12], [235, 6.4.1]; UHF:="uniformly hyperfinite") with the polynomial Q. This shows also that the derivation δ_Q for the case of a nonseparable $\mathfrak{A}^\Pi$ and unbounded X_ξ is described by the same formulas as δ_π is.

6.4 Equilibrium States

6.4.1 Let us consider in this section those states of physical systems which describe the situations corresponding to the *thermodynamic equilibrium* at a given temperature $T \geq 0$. For quantal systems these states are specified usually by the **KMS-condition**, cf. e.g. [54, 106, 235, 271]. We shall investgate here the *KMS states*[3] of systems considered in this chapter, i.e. the systems specified by the triple $(\mathfrak{A}; \sigma(G); \tau^Q)$, cf. also [41]. To avoid possible technical complications, we shall concentrate our attention here on the cases of strongly continuous time evolutions τ^Q including, e.g. the cases described in 6.3.12(iv). Let us use the notation of Theorem 6.3.12, hence $\mathfrak{C} := E_\mathfrak{g}(\mathfrak{C}^G_{bs})$ be the C^*-algebra of (generalized) observables describing the considered system with the dynamics τ^Q. Instead of the above mentioned triple, we shall use also the couple $(\mathfrak{C}; \tau^Q)$ for denoting the system. In most of the analysis of this section an additional structure of the system will be used. Let Π be a locally compact noncompact group and $\pi(\Pi)$ be its representation on $\mathfrak{C}$, i.e. $\pi(p) \in *\text{-}\mathrm{Aut}\,\mathfrak{C}$ for all $p \in \Pi$. Let $\pi(\Pi)$ commutes with τ^Q :

$$\tau_t^Q \circ \pi(p) = \pi(p) \circ \tau_t^Q \quad \text{for all } t \in \mathbb{R}, \ p \in \Pi. \tag{6.4.1}$$

We shall assume usually that $\pi(\Pi)$ has some *asymptotic abelianess* properties. As an example of such a $\pi(\Pi)$ consider the situations described in Sect. 5.1. (i.e. $\mathfrak{A} := \mathfrak{A}^\Pi$ is a tensor product of the mutually commuting 'local algebras' $\mathfrak{A}_p := \mathcal{L}(\mathcal{H}_p)$), where the set $\mathbb{Z}_+ \setminus \{0\}$ is replaced by $\Pi := \mathbb{Z}^r$ (with easy modifications of the whole formalism). Let us write $\pi_p : \mathcal{L}(\mathcal{H}) \to \mathcal{L}(\mathcal{H}_\Pi)$ for the isomorphism defined in (5.1.12), $p \in \Pi$. Now we define $\pi(p) \in *\text{-}\mathrm{Aut}\,\mathfrak{A}^\Pi$ by

$$\pi(p)(\pi_j(A)) := \pi_{j+p}(A), \quad \text{for all } A \in \mathcal{L}(\mathcal{H}), \ p, \ j \in \Pi. \tag{6.4.2}$$

Since the elements $\pi_j(A)$ $(j \in \Pi, A \in \mathcal{L}(\mathcal{H}))$ generate $\mathfrak{A}^\Pi$, (6.4.2) determines an automorphism $\pi(p)$ of $\mathfrak{A}^\Pi$ uniquely. This automorphism can be extended naturally to an (equally denoted) automorphism group $\pi(\Pi)$ of $\mathfrak{C} := E_\mathfrak{g}(\mathfrak{C}^G_{bs})$ by the relation

$$\pi(p)(E_\mathfrak{g}(f)) := \int \pi(p)(f(F))\, E_\mathfrak{g}(dF). \tag{6.4.3}$$

---

[3]KMS is for Kubo, Martin and Schwinger.

The group $\pi(\Pi)$ is *norm-asymptotically abelian*, i.e.

$$\lim_{p \to \infty} \|[\pi(p)(x), y]\| = 0, \quad \text{for all } x, y \in \mathfrak{C}. \tag{6.4.4}$$

In more general cases, the abelianess properties of the action of Π on $\mathfrak{C}$ can be weaker. Systems with this structure **will be denoted**

$$(\mathfrak{C}; \tau^{\varrho}; \pi(\Pi)), \text{ or } (\mathfrak{A}; \sigma(G); \tau^{\varrho}; \pi(\Pi)).$$

We shall use, as usual, $\beta := T^{-1} := (kT)^{-1}$ to denote the inverse *temperature* in convenient units. The following definitions are found e.g. in [54, 5.3.1, 5.3.18, and 5.3.21], and [235, 8.12].

6.4.2 Definition. *Let $(\mathfrak{C}, \tau)$ be a C^*-dynamical system, i.e. the one parameter group $\tau \subset {}^*$-Aut $\mathfrak{C}$ is strongly continuous. The state $\omega \in S(\mathfrak{C})$ is defined to be a τ-KMS state at value $\beta \in \mathbb{R}$, or a (τ, β)-KMS state, if*

$$\omega(x \tau_{i\beta}(y)) = \omega(yx), \quad \text{for all } x, y, \in \mathfrak{C}_\tau^\circ, \tag{6.4.5}$$

where $\mathfrak{C}_\tau^\circ$ is a norm-dense, τ-invariant *-subalgebra of the set $\mathfrak{C}_\tau$ of the entire analytic elements of $\mathfrak{C}$:

$$y \in \mathfrak{C}_\tau^\circ \Leftrightarrow \text{ the function } z \mapsto \tau_z(y) \text{ is analytic for all } z \in \mathbb{C}. \tag{6.4.6}$$

*Let δ_τ be the generator of τ. Then $\omega \in S(\mathfrak{C})$ is called a τ-**ground state** if*

$$-i\,\omega(y^*\delta_\tau(y)) \geq 0, \quad \text{for all } y \in D(\delta_\tau). \tag{6.4.7}$$

In this case, ω is also called a τ-KMS state at value $\beta = \infty$.

6.4.3 Definition. *Let $(\mathfrak{C}; \tau)$ be a C^*-dynamical system with a unital C^*-algebra $\mathfrak{C}$, and let δ_τ be the infinitesimal generator of τ. Then $\omega \in S(\mathfrak{C})$ is said to be a **passive** state if*

$$-i\,\omega(u^*\delta_\tau(u)) \geq 0 \tag{6.4.8}$$

for any $u \in D(\delta_\tau)$ belonging also to the connected component of the identity of the unitary group of $\mathfrak{C}$ in the norm topology.

6.4.4 Let us collect here some important properties of the sets $\mathcal{K}_\beta$ of (τ, β)-KMS states:
Proofs of the listed facts can be found in [54, Chap. 5], or in [275, 4.3]. We shall consider $\beta \in (0, \infty]$, the set $\mathcal{K}_\infty$ being the set of all ground states $\omega \in S(\mathfrak{C})$. Let $(\mathfrak{C}, \tau)$ be a C^*-dynamical system. Then:

(0) Any state $\omega \in \mathcal{K}_\beta$ is τ-invariant: $\omega \circ \tau_t = \omega$ ($t \in \mathbb{R}$).

(i) Any $\mathcal{K}_\beta$ is a *convex W^*-compact* subset of $\mathcal{S}(\mathfrak{C})$.

(ii-a) For $\beta \neq \infty$, $\mathcal{K}_\beta$ is a *simplex* in $\mathcal{S}(\mathfrak{C})$.

(ii-b) $\mathcal{K}_\infty$ is a *face* in $\mathcal{S}(\mathfrak{C})$.

(iii-a) The set $\mathcal{E}\mathcal{K}_\beta$ of extremal points $\omega \in \mathcal{K}_\beta$ ($\beta \neq \infty$) consists of *factor states*: The centers of $\pi_\omega(\mathfrak{C})''$ are trivial.

(iii-b) The extremal points $\omega \in \mathcal{K}_\infty$, i.e. $\omega \in \mathcal{E}\mathcal{K}_\infty$, are *pure states*: $\omega \in \mathcal{E}\mathcal{S}(\mathfrak{C})$, i.e. $\pi_\omega(\mathfrak{C})'' = \mathcal{L}(\mathcal{H}_\omega)$.

(iv) $\omega_j \in \mathcal{E}\mathcal{K}_\beta$ ($\beta \neq \infty$, $j = 1, 2$) implies either $\omega_1 = \omega_2$, or $\omega_1 \perp \omega_2$, i.e. ω_1 and ω_2 are mutually *disjoint*, i.e. the *central covers* s_{ω_1} and s_{ω_2} of the corresponding GNS-representations are mutually orthogonal.

(v) The *extremal decomposition* of $\omega \in \mathcal{K}_\beta$ ($\beta \neq \infty$) coincides with its *central decomposition*, cf. [53, Chap. 4], [235, Chap. 4]. The corresponding probability measure μ_ω^c on $\mathcal{S}(\mathfrak{C})$ is *pseudosupported* (cf. [54, Chap. 6]) by $\mathcal{E}\mathcal{K}_\beta$ and if the Hilbert space of the GNS-representation $\mathcal{H}_\omega$ is separable, then μ_ω^c is supported by $\mathcal{E}\mathcal{K}_\beta$: $\mu_\omega^c(\mathcal{E}\mathcal{K}_\beta) = \mu_\omega^c(\mathcal{S}(\mathfrak{C})) = 1$.

6.4.5 Lemma. *Let $\omega \in \mathcal{S}(\mathfrak{C})$ be a τ-ground state. Let $(\pi_\omega, \mathcal{H}_\omega, \Omega_\omega)$ be the corresponding GNS representation. Then for the unique selfadjoint operator Q_ω on $\mathcal{H}_\omega$ determined by the relation:*

$$\exp(it\, Q_\omega)\, \pi_\omega(y)\Omega_\omega := \pi_\omega(\tau_t(y))\Omega_\omega, \quad \forall t \in \mathbb{R}, \tag{6.4.9}$$

the following is valid:

$$Q_\omega \geq 0, \text{ and for all } t \in \mathbb{R} \text{ one has } \exp(it\, Q_\omega) \in \pi_\omega(\mathfrak{C})''. \tag{6.4.10}$$

Proof. See [54, 5.3.19]. □

6.4.6 Any (τ, β)-KMS state, according to 6.4.4(i), can be approximated in the w^*-topology by convex combinations of extremal KMS states at the same temperature β^{-1}. The set $\mathcal{K}_\beta$ may be void for a general dynamical system and for a given $\beta \in (0, \infty]$. Occurrence of more than one points in $\mathcal{K}_\beta$ means occurrence of several mutually disjoint states in $\mathcal{E}\mathcal{K}_\beta$. Orthogonal central projectors s_1 and s_2 (the central covers of the corresponding GNS representations) are supporting such disjoint states; these $s_j \in \mathfrak{Z}$ (:= the center of $\pi_u(\mathfrak{C})''$) may be interpreted as corresponding to distinct values of a macroscopic (global, classical) quantity for distinct $j = 1, 2$. We interpret this situation as possibility of existence of several mutually different 'phases' of the considered system at the temperature $T = \beta^{-1}$. This interpretation is especially intuitive in cases of quasilocal algebras $\mathfrak{C}$ when the extremal KMS (hence factor) states have *short range correlations* (cf. e.g. [193])—the necessary property of the states representing pure phases of a spatially extended system [271, 6.5]. We shall investigate general properties of the extremal (τ^Q, β)-KMS states of the systems $(\mathfrak{C}; \tau^Q)$ and $(\mathfrak{C}; \tau^Q; \pi(\Pi))$ representing the generalized mean-field theories.

6.4.7 Proposition. *Let* $\omega \in \mathcal{K}_\beta$ *be an extremal* τ^Q-*KMS state of a generalized mean-field theory* $(\mathfrak{A}; \sigma(G); \tau^Q)$. *Then there is an element* $F_\omega \in \operatorname{supp} E_\mathfrak{g}$ *such that the central support* $s_\omega \le E_\mathfrak{g}(B)$ *for any open* $B \subset \mathfrak{g}^*$ *containing* F_ω : $F_\omega \in B$. **The point** F_ω *is a fixed point of the* **classical flow** φ^Q *on* $\mathfrak{g}^*$. *The state* ω *is invariant with respect to the one parameter* **subgroup** *of* $\sigma(G)$ **generated by the element** $\beta^Q_{F_\omega} \in \mathfrak{g}$, *(6.1.17), and the* **generator** Q_ω *of* τ^Q *in* $\pi_\omega(\mathfrak{A})$ **implements** *this subgroup in the sense that*

$$\pi_\omega\left(\sigma(\exp(-\beta^Q_{F_\omega}t))(x)\right)\Omega_\omega = \exp(itQ_\omega)\pi_\omega(x)\Omega_\omega, \quad t \in \mathbb{R}, \ x \in \mathfrak{A}. \qquad (6.4.11)$$

The image $\pi_\omega(\mathfrak{C})$ *of* $\mathfrak{C} := E_\mathfrak{g}(\mathfrak{C}^G_{bs})$ *coincides with* $\pi_\omega(\mathfrak{A})$, $\mathfrak{A} = E_\mathfrak{g}(\mathfrak{A})$ $(\mathfrak{A} \subset \mathfrak{C}^G_{bs}$ *represents here* $\mathfrak{A}$-*valued constant functions).*

Assume that the whole group $\sigma(G)$ is unitarily implemented in the representation $(\pi_\omega, \mathcal{H}_\omega, \Omega_\omega)$. Then we can choose the generators $X_\omega(\xi)$ of the one parameter subgroups $\exp(t\xi)$ in such a way that

$$Q_\omega = X_\omega(\beta^Q_{F_\omega}) = \sum_{j=1}^n \partial_j Q(F_\omega) X_\omega(\xi_j) \qquad (6.4.12)$$

for any basis $\{\xi_j : j = 1, 2, \ldots n\}$ *in* $\mathfrak{g}$.

Proof. The factor state ω is projected by p_M onto a pure state on $\mathfrak{N}_G$, 5.1.35, hence the decomposition of ω in (6.3.16) is concentrated on a one point set $F_\omega \in \operatorname{supp} E_\mathfrak{g}$. Let f_j $(j = 1, 2)$ be any such elements of $\mathfrak{C}^G_{bs}$ that $f_1(F_\omega) = f_2(F_\omega)$. Then

$$\omega(E_\mathfrak{g}(f_1)) = \omega(f_1(F_\omega)) = \omega(f_2(F_\omega)) = \omega(E_\mathfrak{g}(f_2)). \qquad (6.4.13)$$

This proves that $\pi_\omega(\mathfrak{C}) = \pi_\omega(\mathfrak{A})$. The state $\omega \circ \tau^Q_t \equiv \omega$ is then concentrated (in the above described sense) on $\varphi^Q_t(F_\omega)$, and states ω_1 and ω_2 concentrated on $F_1 \ne F_2$ are disjoint: $\omega_1 \perp \omega_2$. Hence, $\varphi^Q_t(F_\omega) = F_\omega$ for all $t \in \mathbb{R}$. This means, however, that the classical Poisson bracket $\{Q, f\}(F_\omega) = 0$ for any function f. It follows that for the generator δ_Q, (6.3.43), in the representation π_ω, one has:

$$\omega(x\delta_Q(E_\mathfrak{g}(f))y) = -\sum_{j=1}^n \partial_j Q(F_\omega)\,\omega(x\delta_{\xi_j}(f(F_\omega))y), \quad x, y \in \mathfrak{A}. \qquad (6.4.14)$$

The definition of the time evolution in Proposition 6.3.8 and the φ^Q-invariance of F_ω shows the identity of the time evolution of $\pi_\omega(\mathfrak{A})$ with the action of the one-parameter group $g_Q^{-1}(t, F_\omega)$, cf. (6.1.13), with the generator $-\beta_{F_\omega}^Q$, cf. (6.1.10). According to (6.1.17) and (6.1.18), we obtain the remaining assertions of the proposition. $\qquad\square$

6.4.8 Note. *The generator of the mean-field time evolution τ^Q of local perturbations of an extremal equilibrium state ω given in (6.4.12) is usually called the Bogoliubov-Haag Hamiltonian, cf.* [23, 140, 312].

6.4.9 We shall assume in the following that $\mathfrak{A}$ is a quasilocal C^*-algebra generated by a net $\{\mathfrak{A}^J : J \subset \Pi, J \text{ finite}\}$ of local subalgebras $\mathfrak{A}^J$ commuting with each other for disjoint J's:

$$x \in \mathfrak{A}^J, \; y \in \mathfrak{A}^{J'}, \; J \cap J' = \emptyset \Rightarrow [x, y] = 0. \tag{6.4.15}$$

Here Π is a countable infinite commutative group acting on $\mathfrak{A}$ by the representation $\pi : \pi_p \in {}^*\text{-}\mathrm{Aut}\,\mathfrak{A}$, in such a way that $\pi_p : \mathfrak{A}^J \to \mathfrak{A}^{J+p}$ is an isomorphism for any $J \subset \Pi$. This is the situation from (6.4.2), where $\mathcal{L}(\mathcal{H})$ is identified with $\mathcal{L}(\mathcal{H}_u), \pi_0 = \pi(0) = \mathrm{id}_{\mathcal{L}(\mathfrak{A})}$ (0 is here the identity of the group Π), hence $\pi(p) = \pi_p$ ($p \in \Pi$).

It will be assumed in the following that each $\mathfrak{A}^J$ ($J \subset \Pi$) is $\sigma(G)$-invariant, and that the action of $\sigma(G)$ commutes with $\pi(\Pi)$. Then also (6.4.1) will be fulfilled ($\pi(\Pi)$ is naturally extended to the equally denoted automorphism groups of $\mathfrak{C}$ and of $\mathfrak{A}^{**}$).

In this situation, let $\omega \in \mathcal{S}(\mathfrak{A})$ be a factor state which is invariant with respect to the action of $\pi(\Pi)$:

$$\omega(\pi_p(x)) = \omega(x), \text{ for all } x \in \mathfrak{A}, \; p \in \Pi. \tag{6.4.16}$$

The locally normal factor states have *short range correlations*, [193], [53, Theorem 2.6.10], hence they are *weakly $\pi(\Pi)$-clustering*, and

$$\lim_{p \to \infty} \omega(\pi_p(x)y) = \omega(x)\omega(y), \text{ for all } x, y, \in \mathfrak{A}. \tag{6.4.17}$$

If $\mathfrak{A}^J$ are faithfully represented in Hilbert spaces $\mathcal{H}_J$, as it was the case of Sect. 5.1, then π_p will be used also for translations of unbounded operators acting on $\mathcal{H}_J$ to unitarily equivalent operators acting on $\mathcal{H}_{J+p}$ (e.g. by translating their spectral projectors belonging to $\mathfrak{A}^J$); this can be done if the isomorphisms of $\mathfrak{A}^J \subset \mathcal{L}(\mathcal{H}_J)$ with $\mathfrak{A}^{J+p} \subset \mathcal{L}(\mathcal{H}_{J+p})$ ($J \subset \Pi, \; p \in \Pi$) are spatial. We shall write also $\mathfrak{A}_p := \mathfrak{A}^J$ with $J := \{p\} :=$ the one-point set, $p \in \Pi$. Let all the $\mathfrak{A}^J$ ($J \subset \Pi$) have common unit and let the C^*-algebras $\mathfrak{A}_p$ with $p \in J$ generate $\mathfrak{A}^J$ ($J \subset \Pi$).

With the introduced notation and assumptions, we shall prove now the following:

6.4.10 Theorem. *Let us consider a system* $(\mathfrak{A}; \sigma(G); \tau^Q; \pi(\Pi))$ *with simple* C^*-*algebra* $\mathfrak{A}$ *and 'local' subalgebras* $\mathfrak{A}^J \subset \mathfrak{A}$ *being factors for all finite* J. *Let* $\omega \in \mathcal{S}(\mathfrak{A})$ *and let* ω^0 *be the restriction of* ω *to the subalgebra* $\mathfrak{A}_0$ $(:= \mathfrak{A}^J$ *with the one-point set* J *containing the identity* $0 \in \Pi)$. *Then the following two statements are equivalent:*

(i) ω *is a locally normal extremal* τ^Q-*KMS state at a positive temperature* $\beta^{-1} > 0$.

(ii) $\omega = \overline{\omega}$, *where* $\overline{\omega}$ *is the* $\pi(\Pi)$-*invariant product state determined by the relation*

$$\overline{\omega}(\pi_{p_1}(x_1)\pi_{p_2}(x_2)\ldots\pi_{p_m}(x_m)) = \prod_{j=1}^{m} \omega^0(x_j), \qquad (6.4.18)$$

with $x_j \in \mathfrak{A}_0$, $p_j \in \Pi$ $(p_j \neq p_k$ *for* $j \neq k)$, $j = 1, 2, \ldots m$, $\forall m \in \mathbb{N}$ *and* ω^0 *is the faithful normal KMS-state at* β *on* $\mathfrak{A}_0$ *corresponding to the one-parameter subgroup* $\{\sigma(\exp(-t\beta_{F_\omega}^Q)) : t \in \mathbb{R}\}$ *of* *-$\mathrm{Aut}\,\mathfrak{A}_0$ *with*

$$\varphi_t^Q(F_\omega) = F_\omega, \text{ for all } t \in \mathbb{R} \qquad (6.4.19)$$

for some element $F_\omega \in \mathfrak{g}^*$. *Moreover, the 'consistency condition'*

$$\omega(E_{\mathfrak{g}}(f_\xi)) = F_\omega(\xi), \quad (\xi \in \mathfrak{g}, \ f_\xi(F) := F(\xi) \text{ for } F \in \mathfrak{g}^*) \qquad (6.4.20)$$

is fulfilled.[4]

Proof. (i) implies $\pi_\omega(\tau_t^Q(x)) = \pi_\omega(\sigma(\exp(-t\beta_{F_\omega}^Q))(x))$ according to (6.4.11). Hence ω satisfies the KMS-condition with respect to the group $\sigma(\exp(-t\beta_{F_\omega}^Q))$ at T^{-1} and the same is true for ω^0, since $\sigma(G)(\mathfrak{A}_0) = \mathfrak{A}_0$. Let $X(\beta_{F_\omega}^Q)$ be the restriction of $X_\omega(\beta_{F_\omega}^Q)$ onto $\overline{\pi_\omega(\mathfrak{A}_0)\Omega_\omega}$. ω is faithful on $\mathfrak{A}$ ($\mathfrak{A}$ is simple) and the cyclic vector Ω_ω is separating for $\pi_\omega(\mathfrak{A})''$, cf. [54, 5.3.9]. Hence $\omega(x^*x) \neq 0$ for $x \neq 0$, and ω^0 is faithful on $\mathfrak{A}_0$. The local normality of ω implies normality of ω^0. According to the *Takesaki's theorem* [54, 5.3.10], the one-parameter automorphism group of $\pi_\omega(\mathfrak{A}_0)$:

$$t \mapsto \exp(itX(\beta_{F_\omega}^Q))\pi_\omega(x)\exp(-itX(\beta_{F_\omega}^Q)), \quad x \in \mathfrak{A}_0, \qquad (6.4.21)$$

coincides with the corresponding modular automorphism group of $\pi_\omega(\mathfrak{A}_0)$ determined by the state ω^0 (up to a rescaling of time t). According to [54, 5.3.29], the KMS state at $\beta := T^{-1} \in \mathbb{R}$ on the factor $\mathfrak{A}_0$ corresponding to its automorphism group $\sigma(\exp(-t\beta_{F_\omega}^Q))$ is uniquely determined faithful normal state on $\mathfrak{A}_0$.

---

[4]The stationarity (6.4.19) is a consequence of the "consistency condition" (6.4.29), i.e. of (6.4.20); hence (6.4.19), and (6.4.20) can be replaced by (6.4.29).

We have to prove that ω is a $\pi(\Pi)$-invariant product state on $\mathfrak{A}$, i.e. that (6.4.18) (with $\omega \hookrightarrow \overline{\omega}$) is satisfied. Let $y := \pi_p(x)$ for some $x \in \mathfrak{A}_0$, $p \in \Pi$. From the commutativity of $\pi(\Pi)$ with $\sigma(G)$ we have for $y' := \pi_p(x')$:

$$\omega(\tau_t^Q(y)y') = \omega \circ \pi_p(\tau_t^Q(x)x'), \text{ for all } x, x' \in \mathfrak{A}_0, \ t \in \mathbb{R}. \tag{6.4.22}$$

We can write here $\omega^p \in \mathcal{S}(\mathfrak{A}_p)$ instead of ω. The state ω^p is a KMS-state, hence $\omega^p \circ \pi_p \in \mathcal{S}(\mathfrak{A}_0)$ is the unique KMS state ω^0:

$$\omega^p \circ \pi_p = \omega^0, \text{ for all } p \in \Pi. \tag{6.4.23}$$

Since all the $\mathfrak{A}^J$ are factors (J finite), we can repeat the above considerations for the restrictions ω^J of ω to $\mathfrak{A}^J$ (with J replacing the one point set $\{0\} \subset \Pi$) : ω^J is the unique KMS state at T^{-1} of $\mathfrak{A}^J$ corresponding to the group $\sigma(\exp(-t\beta_{F_\omega}^Q)) \in$ *-Aut $\mathfrak{A}^J$, and

$$\omega^{J+p} \circ \pi_p = \omega^J \text{ for all finite } J \subset \Pi, \ p \in \Pi. \tag{6.4.24}$$

For an arbitrary local element $x \in \mathfrak{A}^J$ one obtains:

$$\omega \circ \pi_p(x) = \omega^{J+p} \circ \pi_p(x) = \omega^J(x) = \omega(x), \tag{6.4.25}$$

hence we have the translation invariance $\omega \circ \pi_p = \omega$ of the extremal τ^Q-KMS state ω at positive temperature T.

The restriction to $\mathfrak{A}^J$ of the product state $\overline{\omega}$ on the right hand side of (6.4.18) satisfies the KMS condition at T^{-1} with respect to the one parameter group $\{\sigma(\exp(-t\beta_{F_\omega}^Q)) : t \in \mathbb{R}\} \subset$ *-Aut $\mathfrak{A}^J$, since for all $x_j, \ y_j \in \mathfrak{A}_0, \ j = 1, 2, \ldots m$, one has the identity

$$\overline{\omega}\left(\pi_{p_1}(x_1)\pi_{p_2}(x_2)\ldots\pi_{p_m}(x_m)\tau_t(\pi_{p_1}(y_1)\pi_{p_2}(y_2)\ldots\pi_{p_m}(y_m))\right) =$$
$$\overline{\omega}\left(\pi_{p_1}(x_1\tau_t(y_1))\pi_{p_2}(x_2\tau_t(y_2))\ldots\pi_{p_m}(x_m\tau_t(y_m))\right) = \tag{6.4.26}$$
$$\prod_{j=1}^m \omega^0(x_j\tau_t(y_j)), \text{ for all } m\text{-tuples } \{p_1, p_2, \ldots p_m\} \subset \Pi, \ m = 1, 2, \ldots,$$

where $\tau_t \in$ *-Aut $\mathfrak{A}$ leaves all $\mathfrak{A}^J$ invariant: $\tau_t(\mathfrak{A}^J) = \mathfrak{A}^J$, $J \subset \Pi$. Setting $\tau_t := \sigma(\exp(-t\beta_{F_\omega}^Q))$, we obtain the KMS-property of $\overline{\omega}$ from the proved KMS-property of the state ω^0, since the finite linear combinations of the products

$$\pi_{p_1}(x_1)\pi_{p_2}(x_2)\ldots\pi_{p_m}(x_m), \ x_j \in \mathfrak{A}_0, \ p_j \in \Pi, \ m \in \mathbb{Z}_+ \setminus \{0\}, \tag{6.4.27}$$

form such a subset $\mathfrak{A}_L^0$ of $\mathfrak{A}$, that the values

$$\overline{\omega}(y) \in \mathbb{C}, \ y \in \mathfrak{A}_L^0, \tag{6.4.28}$$

determine any locally normal state $\overline{\omega} \in \mathcal{S}(\mathfrak{A})$ uniquely. The uniqueness of the KMS-states on $\mathfrak{A}^J$ (J finite) gives the restrictions of ω to all the $\mathfrak{A}^J$, hence we have equality

$\overline{\omega} = \omega$ of the states on $\mathfrak{A}$, hence the relation (6.4.18). [Warning: This does not imply uniqueness of the τ^Q-KMS states on $\mathfrak{C}$, but we have proved uniqueness of the KMS states on $\mathfrak{C}$ with respect to one parameter groups $\sigma(\exp(t\xi)) =: \sigma_\xi(t)$. Different extremal τ^Q-KMS states at the same temperature T give different values of F_ω and of $\beta_{F_\omega}^Q$, hence lead to different one parameter groups σ_ξ ($\xi := -\beta_{F_\omega}^Q$).]

Let now $\omega^0 \in \mathcal{S}(\mathfrak{A}_0)$ be a given faithful normal KMS-state at the temperature $T > 0$ corresponding to the group σ_ξ with $\xi := -\beta_{F_\omega}^Q$, where $F_\omega \in \mathfrak{g}^*$ satisfies (6.4.19). Then the product state $\overline{\omega}$ from (6.4.18) is locally normal, since the finite product of normal states is a normal state on the tensor product of W^*-algebras, [306, Sect. IV.5]. The factoriality is trivial for product states, [53, 2.6.10]. According to the Pusz-Woronowicz theorem, [54, 5.3.22], ω^0 satisfies the passivity condition (6.4.8) with $\tau := \sigma_\xi$ ($\xi := -\beta_{F_\omega}^Q$). This implies the satisfaction of (6.4.8) with respect to the same group by the state $\overline{\omega}$. The cluster property of the product state gives now the KMS-property of $\overline{\omega}$ with respect to the σ_ξ. Since ω^0 satisfies σ_ξ-KMS condition with $T \neq 0$ positive, the same is true for $\overline{\omega}$. Since F_ω is a fixed point of φ^Q, the derivations of the σ_ξ and of τ^Q coincide in the GNS-representations corresponding to the states supported by $E_\mathfrak{g}(F_\omega)$, cf. (6.4.14). The assumption (6.4.20) ensures, that the macroscopic limit of the product state $\overline{\omega}$ from (6.4.18) is concentrated on F_ω, hence the evolutions of proposition 6.4.7 τ^Q and σ_ξ ($\xi := -\beta_{F_\omega}^Q$) coincide in the representation π_ω corresponding to the state $\omega := \overline{\omega}$ from (6.4.18). $\square$

6.4.11 Corollary. *Let $\mathfrak{A} := \mathfrak{A}^\Pi$ and the system $(\mathfrak{A}; \sigma(G); \pi(\Pi))$ be defined according to Sect. 5.1, i.e. the G-measure $E_\mathfrak{g}$ is given by 5.1.33 and $\sigma(G)$ is locally implementable in states $\omega \in \mathcal{S}_\mathfrak{g}$. Let, with the assumptions of Theorem 6.4.10, ω be locally normal extremal τ^Q-KMS state at $T > 0$. Let X_ξ ($\xi \in \mathfrak{g}$) be the generators of the $(\sigma(G)$-defining) representation $U(G)$ on $\mathcal{H}_0 := \mathcal{H}$, $\mathfrak{A}_0 = \mathcal{L}(\mathcal{H})$, $\sigma(\exp(t\xi))(y) := \exp(-itX_\xi)\, y\, \exp(itX_\xi)$ for all $y \in \mathfrak{A}_0$. Then*

$$\omega^0(exp(itX_\xi)) = \exp(it F_\omega(\xi)), \ \forall \xi \in \mathfrak{g}, \tag{6.4.29}$$

where F_ω is given by the (trivially fulfilled) 'consistency condition'

$$\omega(E_\mathfrak{g}(f_\xi)) = F_\omega(\xi), \ \xi \in \mathfrak{g}. \tag{6.4.30}$$

Proof. Since $\exp(itX_\xi) \in \mathfrak{A}_0$, the generators of the restriction of $\sigma(G)$ onto $\mathfrak{A}_p := \pi_p(\mathfrak{A}_0)$ are $\pi_p(X_\xi)$, where

$$\exp(it\pi_p(X_\xi)) := \pi_p(\exp(itX_\xi)). \tag{6.4.31}$$

The generators of the restriction of $\sigma(G)$ onto $\mathfrak{A}^J$ (finite $J \subset \Pi$) are $X_\xi^J := \sum_{p\in J} \pi_p(X_\xi)$,

$$\exp(itX_\xi^J) := \prod_{p\in J} \exp(it\pi_p(X_\xi)) \in \mathfrak{A}^J. \tag{6.4.32}$$

ω is expressed by (6.4.18), hence according to (5.1.120):

$$\exp(it\,F_\omega(\xi)) = \omega(\exp(it\,X_{\xi\Pi})) = \lim_J \omega(\exp(it\frac{1}{|J|}\,X_\xi^J)) =$$

$$= \lim_J \prod_{p\in J} \omega^0(\exp(\frac{it}{|J|}\,X_\xi)) = \lim_{|J|\to\infty} [\omega^0(\exp(\frac{it}{|J|}\,X_\xi))]^{|J|}. \quad (6.4.33)$$

The result (6.4.29) is now obtained from (6.4.33) by the 'law of large numbers' ([112, II.Chap. XVII.1. Theorem 1]) applied to the arithmetic means of $|J|$ copies of independent real-valued variables with equal distributions μ_ξ^0. The probability measure μ_ξ^0 on $\mathbb{R}$ is given here by the projection-valued spectral measure P_ξ of X_ξ:

$$X_\xi = \int_{\mathbb{R}} \lambda\, P_\xi(d\lambda). \quad (6.4.34)$$

Then we set

$$\mu_\xi^0(d\lambda) := \omega^0(P_\xi(d\lambda)), \quad (6.4.35)$$

and we can write:

$$[\omega^0(\exp(\frac{it}{|J|}\,X_\xi))]^{|J|} = \int_{\mathbb{R}^{|J|}} \exp\left(\frac{it}{|J|}\sum_{p\in J}\lambda_p\right) \bigotimes_{m\in J} \mu_\xi^0(d\lambda_m), \quad (6.4.36)$$

where $\otimes_{m\in J}\mu_\xi^0(d\lambda_m)$ is the tensor product of $|J|$ copies of the measures (6.4.35) describing the simultaneous probability distribution of the $|J|$ independent random variables. Combining (6.4.33) and (6.4.36) gives the wanted result (6.4.29). □

6.4.12 Proposition. *Let us consider the system* $(\mathfrak{A}; \sigma(G); \tau^Q; \pi(\Pi))$ *as in Theorem 6.4.10. Assume that* ω^p ($p \in \Pi$) *are ground states for the restriction of the group* $\sigma(\exp(-t\beta_{F_\omega}^Q))$ *to the subalgebras* $\mathfrak{A}_p$. *Let the product-state*

$$\omega := \bigotimes_{p\in\Pi} \omega^p \quad (6.4.37)$$

satisfy the 'consistency condition'

$$\omega(E_\mathfrak{g}(f_\xi)) = F_\omega(\xi), \quad for\ all\ \xi \in \mathfrak{g}. \quad (6.4.38)$$

Then ω *is a factor ground state of the evolution* τ^Q. *If all the* ω^p *are pure, then* ω *is an extremal* τ^Q-*ground state.*

Proof. The factoriality of ω is a consequence of cluster properties, cf. e.g. [53, 54]. The condition (6.4.7) is fulfilled for $\tau_t := \sigma(\exp(-\beta_{F_\omega}^Q t))$. An application of

Proposition 6.4.7 shows the fulfillment of the ground state condition also for $\tau := \tau^{Q}$. The validity of the remaining assertions is clear. □

6.4.13 Note. *A brief version of the here presented theory together with applications to models of BCS theory and of Josephson junction was published in [40, 41]. Cf. also the next section.*

6.5 An Example: The B.C.S. Model of Superconductivity

6.5.1 We shall illustrate in this section the above developed theory by description and analysis of a perhaps simplest nontrivial and physically interesting mathematical model: The strong coupling version of the *Bardeen-Cooper-Schrieffer model* of the phenomenon of superconductivity in the *quasi spin formulation*; it was formulated and analyzed in [168, 311, 312], in the framework of the traditional QM formalism. It can be presented, completed, and solved in the framework of the constructions of the present work as follows:

It is a tensor product type model of Sect. 5.1 with $G := SU(2)$, $\mathcal{H} := \mathcal{H}_0 := \mathbb{C}^2$, $\Pi := \mathbb{Z}$, the generators of $U(G)$ in $\mathbb{C}^2$ are

$$X_{\xi_j} := i \left. \frac{\mathrm{d}}{\mathrm{d}t} \right|_{t=0} U(\exp(t\xi_j)) = \frac{1}{2}\sigma_j, \ j = 1, 2, 3, \tag{6.5.1}$$

where σ_j are the Pauli matrices and the elements $\xi_j \in \mathfrak{g}$ of the chosen basis satisfy the relations

$$[\xi_j, \xi_k] = \varepsilon_{jkm}\xi_m, \ j, k, (m) = 1, 2, 3. \tag{6.5.2}$$

Let $F_j := F(\xi_j)$ be used for the functions f_{ξ_j} on $\mathfrak{g}^* \ni F$ as well as for their numerical values in the points $F \in \mathfrak{g}^*$. The dynamics of the system is specified by the function Q on $\mathfrak{g}^*$:

$$Q(F) = -2\varepsilon F_3 - \lambda(F_1^2 + F_2^2), \ \ \varepsilon, \lambda \text{ are some positive numbers.} \tag{6.5.3}$$

This specifies the model completely.

6.5.2 The Poisson structure on $\mathfrak{g}^* = su(2)^*$ is determined by the Poisson brackets

$$\{F_j, F_k\} = -\varepsilon_{jkm}F_m, \ j, k, (m) = 1, 2, 3, \tag{6.5.4}$$

which are obtained from (6.5.2) according to (5.1.145). The classical dynamics corresponding to the given Hamiltonian function $Q \in C^{\infty}(su(2)^*, \mathbb{R})$ is then described by the flow φ^Q on $su(2)^*$ which is determined by the Hamilton equations

$$\dot{F}_j(\varphi_t^Q F) := \frac{\mathrm{d}}{\mathrm{d}t}F_j(\varphi_t^Q F) = \{Q, F_j\}(\varphi_t^Q F), \ t \in \mathbb{R}, \ j = 1, 2, 3. \tag{6.5.5}$$

We see from (6.5.4) that φ^Q is nontrivial for a general Q, hence the symplectic (even dimensional) Ad^*-orbits in $su(2)^*$ (which is 3-dimensional) are two-dimensional (with the exception of a zero-dimensional orbit consisting of the point $F = 0$). Since $SU(2)$ is a compact group, orbits are compact orientable two-dimensional manifolds in $su(2)^*$. They are submanifolds of the spheres S_r^2:

$$F^2 := F_1^2 + F_2^2 + F_3^2 = r^2, \tag{6.5.6}$$

because

$$\{F^2, F_j\} = 0 \text{ for } j = 1, 2, 3. \tag{6.5.7}$$

Hence the $Ad^*(SU(2))$-orbits are the spheres S_r^2. The equations of motion with Q from (6.5.3) are

$$\dot{F}_j = \{Q, F_j\} = -2\varepsilon\{F_3, F_j\} - 2\lambda(F_1\{F_1, F_j\} + F_2\{F_2, F_j\}), \tag{6.5.8}$$

that is

$$\dot{F}_1 = 2(\varepsilon - \lambda F_3)F_2, \tag{6.5.9a}$$

$$\dot{F}_2 = -2(\varepsilon - \lambda F_3)F_1, \tag{6.5.9b}$$

$$\dot{F}_3 = 0. \tag{6.5.9c}$$

The solution is elementary: With

$$F_{\pm} := F_1 \pm i F_2, \tag{6.5.10}$$

one has the flow φ^Q determined by the equations

$$F_3(t) = F_3 \equiv F_3(0), \ t \in \mathbb{R}, \tag{6.5.11a}$$

$$F_+(t) = F_+(0) \exp(-i2(\varepsilon - \lambda F_3)t). \tag{6.5.11b}$$

We shall assume $\lambda \neq 0$. The set of all *stationary points* $F \in su(2)^*$ of the flow φ^Q consists of points satisfying the conditions:
Either

$$F_+ = 0, \text{ and } F_3 = \text{arbitrary real number}, \tag{6.5.12a}$$

or

$$F_3 = \frac{\varepsilon}{\lambda}, \text{ and } F_+ = \text{arbitrary complex number}. \tag{6.5.12b}$$

The 'physical region' for the values F of the considered quantum mechanical system consists, however, of the points $F \in \text{supp } E_{\mathfrak{g}} \subset su(2)^*$.

6.5.3 Lemma. supp $E_{\mathfrak{g}} = \{F \in su(2)^* : F^2 \leq \frac{1}{4}\}$.

Proof. The spectra of the generators X_{ξ_j} $(j = 1, 2, 3)$ are the two-point sets $\{\lambda = \pm\frac{1}{2}\}$. According to the proof of Lemma 6.2.17, supp $E_{\mathfrak{g}} = \{F \in \mathfrak{g}^* : F(\xi) \in \text{conv}(sp(X_\xi)) \ \forall \xi \in \mathfrak{g}\}$. Since supp $E_{\mathfrak{g}}$ is Ad^*-invariant and the Ad^*-orbits are spheres S_r^2, the set supp $E_{\mathfrak{g}}$ is the ball $\{F : F \in S_r^2, 0 \leq r \leq \frac{1}{2}\}$.   □

6.5.4 The quantum evolution τ^Q is determined according to Proposition 6.3.8 and 6.3.11 by φ^Q as well as by the cocycle $\sigma(g_Q^{-1}(t, F)) \in *$-Aut $\mathfrak{A}$, where $\mathfrak{A}$ is the quasilocal algebra of our spin system. The action of this cocycle on the local algebra $\mathfrak{A}_0$ (:= the algebra of the $\frac{1}{2}$-spin sitting at the site $0 \in \Pi$) is given by the unitary family $U(g_Q(t, F))$ satisfying the Schrödinger-type evolution equation

$$i\frac{d}{dt} U(g_Q(t, F)) = X(\beta_{F(t)}^Q)U(g_Q(t, F)), \quad F(t) := \varphi_t^Q(F), \qquad (6.5.13)$$

as can be seen from (6.1.16). The elements $\beta_F^Q \in su(2)$ are defined by (6.1.17), i.e.

$$\beta_F^Q := d_F Q = -2\varepsilon\xi_3 - 2\lambda(F_1\xi_1 + F_2\xi_2). \qquad (6.5.14)$$

In the representation $g \mapsto U(g)$ one has

$$X(\beta_F^Q) = -\varepsilon\sigma_3 - \lambda(F_1\sigma_1 + F_2\sigma_2) = -a(F)\,\mathbf{n}(F) \cdot \boldsymbol{\sigma}, \qquad (6.5.15)$$

where $\boldsymbol{\sigma} := \{\sigma_1, \sigma_2, \sigma_3\}$ is the 3-vector of σ-matrices,

$$a(F) := \sqrt{\varepsilon^2 + \lambda^2 F_+ F_-}, \qquad (6.5.16)$$

and $\mathbf{n}(F) := \{n_1, n_2, n_3\}$ with

$$n_1 := \frac{\lambda F_1}{a(F)}, \quad n_2 := \frac{\lambda F_2}{a(F)}, \quad n_3 := \frac{\varepsilon}{a(F)}, \qquad (6.5.17)$$

and $\mathbf{n} \cdot \boldsymbol{\sigma} := n_j\sigma_j$ is the scalar product.

If $F \in su(2)^*$ is one of the stationary points (6.5.12), then the function $t \mapsto g_Q(t, F)$ will be a one-parameter subgroup of $SU(2)$ with the generator β_F^Q. This subgroup is the stability subgroup of F with respect to the $Ad^*(SU(2))$-representation (for $F \neq 0$). The time evolution τ^Q in those states ω the classical projection of which is concentrated on $F_\omega = F$ is now identical with the evolution according to the subgroup of $\sigma(SU(2))$ specified by the element $\beta_F^Q \in \mathfrak{g}$. The generator Q_ω of this evolution in the representation π_ω can be expressed by its commutators with $\pi_\omega(y)$, $y \in \mathfrak{A}^J$ (finite $J \subset \Pi$):

$$[Q_\omega, \pi_\omega(y)] = [\pi_\omega(X^J(\beta_F^Q)), \pi_\omega(y)] \text{ for } y \in \mathfrak{A}^J, \ J := \{p_1, \ldots p_m\}, \quad (6.5.18)$$

where the usual notation $X^J(\xi) := \sum_{p \in J} \pi_p(X(\xi))$ was used, cf. also (6.4.12). The generator Q_ω is a well defined selfadjoint operator on the space $\mathcal{H}_\omega$ of the representation π_ω chosen so that $Q_\omega \Omega_\omega = 0$ on the cyclic vector Ω_ω. This is the meaning of the Bogoliubov-Haag Hamiltonian operator Q_ω in the GNS-representations of *macroscopically pure* and *macroscopically stationary* states of the system.

6.5.5 The KMS-states of $(\mathfrak{A}; \tau^Q)$ at positive temperature $T > 0$:

The algebra $\mathfrak{A}$ is separable, hence the representation space $\mathcal{H}_\omega$ of any cyclic representation is separable and the *KMS-states* ω of this system are supported by the *extremal KMS states*. This means, roughly speaking, that any KMS-state can be constructed as an integral of the extremal KMS states at the same temperature T. Hence, the evaluation of all extremal KMS states is sufficient to characterization of all KMS states of the system. Let us consider now the extremal KMS states.

Any *extremal τ^Q-KMS state* at $T > 0$ (hence at $\beta := T^{-1} \neq \infty$) is determined uniquely by its restriction ω^0 to $\mathfrak{A}_0$, cf. Theorem 6.4.10 (remember that all states on the UHF-algebra $\mathfrak{A}$ are locally normal). Let $F_\omega \in \mathfrak{g}^*$ be the classical phase point corresponding to a given extremal τ^Q-KMS state on $\mathfrak{A}$. Then the strong version (6.4.29) of the 'consistency condition' is valid, i.e.

$$\omega^0(X_\xi) = F_\omega(\xi) \text{ for all } \xi \in \mathfrak{g}. \tag{6.5.19}$$

Here ω^0 is the (unique, if it exists) KMS-state on $\mathfrak{A}_0$ at the same temperature T as the state $\omega \in \mathcal{S}(\mathfrak{A})$, corresponding to the evolution given by the generator $-X(\beta_{F_\omega}^Q)$. There is one-one correspondence between the extremal τ^Q-KMS states of the infinite system and the states ω^0 satisfying the above listed conditions for some stationary point F_ω of the classical equations lying in the physical domain, $F_\omega \in \text{supp } E_\mathfrak{g}$.

Let a stationary point $F_\omega \in \text{supp } E_\mathfrak{g}$ be given. Then any $\sigma(\exp(-t\beta_{F_\omega}^Q))$-KMS state ω^0 on $\mathfrak{A}_0$ coincides with the Gibbs state ω_T^0 at some temperature T. The state ω_T^0 is given by:

$$\omega_T^0(y) := \left(Tr \, \exp\left(\frac{a(F_\omega)}{T} \, \boldsymbol{n}(F_\omega) \cdot \boldsymbol{\sigma} \right) \right)^{-1} Tr \left(\exp\left(\frac{a(F_\omega)}{T} \, \boldsymbol{n}(F_\omega) \cdot \boldsymbol{\sigma} \right) y \right),$$
$$\tag{6.5.20}$$

for all $y \in \mathfrak{A}_0$. It is sufficient to calculate (6.5.20) for $y = \sigma_j$, $j = 1.2.3$. We obtain

$$\omega_T^0(\sigma_j) = n_j(F_\omega) \tanh(T^{-1} a(F_\omega)), \ j = 1, 2, 3, \tag{6.5.21}$$

and the consistency condition (6.5.19) means:

$$n_j(F_\omega) \tanh(T^{-1} a(F_\omega)) = 2 F_\omega(\xi_j), \ j = 1, 2, 3, \tag{6.5.22}$$

which is equivalent to the following conditions:

$$\frac{\lambda F_\omega(\xi_j)}{a(F_\omega)} \tanh(T^{-1}a(F_\omega)) = 2F_\omega(\xi_j), \quad j = 1, 2;$$ (6.5.23a)

$$\frac{\varepsilon}{a(F_\omega)} \tanh(T^{-1}a(F_\omega)) = 2F_\omega(\xi_3).$$ (6.5.23b)

These conditions are satisfied by $F_\omega = F$, where

(i) either (in the cases of arbitrary positive ε and λ)

$$F_1 = F_2 = 0, \text{ and } F(\xi_3) := F_3 = \frac{1}{2} \tanh\left(\frac{\varepsilon}{T}\right), \quad T > 0,$$ (6.5.24)

(ii) or (in the cases with $0 < 2\varepsilon < \lambda$)

$$F(\xi_3) = \frac{\varepsilon}{\lambda}, \ 2a(F) = \lambda \tanh(T^{-1}a(F)), \ 0 < T < T_c := \varepsilon\left(\tanh^{-1}\left(\frac{2\varepsilon}{\lambda}\right)\right)^{-1}.$$ (6.5.25)

Note that the condition (6.5.25) can be fulfilled with $F_+ \neq 0$ only, hence the sets of values $F \in \mathfrak{g}^*$ determined by the two conditions (6.5.24) and (6.5.25) are mutually disjoint. These relations allow us to give the list of all F_ω corresponding to extremal τ^Q-KMS states at a given temperature $T > 0$:

(i) $T \geq T_c$; in this case $F_\omega(\xi_1) = F_\omega(\xi_2) = 0$, $F_\omega(\xi_3) = \frac{1}{2} \tanh\left(\frac{\varepsilon}{T}\right)$.
(ii) $0 < T < T_c$; here one has a state with F_ω described in (i) above, and, if $0 < 2\varepsilon < \lambda$, one has, moreover, a one-parameter family of possible $F_\omega \in su(2)^*$ such that:

$$F(\xi_3) = \frac{\varepsilon}{\lambda}, \ 2a(F_\omega) = \lambda \tanh(T^{-1}a(F_\omega)).$$

There is one-one correspondence between the elements F_ω corresponding to a given value of $T > 0$ in this list and the *extremal* $(\tau^Q, \beta := T^{-1})$-*KMS states* of the infinite quantal system.

We see that, in the considered model, a KMS-state exists at any positive T, and for $T \geq T_c$ this state is unique. For $0 < T < T_c$, except of the 'trivial possibility' (6.5.24), there is a circle of points $F_\omega \in \mathfrak{g}^*$ numbering the elements of pairwise *mutually disjoint* extremal KMS states at the same temperature. If we call the subgroup $\exp(t\xi_3)$ the 'gauge group', then the *gauge-invariant KMS-states* exist at all $T > 0$ (the trivial possibilities (6.5.24) are gauge invariant); the extremal KMS states for temperatures $0 < T < T_c$ are not invariant with respect to the gauge group and they are transformed by the group actions into one another: here appears the **spontaneous symmetry breaking** phenomenon. For $0 < T < T_c$, there is another gauge invariant state $w_T^s \in K_\beta \subset \mathcal{S}(\mathfrak{A})$, $\beta := T^{-1}$, given by the integral of the states $w_T^F \in \mathcal{E}K_\beta$ corresponding to the values F from (6.5.25):

$$w_T^s(y) = \frac{1}{2\pi} \int_0^{2\pi} w_T^F(\sigma(\exp(\iota\xi_3))(y)) \, d\iota, \ 0 < T < T_c. \tag{6.5.26}$$

Let us denote by w_T^n the (extremal) KMS-state at $\beta = T^{-1}$ corresponding to the values (6.5.24) of $F_w = F$. The states w_T^n ($T > 0$) are interpreted as describing the 'normal conducting phase', and the states w_T^s ($T < T_c$) represent the 'superconducting phase'. The mixtures $w_T := \lambda w_T^s + (1 - \lambda)w_T^n$ are also $(\tau^Q, \beta = T^{-1})$-KMS states at $0 < T < T_c$, $0 \le \lambda \le 1$. The equilibrium states of the considered system can be defined as the thermodynamic limits of the (unique) Gibbs states of local systems $(\mathfrak{A}^J; \tau^J)$, $|J| < \infty$, where $\tau_t^J \in *$-Aut $\mathfrak{A}^J$ is generated by the local Hamiltonians Q^J defined in (6.1.1). According to [168], these thermodynamic limits coincide with w_T^n for $T \ge T_c$, whereas for $0 < T < T_c$ the limit $J \to \Pi$ leads to the state w_T^s.

6.5.6 The ground states of $(\mathfrak{A}; \tau^Q)$:

Let us consider now an *extremal τ^Q-ground state* w of our system, $w \in \mathcal{E}\mathcal{K}_\infty$. Let F_w be the corresponding classical stationary point in supp $E_\mathfrak{g}$. The restriction w^0 of w to the subalgebra $\mathfrak{A}_0$ is the unique ground state of the generator $X(\beta_{F_w}^Q)$, (6.5.15), corresponding to its eigenvector $\chi(F_w) \in \mathbb{C}^2$ with the minimal eigenvalue:

$$\mathbf{n}(F) \cdot \sigma \, \chi(F) = \chi(F), \ F \in su(2)^*. \tag{6.5.27}$$

Due to the uniqueness of the ground state $w^0 \in S(\mathfrak{A}_0)$ corresponding to a given $F_w \in su(2)^*$, any *extremal τ^Q-ground state* is an $\pi(\Pi)$-invariant product state. Conversely, the $\pi(\Pi)$-invariant product state constructed from a vector $\chi(F)$ defined in (6.5.27) will be a pure ground state of $(\mathfrak{A}; \tau^Q)$ if the 'consistency condition' $[(\chi_1, \chi_2)$ is here the scalar product in $\mathbb{C}^2]$

$$(\chi(F), X(\xi)\chi(F)) = F(\xi), \ \xi \in su(2), \tag{6.5.28}$$

will be satisfied. This is a consequence of the considerations in Sect. 6.4. Let us solve (6.5.28) for F. For $\xi := \xi_j$ ($j = 1, 2, 3$) one has

$$(\chi(F), X(\xi_j)\chi(F)) = \frac{1}{2}n_j(F), \ j = 1, 2, 3, \tag{6.5.29}$$

where $n_j(F)$ is defined in (6.5.17). The obtained condition

$$n_j(F) = 2F(\xi_j), \ j = 1, 2, 3, \tag{6.5.30}$$

leads to the following possibilities for $F = F_w$, $w \in \mathcal{E}\mathcal{K}_\infty$:

(i) if ε and λ are arbitrary positive, then one can have:

$$F_1 = F_2 = 0, \quad F_3 = \frac{1}{2}; \tag{6.5.31}$$

(ii) for $0 < 2\varepsilon < \lambda$, one has, moreover, the possibilities:

$$F_1^2 + F_2^2 = \frac{1}{4} - \left(\frac{\varepsilon}{\lambda}\right)^2, \quad F_3 = \frac{\varepsilon}{\lambda}. \tag{6.5.32}$$

Hence, in the case $0 < 2\varepsilon < \lambda$, the set of ground states has similar classical picture in $su(2)^*$ as the set of τ^{ϱ}-KMS states with temperatures lying under the 'critical temperature' T_c. Let $\omega_0^n \in \mathcal{K}_\infty \subset \mathcal{S}(\mathfrak{A})$ corresponds to the value F_ω from (6.5.31), and let ω_0^s be given by (6.5.26) with $T = 0$ and with $\omega_0^F \in \mathcal{E}\mathcal{K}_\infty$ corresponding to any value of F given in (6.5.32). According to [168], the thermodynamic limit of the (unique) local ground states on $\mathfrak{A}^J$ corresponding to the Hamiltonians Q^J coincides with ω_0^s.

Chapter 7
Some Models of "Quantum Measurement"

7.1 Introductory Notes

7.1.1 The interactions in the models of large quantal systems described in Chap. 6 were of specific long-range type. All the elementary subsystems ("particles" or "spins") mutually interacted with each other 'in the same way' as if all the subsystems were not distinguishable from each other, i.e. the multi-particle interaction was invariant with respect to permutations of the particles independent of their positions in the lattice Π, as specified by (6.1.1). Such interactions led in infinite limit of the number N of the subsystems to the dynamics of "mean-field type", i.e. to such a dynamics that each individual subsystem moved as if it was immersed in an external (in general time dependent) field produced by the whole collection of the infinite number of all the subsystems and independent of any changes of the state of any of these subsystems. The resulting dynamics was such that macroscopic (classical) parameters of the infinite system were varying in time according to the dynamics of some classical mechanical Hamiltonian system.

In this chapter, we shall describe several specific models of large quantal systems whose elementary subsystems interact by short range interactions. The macroscopic, or "classical", variables of the infinite systems will change now just in the limit $t \to \infty$, because the short range interaction results in finite velocity of spreading of local changes across the infinite system, hence in finite times only local variables corresponding to changes of finite subsystems are changed.

7.1.2 We shall briefly describe here a few quantum-mechanical model systems describing interactions of a 'microscopic system' with a 'macroscopic system' leading to a 'macroscopic change' in the second system. This means that such systems describe schemes modeling dynamics of processes like '*quantum measurement*' as a process ascribing a classical probability distribution of 'measurement results' (given by macroscopically distinct states of the 'macroscopic system' which plays the role of the '*measuring apparatus*') to the corresponding (according to the 'measured observable') quantum-mechanical linear decomposition of the wave function of the 'microscopic system'. Construction of these models was inspired mainly by the classical

© Springer Nature Switzerland AG 2020, corrected publication 2020

P. Bóna, *Classical Systems in Quantum Mechanics*,
https://doi.org/10.1007/978-3-030-45070-0_7

paper by *Klaus Hepp* [153], cf. also [253, 254]. According to the previous chapters, we are able to describe in QM in a mathematically clear way macroscopic systems (with coordinates undergoing classical behaviour) by models of infinite quantal systems only. Of course, the infinity of the number of degrees of freedom should be considered as a convenient approximation to large but finite systems. Also infinite time duration of the processes of changing macroscopic parameters corresponding to considered microscopic influences is connected with this infinity. In this connection, it is relevant to be interested in the speed of the corresponding macroscopic changes. In the 'infinite models' presented here the convergence to a macroscopic change is very slow.

A much larger speed of convergence is reached in the model of finite (arbitrary long) 'Quantum Domino'—spin chain (cf. 7.1.3 and Sect. 7.3) interacting with fermion field in such a way that after all the spins in the chain changed their orientations into the opposite ones the chain emits a fermion. In this case the speed of convergence to final stationary state is 'almost exponential'. The model is described in [39]. Its interpretation as "a model of quantum measurement" is, however, questionable: Due to its finite dimension a definition of "macroscopic difference" is ambiguous and it would need probably a longer discussion. Cf. notes on this problem in the original Hepp's work [153], and also in our Sect. 7.7.

It should be stressed that we do not intend to present the described models of micro-macro interaction as a definitive solution of the 'measurement problem in QM', cf. Sect. 7.7. They could be considered rather as an illustration of possibilities of the standard quantum mechanical formalism to include, by using this specific way of description of macroscopic observables, some descriptions of possible responses of large systems (hence changes of their 'macroscopic variables') to some of their interactions with microsystems. It is shown how can various states of a microsystem interacting with a macrosystem lead in QM to various 'corresponding changes' of values of their macroscopic (resp. 'classical') observables.

7.1.3 We present here four models, the second of which is based on the first one, the "*Quantum Domino*" (*QD*), published originally in [36]. The idea of the third model is similar to that of QD, but it is based on the known X-Y model of the spin chain [201]. QD is a model of an infinite quantum system—an infinite (or semiinfinite) spin chain with a short range interaction—in which any local (microscopic) change of a specific stationary state leads to subsequent evolution (with time $t \to \infty$) to a new, macroscopically different stationary state. The initial local changes of these stationary states of this model are realized "by hand", i.e. a locally perturbed stationary state is chosen as an initial condition for the forthcoming time-evolved states of that system. This local perturbation can be realized by a change of quantum state of a single spin (say the first one in the semiinfinite chain), and this spin can be considered, e.g. as an additional microsystem (the 'measured system') interacting with the infinite rest of the chain.[1]

---

[1] In the case of some different choices of (locally perturbed stationary) initial states in this model, the subsequent time evolutions of the chain could be different: e.g., an initial segment could move quasiperiodically and the infinite rest of the chain will converge to a macroscopically different state.

The second model consists in the composition of two systems: of the previous (QD) one and of a point particle scattered on it; the QD-spin chain occurs initially in its specific stationary state. The scalar particle (moving in the configuration space $\mathbb{R}^3$) perturbs locally the infinite system (by scattering on its 'first two' spins) and the chain develops then (after $t \to \infty$) with some probability to a macroscopically different state. This process can be interpreted as modeling detection of the particle by a macroscopic detector. The model is interesting by that it does not correspond to an "*ideal measurement*" the results of which are described usually by a *projector valued measure (PVM)* realizing, e.g., the spectral decomposition of some selfadjoint operator—the 'measured observable'. In our case, however, we obtain a *positive operator valued measure (POVM)* describing the probabilities of responses to incoming states of the particle; this expresses the technical characteristics of the detector with less than 100% efficiency. This model is presented here in detail, since it is presented here for the first time—it is a more complex and more complete version of an older model. The original version of this model was published in [38].

The third presented model is based on an 'X-Y modification' of the Heisenberg spin-chain models, cf. e.g. [267]. This "model of quantum measurement" consists of the 1/2-spin chain with a nearest neighbourhood interaction, which is interrupted in one link, and in the point of the interruption an additional 1/2-spin modeling a simplest possible "measured microsystem" is included (together with its interaction with the rest of the chain).

The fourth model consists of a finite portion of QD of the length $N \gg 1$ coupled to Fermi field and working so that in the initial state "all the N spins are pointing down", but after reversing the first spin the chain moves until all the spins are "pointing up" and, after reversing the last N-th spin, the chain emits a Fermi particle. With the time $t \to \infty$ the particle moves freely to infinity and the chain remains in a new stationary state with "all spins pointing up". The finite length of the chain needs a different interpretation as a "measuring device" in comparison with the preceding three infinite models.

7.2 On 'Philosophy' of "Models"

The term "*model*" is used repeatedly in this Chapter, as well as in science in general. This word is generally used in various connections and meanings. It is usually considered as denoting human constructs (material or mental) approximating in some way an aspect of a considered 'part of reality'. But, can we determine where there is a borderline between 'only approximation' and 'full picture of truth'? What is the 'reality'? What is the *meaning of 'the truth'* (as it was asked also by *Pontius Pilate* very appropriately in Bible—New Testament: John 18:38)?

Let us consider (not only here) any human symbolic formulation of any *knowledge as a "model"*. Hence, also our laws of nature including the whole physics are models—they are provisional and waiting for further completions and/or reformulations.

It is motivating and orientating for researchers to believe in the existence of some 'final truth'. It is an important psychological aspect of scientific progress. The faith in our 'reliably verified knowledge' is perhaps necessary also for the success of our practical life. But if a theory is completed (i.e. if it is in agreement with all available 'trustworthy' experimental results), it can be (and eventually should be) challenged in science.

Any theory, as well as *any concept* appearing in our consciousness or/and used in our communication **is a human construction**. Hence it is dependent on human interests and activities, and these activities are perpetually evolving–sometimes even substantially changing. Hence, also our attention and interests are changing. This implies that the motives for our intellectual activity are perpetually developing. The resulting our '*pictures of the world*', either global, or various special, are correspondingly changing along with these other changes. And, people also look then on 'the same things' by different ways and from different points of view than before.

The 'models' presented in this chapter are just very simple abstractions imitating certain features of mutual interactions of general classes of physical systems: microsystems described adequately by QM, and macroscopic systems (usually described by CM) consisting of a large number of microsystems. We tried to be mathematically rigorous in proceeding from basic axioms of QM to definitions of introduced concepts and constructions of the mathematical models, as well as to description and obtaining the consequences of the used dynamics. This emphasis on mathematical rigor was motivated by our desire to show clearly that the obtained results are exact consequences of the currently generally accepted formal theory of QM.

7.3   Quantum Domino

7.3.1 We shall describe here briefly (for more details we refer to [35, 36]) the model of infinite spin chain which we call, due to the character of its time evolution, **Quantum Domino** (*QD*). The 1/2 spins are ordered by the values of the index $i \in \mathbb{Z}$ and the Hamiltonian produces a local nearest three body interaction. This interaction can be described easily as follows: If the hamiltonian acts on the state with the i-th spin "pointing up" and the $(i + 2)$-nd spin "pointing down", then the $(i + 1)$-st spin changes its orientation to the opposite one. The dynamics of the two sided infinite spin-1/2 quantum chain has spin configurations "all spins pointing up", and "all spins pointing down" as stationary states, which are unstable: If we reverse the direction of one of the spins in these states, the new state will develop in the limit $t \to \infty$ into another stationary (and 'macroscopically' stable) state, in which all the spins lying on one side of the reversed spin are also reversed, and all the spins lying on the other side of that spin stay unchanged. Since this evolution leads to the change of the value of a macroscopic observable of the chain, it can be used as a model for 'quantum measurement' of microscopic observables of a single spin of the chain. We shall show in this section how such model works.

7.3.2 Let the C^*-algebra of observables $\mathfrak{A}$ be the C^*-tensor product of count-ably infinite set of copies of the algebra of complex 2×2 matrices generated by the spin creation and annihilation operators $a_j^*, a_j, \ j \in \mathbb{Z}$ satisfying the following (anti)commutation relations

$$a_i a_j - a_j a_i =: [a_i, a_j] = [a_i^*, a_j] = 0, \ \ i \neq j \tag{7.3.1}$$

$$a_i a_i = 0, \ \ a_i^* a_i + a_i a_i^* = 1,$$

for all $i, j \in \mathbb{Z}$. The algebra $\mathfrak{A}$ is simple, hence each its nonzero representation is faithful. We shall describe the dynamics in $\mathfrak{A}$ in the "vacuum" representation, i.e. in the GNS representation corresponding to the "vacuum state" $\omega_0 \in \mathfrak{A}_{+1}^* \equiv \mathcal{S}(\mathfrak{A})$ that is given by the relation

$$\omega_0(a_j^* a_j) = 0, \ \text{for all } j \in \mathbb{Z}. \tag{7.3.2}$$

This state is pure, hence the GNS representation is irreducible. We shall call the spins in this state to be "pointing down", to be specific in verbal expression. Let the **cyclic vector ("vacuum" in the lattice gas terminology) of this representation be denoted by** Ω_0, i.e. for all elements $x \in \mathfrak{A}$ it is

$$\omega_0(x) = \langle \Omega_0 | x | \Omega_0 \rangle, \ \text{for all } x \in \mathfrak{A}. \tag{7.3.3}$$

Here and in the following we shall denote the elements of $\mathfrak{A}$ and their operator representatives in the considered irreducible Hilbert space representation by the same symbols. Let us **denote this Hilbert space by** $\mathcal{H}_{vac}$.

Let us define a "finite-subchain Hamiltonian" $H_{(j,k)}$:

$$H_{(j,k)} := \sum_{n=j+1}^{k-2} a_n^* a_n (a_{n+1}^* + a_{n+1}) a_{n+2} a_{n+2}^*. \tag{7.3.4}$$

Local time evolution automorphisms of $\mathfrak{A}$ are given by

$$\tau_t^n(x) := \exp(it H_{(-n,n)}) \, x \, \exp(-it H_{(-n,n)}), \tag{7.3.5}$$

and the norm limits

$$\tau_t(x) := \text{norm-} \lim_{n \to \infty} \tau_t^n(x) \tag{7.3.6}$$

determine the time evolution in $\mathfrak{A}$ (in the "Heisenberg picture").

In our vacuum representation, this evolution is determined by a selfadjoint Hamiltonian H,

$$\tau_t(x) = e^{it H} x \, e^{-it H}. \tag{7.3.7}$$

Here, the (unbounded) operator H can be written in the evident form (its obvious definition and a proof of selfadjointness is given in [36, Proposition II.1])

$$H := \sum_{n \in \mathbb{Z}} a_n^* a_n (a_{n+1}^* + a_{n+1}) a_{n+2} a_{n+2}^*. \tag{7.3.8}$$

This evolution is *time-reflection invariant*, but it is not invariant with respect to the *space reflection* $n \mapsto -n$. Let us introduce the operators

$$g_j := a_j a_j^* a_{j+1}^* a_{j+1} .$$

These quantities are integrals of motion. One can also prove that the Hilbert space $\mathcal{H}_{vac}$ can be decomposed into H-invariant orthogonal subspaces and on each of them the restriction of the Hamiltonian H is a bounded operator.

Let $X \subset \mathbb{Z}$ be of finite cardinality, and **let** $\Omega_X := \prod_{j \in X} a_j^* \, \Omega_0$. The vectors Ω_X with all mutually distinct finite $X \subset \mathbb{Z}$, **with** $\Omega_\emptyset := \Omega_0$, form an orthonormal basis in $\mathcal{H}_{vac}$. Each finite $X \subset \mathbb{Z}$ is of the form $Y_1 \cup Y_2 \cup \ldots \cup Y_r$, where all $Y_k \subset \mathbb{Z}$ are nonempty finite, mutually disjoint and of the form $\{j_k + 1, j_k + 2, \ldots, j_k + m_k\}$, with $j_{k+1} > j_k + m_k$, $|Y_k| \equiv m_k$, i.e the sets $Y_k \subset X$ $(k = 1, 2, \ldots, r)$ form mutually separated "connected islands" consisting of "pointing up" spins. All the vectors Ω_X are eigenvectors of all the operators g_j. For the set X of the just described structure we have

$$g_j \Omega_X = \begin{cases} \Omega_X & \text{for } j = j_k, \quad k = 1, 2, \ldots, r \\ 0 & \text{otherwise.} \end{cases} \tag{7.3.9}$$

This implies that the time evolution of the vectors Ω_X conserves the number of islands, leaving the initial ("left") points $j_k + 1$ of each Y_k $(k = 1, 2, \ldots, r)$ unchanged ("occupied", or "pointing up"), and the places j_k, $k = 2, 3, \ldots, r$ as well as $j_1 - n$ $(n \in \mathbb{Z}_+)$ remain all the time "unoccupied" (i.e. spins are there "pointing down"). Hence, the subspaces $\mathcal{H}_{\{j\}}$ spanned by all such vectors with a fixed set $\{j\} := \{j_1, j_2, \ldots, j_r\}$ are left invariant with respect to the action of the Hamiltonian H. Then the space $\mathcal{H}_{vac}$ decomposes as

$$\mathcal{H}_{vac} = \bigoplus_{\{j\}} \mathcal{H}_{\{j\}}, \tag{7.3.10}$$

where the orthogonal sum is taken over all mutually different $\{j\}$; note that **the stationary subspace $\mathcal{H}_{\{\emptyset\}} := \{\lambda \Omega_0 : \lambda \in \mathbb{C}\}$ is one dimensional.**

The structure of the Hamiltonian H shows, moreover, that each $\mathcal{H}_{\{j\}}$ can be written as (i.e. it is isomorphic to) the tensor product of a vector (resp. of a one-dimensional subspace) and a finite number of Hilbert spaces corresponding to restricted subchains of spins:

$$\mathcal{H}_{\{j\}} = \Omega_{(-\infty, j_1]}^0 \otimes \mathcal{H}_{(j_1, j_2)} \otimes \mathcal{H}_{(j_2, j_3)} \otimes \cdots \otimes \mathcal{H}_{(j_r, +\infty)}, \tag{7.3.11}$$

where $\Omega^0_{(-\infty, j_1]}$ is one-dimensional space containing the vector with all spins numbered by $j \leq j_1$ "pointing down", and the spaces $\mathcal{H}_{(j_k, j_{k+1})}$ are spanned by $j_{k+1} - j_k - 1$ vectors corresponding to the "islands" Y_k of all permitted lengths $1 \leq |Y_k| < j_{k+1} - j_k$. Here we understand that $j_{r+1} \equiv +\infty$. We see from the form of the Hamiltonian that the time evolution of vectors in the subspaces $\mathcal{H}_{\{j\}}$ is described by ('mutually independent') evolutions in each $\mathcal{H}_{(j_k, j_{k+1})}$ determined by the Hamiltonians $H_{(j_k, j_{k+1})}$, cf. (7.3.4) ; for more details see [35, 36].

7.3.3 The result of these considerations is that the evolution of general vectors of our representation (hence also the evolution of any states from $\mathcal{S}(\mathfrak{A})$) can be described by two simpler kinds of evolution, namely, the evolutions in finite chains described by Hilbert spaces $\mathcal{H}_{(j_k, j_{k+1})}$, as well as in the Hilbert spaces $\mathcal{H}_{(j_r, +\infty)}$ spanned by vectors of arbitrary one-sidedly unrestricted lengths. Because the interaction in our infinite chain is translation invariant, we can describe these two possibilities as[2]

(1) the evolution in the finite-dimensional Hilbert space $\mathcal{H}_{(0, N+1)}$ spanned by the vectors

$$|m\rangle := a_1^* a_2^* \ldots a_m^* \Omega_0 \ (m = 1, 2, \ldots, N) \tag{7.3.12a}$$

by the unitary evolution group $U_N(t) := e^{-it H_N}$ with the Hamiltonian $H_N := H_{(0, N+1)}$ from (7.3.4), and

(2) the evolution in the infinite-dimensional Hilbert space $\mathcal{H}_{(0, \infty)}$ spanned by the vectors

$$|m\rangle := a_1^* a_2^* \ldots a_m^* \Omega_0 \ (m \in \mathbb{Z}, m \geq 1) \tag{7.3.12b}$$

by the unitary evolution operators $U_\infty(t) := e^{-it H}$ with the Hamiltonian $H := H_{(0, +\infty)}$.

Let us express these two instances of dynamics by the matrix elements $\langle n|U(t)|m\rangle$. The result can be obtained by explicitly solving the eigenvalue problem for H_N. The action of H_N is:

$$H_N|1\rangle = |2\rangle, \tag{7.3.13a}$$

$$H_N|m\rangle = |m - 1\rangle + |m + 1\rangle, \ m = 2, 3, \ldots, N - 1, \tag{7.3.13b}$$

$$H_N|N\rangle = |N - 1\rangle, \tag{7.3.13c}$$

$$H_N|k\rangle = 0 \quad \text{for } k > N. \tag{7.3.13d}$$

For the eigenvectors $\psi_E : H_N \psi_E = E \psi_E$ written in the basis of vectors $|m\rangle$:

$$\psi_E = \sum_{m=1}^{N} c_m(E)|m\rangle \tag{7.3.14}$$

---

[2]We shall use here the Dirac bra-ket notation for convenience.

we obtain the eigenvalue problem in the form:

$$Ec_1(E) = c_2(E), \tag{7.3.15a}$$

$$Ec_m(E) = c_{m-1}(E) + c_{m+1}(E), \ m = 2, 3, \ldots N - 1, \tag{7.3.15b}$$

$$Ec_N(E) = c_{N-1}(E). \tag{7.3.15c}$$

The equations (7.3.15) lead to

$$c_m(E) = U_{m-1}(E/2)c_1(E), \tag{7.3.16}$$

where

$$U_{m-1}(z) := \frac{\sin(m \arccos z)}{\sin(\arccos z)} \tag{7.3.17}$$

are the **Tshebyshev polynomials** of the second kind [129, 8.940]. This is seen from the recurrent relations for U_n following from (7.3.15), cf. [36, III.(27)]:

$$U_{n+1}(z) = 2z U_n(z) - U_{n-1}(z), \ U_0(z) = 1, \ U_1(z) = 2z.$$

The equation (7.3.15c) has now the form

$$U_N(E/2) = 0, \tag{7.3.18}$$

which is the secular equation corresponding to our eigenvalue problem. Its solutions are

$$E_j = 2 \cos\left(\frac{j\pi}{N+1}\right), \ j = 1.2.\ldots N, \tag{7.3.19}$$

hence we have the expressions

$$c_m(E_j) = \left[\frac{2}{N+1}\right]^{1/2} \sin\left[\frac{jm\pi}{N+1}\right]. \tag{7.3.20}$$

We shall need also the following definition:

$$J_n^{(N)}(z) := \frac{i^n}{N+1} \sum_{j=1}^{N} \exp\left[-iz \cos\left(\frac{j\pi}{N+1}\right)\right] \cos\left(n\frac{j\pi}{N+1}\right). \tag{7.3.21}$$

This is an integral sum of *Sommerfeld integral representation* of the *Bessel function* $J_n(z)$, see also [129, 8.41]:

$$J_n(z) = \frac{i^n}{\pi} \int_0^\pi e^{-iz \cos \alpha} \cos(n\alpha) d\alpha. \tag{7.3.22}$$

We can now write the desired expression for the Green function of a finite chain:

$$\langle n|U_N(t)|m\rangle = (-i)^{n-m} J_{n-m}^{(N)}(2t) - (-i)^{n+m} J_{n+m}^{(N)}(2t), \tag{7.3.23}$$

what can be obtained by a standard way using the completeness of the orthonormal system of vectors (7.3.14) in $\mathcal{H}_{(0,N+1)}$.
This, for an infinite chain with $N \to \infty$, gives:

$$\langle n|U_\infty(t)|m\rangle = (-i)^{n-m} J_{n-m}(2t) - (-i)^{n+m} J_{n+m}(2t). \tag{7.3.24}$$

7.3.4 Let us now consider the local perturbation $\omega_1(x) := \omega_0(a_1 x a_1^*)$ ($x \in \mathfrak{A}$) of the time-invariant vacuum state ω_0. The state ω_1 describes the infinite spin-chain in the state where all the spins except of the one sitting in the site $j = 1$ are pointing down. Its time evolution $\omega_1(\tau_t(x)) \equiv \omega_1^t(x)$ can be expressed in terms of the results given above. Let us, for example, calculate the expectation of "flipping up" of the spin placed in the j-th place at the time t. We have

$$\omega_1^t(a_j^* a_j) = \sum_{m=1}^\infty \langle 1|e^{itH} a_j^* a_j|m\rangle\langle m|e^{-itH}|1\rangle \tag{7.3.25}$$

$$= \sum_{m=j}^\infty \langle 1|e^{itH}|m\rangle\langle m|e^{-itH}|1\rangle = 1 - \sum_{m=1}^{j-1} |\langle m|e^{-itH}|1\rangle|^2,$$

since

$$a_j^* a_j|m\rangle = \begin{cases} 0 & (m < j), \\ |m\rangle & (m \geq j), \end{cases}$$

and the set of vectors $\{|m\rangle : m \in \mathbb{Z}\}$ forms an orthonormal basis in the relevant Hilbert space. From (7.3.24) and from the recurrent formula for Bessel functions

$$J_{p+1}(z) + J_{p-1}(z) = \frac{2p}{z} J_p(z), \tag{7.3.26a}$$

we obtain

$$\omega_1^t(a_j^* a_j) = 1 - \sum_{m=1}^{j-1} \left[\frac{m}{t} J_m(2t)\right]^2. \tag{7.3.26b}$$

Because of the asymptotic behaviour of the Bessel function for large real arguments $|\xi| \to \infty$, given by $J_p(\xi) = O(|\xi|^{-\frac{1}{2}})$, we obtain asymptotic behaviour of our expectation:

$$\omega_1^t(a_j^* a_j) \asymp 1 - \frac{\text{const.}}{|t^3|}, \quad (\forall j \in \mathbb{N}) \text{ for } t \to \infty. \tag{7.3.27}$$

Hence the local perturbation of the state "all spins are pointing down" converges according to (7.3.27) to the state "all spins sitting in sites with $j > 0$ are pointing up". For more details see also [35, 36].

7.3.5 This can be used for construction of models imitating the 'quantum measurement process'. For instance, let the infinite chain without the spin sitting in the site $j = 0$ model an "apparatus" and the spin at $j = 0$ serve as a "measured microsystem". If the apparatus is initially in the state $\omega_\downarrow$ with all its spins pointing down, and the measured spin in a superposition $\varphi := c_\downarrow |\downarrow\rangle + c_\uparrow |\uparrow\rangle$, then the compound system "measured microsystem + apparatus" is in the time $t = 0$ in the state described by the state-vector $c_\downarrow \Omega_0 + c_\uparrow a_0^* \Omega_0$, which is a coherent superposition of vectors in the 'vacuum representation' of the algebra of observables of the compound system. Then the final state of the chain (at $t = \infty$) will be (as a state on the algebra $\mathfrak{A}$ of the compound system "measured system + apparatus")[3] in an incoherent genuine mixture ω_f according to the above described dynamics: $\omega_f = |c_\downarrow|^2 \omega_0 + |c_\uparrow|^2 \omega_\uparrow$, where the state $\omega_\uparrow$ means that all spins of the compound system lying in sites $j \geq 0$ are pointing up, whereas the spins lying in sites $j < 0$ remain pointing down. The states ω_0 and $\omega_\uparrow$ on $\mathfrak{A}$ are mutually *disjoint*; this is interpreted here as "macroscopic difference" of these states. Also, the states ω_0 and $\omega_\uparrow$ define two representations of the algebra of quasi-local observables (see also [53, 54, 274, 275] for further details) which are not unitary equivalent, and can be distinguished by a measurement of a macroscopic observable.

As the macroscopic observable distinguishing these states could be chosen, e.g., the weak limit $\gamma \in \mathfrak{A}^{**}$ for $n \to \infty$ of the sequence

$$\gamma_n := \frac{1}{2n + 1} \sum_{j=-n}^{n} a_j^* a_j, \tag{7.3.28}$$

and for the states $\omega_0, \omega_\uparrow$ (now considered as being extended to normal states on the von Neumann algebra $\mathfrak{A}^{**}$) we obtain: $\omega_0(\gamma) = 0$, $\omega_\uparrow(\gamma) = \frac{1}{2}$. This is an example in the spirit of the models proposed in the classical paper by Hepp [153] for modeling the "quantum measurement process".

7.3.6 Observable quantities in QM, or "observables", are described usually by self-adjoint operators A acting on the Hilbert space where the "observed" states of a

---

[3]We consider here, for the sake of simplicity, the measured system after the measurement as a part of the apparatus, what makes no difference for observing results of measurements via various macrostates—the macroobservables of the compound system are identical with those of the measuring apparatus alone. See however the Sect. 7.3.7 below.

considered physical system appear. In another setting, we can speak instead of a selfadjoint operator A about its **projection-valued measure** ($\equiv$ *projector-valued measure*) (PM) $\Lambda \mapsto E_A(\Lambda)$ for $\Lambda \subseteq \Gamma \equiv$ the set (with a given σ-algebra structure) of possible values of the observable (specifying the operator uniquely); here $E_A(\Lambda)$ are mutually commuting orthogonal projectors satisfying σ-additivity with respect to set unions of various disjoint arguments $\Lambda \subset \Gamma$, with $E_A(\Gamma) = I_{\mathcal{H}}$.

More general concept of "**observable**" in QM is again σ-additive **positive operator valued measure** (*POVM*) $\Lambda \mapsto \mathbf{A}(\Lambda)$, with $\mathbf{A}(\Lambda) \in \mathcal{L}(\mathcal{H})$, $0 \le \mathbf{A}(\Lambda) \le \mathbf{A}(\Gamma) = I_{\mathcal{H}}$, $\Lambda_i \cap \Lambda_j = \emptyset$ $(\forall i, j) \Rightarrow \mathbf{A}(\cup_k \Lambda_k) = \sum_k \mathbf{A}(\Lambda_k)$, which also specifies a selfadjoint operator A, but is not specified by it uniquely. The different $\mathbf{A}(\Lambda)$, $\Lambda \subset \Gamma$, need not be now mutually commutative. According to a general 'philosophy' of QM, to each observable corresponds a measuring apparatus (better: a class of equivalent apparatuses) characterized abstractly by the observable, by which it can be measured. Conversely, if we perform a measurement on some quantum-mechanical system, some observable is measured. The results of the measurement of A on the state ϱ is found in the set $\Lambda \subseteq \Gamma$ with the probability $pr_A(\varrho, \Lambda) = Tr(\varrho \mathbf{A}(\Lambda))$. If $\Lambda \mapsto pr(\varrho, \Lambda)$ $(\Lambda \subseteq \Gamma)$ is a probability measure for any ϱ and this mapping depends on ϱ affinely: $pr(\lambda \varrho_1 + (1 - \lambda)\varrho_2, \Lambda) \equiv \lambda\, pr(\varrho_1, \Lambda) + (1 - \lambda) pr(\varrho_2, \Lambda)$, then there is a unique observable A of the measured system such that $pr(\varrho, \Lambda) \equiv Tr(\varrho \mathbf{A}(\Lambda))$. If the distribution of the results of a measurement is expressed in this way by some POVM $\mathbf{A} \ne E_A$, the measurement is often called a **nonideal measurement**. For more complete formulations cf. [84, 149].

We are dealing in this work with infinite quantal systems described by C^*-algebras having many mutually inequivalent representations. Hence, we cannot restrict the concept of observables to operators acting e.g. on a Hilbert space $\mathcal{H}_\omega$ of a specific cyclic representation. If we want stay in a framework of the above presented scheme, we can, and we presently shall, use the universal representation of C^*-algebra $\mathfrak{A}$ in $\mathcal{H}_u$, resp. of its weak closure, which is a W^*-algebra isomorphic to the double dual $\mathfrak{A}^{**}$ of $\mathfrak{A}$. For some comments on this reformulation see e.g. [84, Sect. 2.5].

7.3.7 We can now ask, which observable (in the sense of 7.3.6) was measured by the 'measuring apparatus' modeled by our QD, as it was sketched in 7.3.5. The 'microsystem' being measured consists in the spin sitting in the point $j = 0$ of the infinite spin-chain and the rest of the chain is the 'measuring apparatus'. Let us consider as the apparatus the half-infinite chain with spins sitting in the points numbered by $j = 1, 2, \dots \infty$ only, because the spins sitting in the points with $j < 0$ do not take part in these measurements.[4] An integral part of the characterization of the apparatus is, however, also its initial state 'with all spins pointing down', as well as its dynamics including the interaction with the measured spin. The results of these measurements are read by looking at the final states of the apparatus.[5] There are just two possibilities in this process: The state $\omega_\downarrow$ with all spins pointing down, i.e.

---

[4] In accordance with that, the notation in this section will be changed slightly with respect to the notation in the Sect. 7.3.5.

[5] We are speaking here about the states on the algebra generated by a_j, a_j^* with $j > 0$ only.

$\omega_\downarrow(a_j a_j^*) \equiv 1$, and the state $\omega_\uparrow$ with all spins pointing up, i.e. $\omega_\uparrow(a_j^* a_j) \equiv 1$, which is *disjoint* from the state $\omega_\downarrow$. If these states are (uniquely) extended to normal states on the double dual of the algebra of measuring apparatus, their values can be calculated on the 'macroscopic observable' γ defined now as the weak limit of the sums

$$\gamma_n := \frac{1}{n} \sum_{j=1}^{n} a_j^* a_j. \tag{7.3.29}$$

Then it is $\omega_\downarrow(\gamma) = 0$, $\omega_\uparrow(\gamma) = 1$. The "spectral set" Γ from 7.3.6 consists now of only two points, let us denote them (arbitrarily, but taking into account the actual measurement process) $\pm\frac{1}{2}$, hence $\Gamma := \{\frac{1}{2}, -\frac{1}{2}\}$.

The initial (=measured) state of the 'microsystem' in the example of 7.3.5 was given by the normalized vector $|\varphi\rangle := c_\downarrow |\downarrow\rangle + c_\uparrow |\uparrow\rangle$ corresponding to the density matrix $\varrho = |\varphi\rangle\langle\varphi|$ being just the one-dimensional projector on the pure state $|\varphi\rangle$ of the measured system. The final state of the apparatus was in this case (according to 7.3.5) $w_f := |c_\downarrow|^2 \omega_\downarrow + |c_\uparrow|^2 \omega_\uparrow$, where $|c_\downarrow|^2$, $|c_\uparrow|^2$ are the desired probabilities $pr(\varrho, \mp\frac{1}{2})$. From the linearity of the tensor products, as well as of time evolution, we can see that the extension of the previously introduced function $pr(\varrho, \mp\frac{1}{2})$ to general density matrices ϱ is an affine function of ϱ. Hence, e.g. for convex combination of two 'pure' density matrices,

$$\varrho := \lambda_1 |\varphi_1\rangle\langle\varphi_1| + \lambda_2 |\varphi_2\rangle\langle\varphi_2|, \text{ with } |\varphi_j\rangle := c_{j\downarrow} |\downarrow\rangle + c_{j\uparrow} |\uparrow\rangle, \ j = 1, 2, \tag{7.3.30}$$

we obtain

$$pr(\varrho, -\frac{1}{2}) = \lambda_1 \, pr(|\varphi_1\rangle\langle\varphi_1|, -\frac{1}{2}) + \lambda_2 \, pr(|\varphi_2\rangle\langle\varphi_2|, -\frac{1}{2})$$
$$= \lambda_1 |c_{1\downarrow}|^2 + \lambda_2 |c_{2\downarrow}|^2, \tag{7.3.31}$$

$$pr(\varrho, \frac{1}{2}) = \lambda_1 \, pr(|\varphi_1\rangle\langle\varphi_1|, \frac{1}{2}) + \lambda_2 \, pr(|\varphi_2\rangle\langle\varphi_2|, \frac{1}{2})$$
$$= \lambda_1 |c_{1\uparrow}|^2 + \lambda_2 |c_{2\uparrow}|^2.$$

Let us define the operator $A := \frac{1}{2} |\uparrow\rangle\langle\uparrow| - \frac{1}{2} |\downarrow\rangle\langle\downarrow|$ on the Hilbert state space of the measured system. Its spectral projections are $P_\uparrow := |\uparrow\rangle\langle\uparrow|$ and $P_\downarrow := |\downarrow\rangle\langle\downarrow|$ and the corresponding mutually distinct eigenvalues are chosen to be $\pm\frac{1}{2}$. Then, for our density matrix there holds

$$pr(\varrho, -\frac{1}{2}) = Tr(P_\downarrow \varrho), \quad pr(\varrho, \frac{1}{2}) = Tr(P_\uparrow \varrho). \tag{7.3.32}$$

Hence, our measuring process corresponds to measurement of operators with PM given by the one-dimensional orthogonal projectors $P_{\uparrow,\downarrow}$. Our choice of the values of elements in the set Γ corresponds to the observable describing a component of the $\frac{1}{2}\hbar$-spin, which is usually described in this way. We did not need here a generalized observable determined by a POVM, which will be, however, the case of the following example.

7.4 Particle Detection—A "Nonideal" Measurement

7.4.1 This model describes a compound system of a spin chain A with a particle B; it is a completed version of the model presented originally in [38]. The model of the spin chain is the half-infinite chain of the form described in the Sect. 7.3, and the particle is a nonrelativistic scalar particle.

Let us use (essentially) the notation of Sect. 7.3. Hence, the algebra $\mathfrak{A}$ of the observables of the spin chain is now generated by the elements $a_n^*, a_n, \ n \geq 1$. Let the Hamiltonian of the chain be the operator (cf. (7.3.8))

$$H_A := \sum_{n \geq 1} a_n^* a_n (a_{n+1}^* + a_{n+1}) a_{n+2} a_{n+2}^* \tag{7.4.1}$$

acting in the Hilbert space $\mathcal{H}_{vac}$ of the GNS-representation of $\mathfrak{A}$ with **the cyclic vector** Ω_0 corresponding to the state

$$\omega_{\downarrow}^A (a_j^* a_j) = 0, \ \text{for all } j \geq 1. \tag{7.4.2}$$

The particle B is moving in the 3-dimensional Euclidean space and is described as in elementary QM by operators acting in the space $\mathcal{H}_B := L^2(\mathbb{R}^3, d^3x)$, so that its states are described by vectors (resp. the corresponding unit rays) $\psi \in \mathcal{H}_B$. The free particle's Hamiltonian will be just the kinetic energy (in conveniently chosen units and in the "**x**-representation")

$$H_B := \hat{\mathbf{p}}^2 = -\sum_{j=1}^{3} \frac{\partial^2}{\partial x_j^2}. \tag{7.4.3}$$

The **interaction Hamiltonian** will be V_φ, with

$$V_\varphi := (a_1^* + a_1) a_2 a_2^* \otimes |\varphi\rangle\langle\varphi| \in \mathcal{L}(\mathcal{H}_{vac} \otimes \mathcal{H}_B), \tag{7.4.4}$$

where $\varphi \equiv |\varphi\rangle \in \mathcal{H}_B$ is a conveniently chosen normalized vector, hence $|\varphi\rangle\langle\varphi| \equiv P_\varphi$ is a one-dimensional projector in $\mathcal{H}_B$.

The **total Hamiltonian** H **of the** compound system {spin chain & particle} will be

$$H := H_A + H_B + \gamma V_\varphi, \ \gamma \in \mathbb{R}. \tag{7.4.5}$$

Some restrictions on the interaction constant γ and on the unit vector φ will be specified later.

7.4.2 We want to prove, for conveniently chosen parameters γ and φ of interaction and for suitable initial states $\psi \in \mathcal{H}_B$ of the particle as well as for given initial state of the spin chain with all spins "pointing down", that the compound system will evolve for $t \to \infty$ with positive probability into a convex combination of two mutually *disjoint* (hence 'macroscopically different') states, one of which corresponds to the unchanged initial state of the apparatus and in the other the apparatus has all its spins reversed to the "pointing up" direction. If we denote by $\mathfrak{B} := \mathcal{L}(\mathcal{H}_B)$ the algebra of all bounded operators on $\mathcal{H}_B$, which is the C^*-algebra of the observables of the particle, and **by $\mathfrak{C} := \mathfrak{A} \otimes \mathfrak{B}$ the C^*-algebra of the compound system,** then $\mathcal{S}(\mathfrak{C})$ **will be the state-space** $\mathfrak{C}^*_{+1}$ (i.e. positive normalized elements of the topological dual of $\mathfrak{C}$) of the compound system.

We will prove that **the initial state** $\omega_0^{A\&B} \equiv \omega_\downarrow^A \otimes \omega_\psi^B \in \mathcal{S}(\mathfrak{C})$, **where** $a \mapsto \omega_\psi^B(a) := \langle \psi | a | \psi \rangle$ for $a \in \mathfrak{B}$, will evolve to the state $\overline{\omega} \in \mathcal{S}(\mathfrak{C})$, $\overline{\omega} = (w(\psi) \omega_\uparrow^A + (1 - w(\psi)) \omega_\downarrow^A) \otimes \omega_0^B$, and where $\omega_0^B \in \mathcal{S}(\mathfrak{B})$ is the state without particles, cf. 7.4.3, and $0 < w(\psi) < 1$ for any of the considered initial state-vectors ψ.

If we ask "which observable is measured by this process", the relevant answer is—if we consider only the mathematical expression of the "observable" appearing in the question—in the expression of the **probability** $w(\psi)$ as a diagonal matrix element of a **positive operator** $W \equiv W_\gamma$ between the state vectors of the particle's initial state ψ: $w(\psi) = \langle \psi | W | \psi \rangle$. The operator W, $0 < W < 1$, $W \neq W^2$, replaces here the usual appearance of a projector from the PM of measured selfadjoint operator in the cases of '**ideal measurements**', cf.also [149]. Our simple specific model represents more general instances of measurements: The '**nonideal measurement**' is described by a *POVM* (=positive operator valued measure). Hence, our model illustrates the concept of "generalized observables" introduced in [84, Sect. 3.1], cf. also our 7.3.6 and 7.3.7, and its usefulness. The quantity $w(\psi) = \langle \psi | W | \psi \rangle$ has to be interpreted as the measured probability of one of two possible results of a two-valued observable of the particles prepared at $t = 0$ in the state ψ. A verbal expression of the intuitive physical meaning of "the particle's observable W" might be here just something like "what can be registered by this specific measuring apparatus", with two different pointer values: to be or not to be registered by this specific apparatus.

7.4.3 Notation. *We shall use the following symbols:*

1. *The state without particles could be defined in a standard way, e.g. as the vacuum state in the Fock representation, where the algebra of observables of particles is constructed by creation-annihilation operators, cf. [38]. To avoid this (here unnecessary) complication, we shall define the no-particle state as the normal linear functional $\omega_0^B \in \mathcal{S}(\mathfrak{B})$ on $\mathfrak{B} = \mathcal{L}(\mathcal{H}_B)$, $\omega_0^B : b \mapsto \omega_0^B(b)$ (remember that $\dim \mathcal{H}_B = \infty$) such that*

$$\omega_0^B(b) = 1, \; \text{if } b = I_{\mathcal{H}_B}; \; \omega_0^B(b) = 0, \; \text{if } b = |\psi_1\rangle\langle\psi_2|, \; \psi_j \in \mathcal{H}_B.$$

This will give equivalent results of our considerations to those obtained from the considerations using the formalism of nonrelativistic quantum field theory.

2. *Let us introduce also the symbol* $\mathcal{H}_A$ *for the Hilbert (sub-)space of the chain generated by the vectors* $\{|m\rangle \mid m = 1, 2, \ldots\}$ *introduced in (7.3.12). We shall use also:* $U_t := \exp(-it H)$ *with* H *from (7.4.5), and* $\tau_t c := e^{it H} c \, e^{-it H}$ *for* $c \in \mathfrak{C}$. *The vector* $\Omega_0 = |0\rangle$ *is defined in (7.3.2) and (7.3.3). We shall also use* $\Omega_0^{\chi} := \Omega_0 \otimes \chi, \ \chi \in \mathcal{H}_B$.

3. *Let* $\varphi \in \mathcal{H}_B$, $\|\varphi\|_2 = 1$, *be the vector appearing in the interaction Hamiltonian* V_φ *in (7.4.4), and let* $\psi \in \mathcal{H}_B$ *be the (also normalized) initial state-vector of the particle. We shall* **introduce the symbols** $F^0(t)$, $g(t)$, *and* $F(t)$ *as:*

$$F^0(t) \equiv F^0_\varphi(\psi)(t) := \langle\varphi|e^{-it H_B}|\psi\rangle, \ g(t) := F^0_\varphi(\varphi)(t) \equiv \langle\varphi|e^{-it H_B}|\varphi\rangle.$$
$$(7.4.6a)$$

$$F(t) \equiv F_\varphi(\psi)(t) := \langle\varphi| \otimes \langle 0|e^{-it H}|0\rangle \otimes |\psi\rangle \equiv \langle\Omega_0^\varphi|e^{-it H}|\Omega_0^\psi\rangle,$$
$$(7.4.6b)$$

where $H := H_A + H_B + \gamma V_\varphi$ *is the total Hamiltonian of the compound system (7.4.5).*

The symbols $f_m(t)$, $f(t)$ *will be also useful abbreviations (cf. (7.3.24)):*

$$f_m(t) := \langle m|e^{-it H_A}|1\rangle = (-i)^{m-1} \frac{m}{t} \, J_m(2t), \ m = 1, 2, \ldots \quad (7.4.7)$$

$$f(t) := g(t) f_1(t) = \langle\varphi| \otimes \langle 1|e^{-it(H_A+H_B)}|1\rangle \otimes |\varphi\rangle. \quad (7.4.8)$$

4. *To restrict a function* $t \mapsto h(t)$ *defined on the whole real line* $t \in \mathbb{R}$ *to the positive (resp. negative) values of its argument* $t \in \mathbb{R}_+$ *(resp.* $\mathbb{R}_-$*), we shall use the (Heaviside)* $\theta(t)$*-function equal to zero for* $t < 0$ *and equal to one for* $t \geq 0$. *We shall denote these restrictions as* $h_\pm(t)$:

$$h_+(t) := \theta(t) h(t), \ resp. \ h_-(t) := \theta(-t) h(t), \quad t \in \mathbb{R}. \quad (7.4.9)$$

Such restrictions $f \mapsto f_+$ *will be useful here, e.g., for rewriting certain equations in the convolution form.*

5. *The* **convolution** $f * h(t)$ *of two complex-valued integrable functions is defined by*

$$f * h(t) = \int_{-\infty}^{+\infty} d\tau \, f(t - \tau) h(\tau) = h * f(t). \quad (7.4.10)$$

For more details on existence conditions of convolutions see e.g. [262, IX.4]. The operation $*$ *is not only commutative, but also associative. It can be trivially extended to functions* $t \mapsto h(t)$ *defined for* $t \in \mathbb{R}^n$, *as well as to some other classes of functions and of distributions, see e.g. [262, 324].*

6. *Let us define and denote, for purposes of the present section, to any integrable function* $h \in L^1(\mathbb{R})$, *its Fourier transformed function* $\mathcal{F}(h) \equiv \hat{h}$:

$$\hat{h}(u) \equiv \mathcal{F}(h)(u) := \frac{1}{\sqrt{2\pi}} \int_{-\infty}^{+\infty} e^{-itu} h(t) \, dt, \ u \in \mathbb{R}. \quad (7.4.11a)$$

In the case of higher dimensional arguments of the $\mathbb{C}$*-valued functions* $h \in L^1(\mathbb{R}^n)$ *the analogous formula applies:*

$$\hat{h}(u) \equiv \mathcal{F}(h)(u) \equiv \mathcal{F}(h(t))(u) := (2\pi)^{-\frac{n}{2}} \int_{\mathbb{R}^n} e^{-it \cdot u} h(t) \, d^n t, \ u \in \mathbb{R}^n.$$

(7.4.11b)

The inverse $\mathcal{F}^{-1}$ *of* $\mathcal{F}$ *defined on the image* $\hat{h} = \mathcal{F}(h)$ *has the similarly looking form:*

$$h(t) = \mathcal{F}^{-1}(\hat{h})(t) = \mathcal{F}(\hat{h}(-u))(t) = (2\pi)^{-\frac{n}{2}} \int_{\mathbb{R}^n} e^{it \cdot u} \hat{h}(u) \, d^n u, \ t \in \mathbb{R}^n.$$

(7.4.11c)

Generalizations to various classes of functions h *and also to tempered distributions is very useful in process of solution of various equations. Many important properties of the Fourier transformation can be found, e.g. in [262, 324]. One of the most useful properties of* $\mathcal{F}$ *is the possibility to extend it from* $L^1(\mathbb{R}^n)$ *to a unitary transformation in the Hilbert space* $L^2(\mathbb{R}^n)$—*the Plancherel theorem: The scalar product* $\langle \cdot | \cdot \rangle$ *is invariant with respect to the transformation* $\mathcal{F}$*; for* $\varphi, \psi \in L^2$ *it means:* $\langle \varphi | \psi \rangle = \langle \hat{\varphi} | \hat{\psi} \rangle$*. Moreover, the following important property concerning the interconnection between the convolution and the Fourier transformation is valid:*

$$\mathcal{F}(h_1 * h_2) = (2\pi)^{\frac{n}{2}} \mathcal{F}(h_1) \mathcal{F}(h_2) \equiv (2\pi)^{\frac{n}{2}} \hat{h}_1 \hat{h}_2,$$

(7.4.12)

with the pointwise multiplication of functions.

For the proof of our main result formulated in Theorem 7.4.8, we shall also need several following lemmas. The first one together with its proof can be deduced from [151]:

7.4.4 Lemma. *Let* H *be a lower-bounded selfadjoint operator on a Hilbert space* $\mathcal{H}_v$ *with its spectrum* $sp(H) \geq a$*. Then, for any two nonzero vectors* $\varphi, \psi \in \mathcal{H}$*, it is:*

(a) either $\langle \varphi | e^{-itH} | \psi \rangle \equiv 0, \ \forall t \in \mathbb{R}$,
(b) or $\langle \varphi | e^{-itH} | \psi \rangle \neq 0$ *for* t *in an open dense subset of* $\mathbb{R}$ *of total Lebesgue measure.*

If the above chosen φ *is fixed, then the set of all* $\psi \in \mathcal{H}$ *satisfying* **(a)** *forms a closed linear subspace of* $\mathcal{H}_v$*, hence the open complement in* $\mathcal{H}_v$ *of this set consists of those* $\psi \in \mathcal{H}$ *which satisfy the point* **(b)**.

Proof. Let $\lambda \mapsto E_H(\lambda)$ be the projection measure of H. According to the functional calculus (cf. e.g. [262]) it is

$$\langle \varphi | e^{-itH} | \psi \rangle = \int_a^\infty d\lambda \, e^{-it\lambda} \langle \varphi | E_H(\lambda) | \psi \rangle.$$

(7.4.13)

This function of time $t \in \mathbb{R}$ can be analytically continued to the lower complex half-plain of t, i.e. extended to $t \mapsto t - i\varepsilon =: z, \ \varepsilon \geq 0$:

$$\langle\varphi|e^{-i(t-i\varepsilon)H}|\psi\rangle = \int_a^\infty d\lambda e^{-i(t-i\varepsilon)\lambda}\langle\varphi|E_H(\lambda)|\psi\rangle \equiv \langle\varphi|e^{-izH}|\psi\rangle, \ \text{Im } z \le 0,$$

(7.4.14)

which is analytic in the open lower complex half-plain of z and continuous in the closed lower half-plain, hence also on the real line $z = t - i\varepsilon \to t - i0+$. Assume that $\langle\varphi|e^{-itH}|\psi\rangle \equiv 0$, $\forall t \in I \subset \mathbb{R}$, where I is an interval of positive length. Then, according to the Schwarz reflection principle, the analytic function $z \mapsto \langle\varphi|e^{-izH}|\psi\rangle$ is complex-analytic also on this interval I, hence it is identically zero also in lower complex half-plain. Due to its continuity on $\mathbb{R}$, the function $t \mapsto \langle\varphi|e^{-itH}|\psi\rangle \equiv 0$ ($\forall t \in \mathbb{R}$).

In the other cases, there is no interval of nonzero length $I \subset \mathbb{R}$ on which the function $t \mapsto \langle\varphi|e^{-itH}|\psi\rangle$ identically vanishes. Since it is continuous, it is $\ne 0$ on open intervals composing an open dense subset of $\mathbb{R}$. But union of all these intervals is a set of total Lebesgue measure on $\mathbb{R}$, as is shown in [151]. Hence the function $t \mapsto \langle\varphi|e^{-itH}|\psi\rangle \ne 0$ a.e. with respect to the Lebesgue measure.

Linearity of the set of the ψ's satisfying (a) is clear. That this subspace is closed in $\mathcal{H}_v$ follows from the norm-continuity of the matrix elements $\psi \mapsto \langle\varphi|e^{-itH}|\psi\rangle$; the last assertion follows from the other proved assertions of this Lemma. □

7.4.5 Lemma. *The condition $\varphi \in \mathcal{D}(\mathbb{R}^3)$ for the choice of the vector φ occurring in the definition of the interaction Hamiltonian in (7.4.4), as well as the condition $\psi \in \mathcal{H}_B \cap L^1(\mathbb{R}^3)$ for the choice of the particle's initial vector ψ, both imposed in the Theorem 7.4.8, guarantee the following properties of the functions $t \mapsto F_\varphi^0(\psi)(t)$ (7.4.6a) of the time variable $t \in \mathbb{R}$:*

$$F_\varphi^0(\psi) \in L^2(\mathbb{R}) \cap L^1(\mathbb{R}), \quad g \equiv F_\varphi^0(\varphi) \in L^2(\mathbb{R}) \cap L^1(\mathbb{R}).$$

(7.4.15)

The set $L^1(\mathbb{R}^3) \cap L^2(\mathbb{R}^3) \supset \mathcal{D}(\mathbb{R}^3)$ is dense in $\mathcal{H}_B$ together with $\mathcal{D}(\mathbb{R}^3)$.

Proof. According to the Theorem IX.30 of [262], there is for $\psi \in L^1(\mathbb{R}^3) \cap L^2(\mathbb{R}^3)$:

$$\text{ess sup}_{\mathbf{x}\in\mathbb{R}^3} |e^{-itH_B}\psi(\mathbf{x})| \equiv \|e^{-itH_B}\psi\|_\infty \le |t^{-\frac{3}{2}}|\,\|\psi\|_1.$$

(7.4.16)

The function φ has finite support, say $\varphi(\mathbf{x}) \ne 0 \Rightarrow |\mathbf{x}| < R < \infty$. Let us denote by $B_R \subset \mathbb{R}^3$ the ball of radius R containing the support of φ. Due to the implication $\chi \in L^2(\mathbb{R}^3) \Rightarrow |\chi| \in L^2(\mathbb{R}^3)$, we have

$$|\langle\varphi|e^{-itH_B}\psi\rangle| \le \int_{B_R} d^3\mathbf{x}\,|\varphi(\mathbf{x})| \cdot |e^{-itH_B}\psi(\mathbf{x})| \le$$

$$\int_{B_R} d^3\mathbf{x}\,|\varphi(\mathbf{x})| \cdot \frac{\|\psi\|_1}{|t^{\frac{3}{2}}|} = \frac{\|\varphi\|_1\|\psi\|_1}{|t^{\frac{3}{2}}|}.$$

(7.4.17)

But the matrix element of a unitary operator between two normalized vectors in $\mathcal{H}_B$ is bounded by unity: $|F_\varphi^0(\psi)(t)| \leq 1$, hence we have

$$|\langle \varphi | e^{-itH_B} \psi \rangle| \leq \min\left\{ 1; \frac{\|\varphi\|_1 \|\psi\|_1}{|t^{\frac{3}{2}}|} \right\}, \quad \text{for all } t \in \mathbb{R}, \tag{7.4.18}$$

and the obtained estimate is

$$|F_\varphi^0(\psi)(t)| \leq \theta\left((\|\varphi\|_1\|\psi\|_1)^{\frac{2}{3}} - |t| \right) + \theta\left(|t| - (\|\varphi\|_1\|\psi\|_1)^{\frac{2}{3}} \right) \frac{\|\varphi\|_1\|\psi\|_1}{|t^{\frac{3}{2}}|}. \tag{7.4.19}$$

The function $t \mapsto |t^{-\frac{3}{2}}| \theta(|t| - k)$, $k > 0$, belongs to $L^1(\mathbb{R}) \cap L^2(\mathbb{R})$, hence $F_\varphi^0(\psi)$ also belongs there $\forall \psi \in L^1(\mathbb{R}^3) \cap L^2(\mathbb{R}^3)$. Since also our $\varphi \in \mathcal{D}(\mathbb{R}^3) \subset L^1(\mathbb{R}^3) \cap L^2(\mathbb{R}^3)$, the both relations in (7.4.15) are proved. The density of $\mathcal{D}(\mathbb{R}^3)$ in $L^2(\mathbb{R}^3)$ is easily seen, cf. e.g. [324, I.1.7]. $\qquad\square$

7.4.6 Lemma. *Let* $G \in L^1(\mathbb{R}^n) \cap L_0^\infty(\mathbb{R}^n)$ *and* $G' \in L^p(\mathbb{R}^n) \cap L_0^\infty(\mathbb{R}^n)$ $(1 \leq p \leq \infty)$, *where* L_0^∞ *is the space of (essentially) uniformly bounded functions converging to zero at infinity. Then the convolution* $G * G' \in L^p \cap L_0^\infty$.

Proof. According to the Theorem 1.3. in [298], $\|G * G'\|_p \leq \|G\|_1 \cdot \|G'\|_p$, and also $\|G * G'\|_\infty \leq \|G\|_1 \cdot \|G'\|_\infty$, hence $G * G' \in L^p \cap L^\infty$. It remains to prove the convergence to zero at infinity.

Let us choose $\delta > 0$. For any such δ there is a $T_\delta > 0$ such, that $\forall |\tau| > T_\delta \Rightarrow |G'(\tau)| < \delta$. Then for $|t| > T_\delta$ it is

$$|G * G'(t)| \leq \int_{|\tau| < T_\delta} d^n\tau \, |G(t - \tau)G'(\tau)| + \delta \int_{|\tau| > T_\delta} d^n\tau \, |G(t - \tau)|$$
$$\leq \|G'\|_\infty \, \Omega_n(T_\delta) \sup_{|\eta| \geq ||t| - T_\delta|} |G(\eta)| + \delta \|G\|_1, \tag{7.4.20}$$

where $\Omega_n(T)$ is the Euclidean volume of the n-dimensional ball of radius T. With any fixed δ, the supremum converges to zero with $|t| \to \infty$. Hence, by a convenient choice of $\delta > 0$ and for sufficiently large $|t|$, the right hand side of (7.4.20) can be made arbitrarily small, hence the left hand side converges with $|t| \to \infty$ to zero. $\qquad\square$

A similar useful Lemma for functions of $t \in \mathbb{R}$ restricted to $\mathbb{R}_+$ claims:

7.4.7 Lemma. *For* $h \in L^1(\mathbb{R}) \cap L_0^\infty(\mathbb{R})$ *and* $k \in L^1(\mathbb{R})$ *it is:*

$$h_+ * k_+ \in L^1(\mathbb{R}) \cap L_0^\infty(\mathbb{R}). \tag{7.4.21}$$

Proof. Again from the known L^p-estimate [298] there is $h_+ * k_+ \in L^1(\mathbb{R}) \cap L^\infty(\mathbb{R})$, and also $\|h_+ * k_+\|_p \leq \|h_+\|_p \|k_+\|_1$ for $p = 1, \infty$. Let us prove the convergence to zero. It is

$$h_+ * k_+(t) = \theta(t) \int_0^{\frac{t}{2}} d\tau \, [\, h(t-\tau)k(\tau) + h(\tau)k(t-\tau) \,], \qquad (7.4.22)$$

and the needed estimate is:

$$|h_+ * k_+(t)| \leq \theta(t) \left[\|k\|_1 \sup_{\tau > \frac{t}{2}} |h(\tau)| + \|h\|_\infty \int_{\frac{t}{2}}^{+\infty} d\tau \, |k(\tau)| \right]. \qquad (7.4.23)$$

The first term on the right hand side converges for $t \to +\infty$ to zero because the function h converges to zero. The second term converges to zero due to integrability of $k \in L^1(\mathbb{R})$. This shows that $h_+ * k_+ \in L_0^\infty$. The assertion is proved. $\qquad\square$

We shall give here a proof of the main result of this section:

7.4.8 Theorem. *Let the dynamics of the compound system: nonrelativistic point particle B (as a "detected microsystem") and the one-dimensional spin chain A, described in Sect. 7.3 (as a "detector"), be given by the Hamiltonian (7.4.5) defined in the ground-state representation (corresponding to the state $\omega_\downarrow^A$ of the chain with "all spins pointing down").*

Let the particle's initial normalized state-vector be $\psi \in \mathcal{H}_B \cap L^1(\mathbb{R}^3) \equiv L^2(\mathbb{R}^3) \cap L^1(\mathbb{R}^3)$, and the initial state of our half-infinite chain be $\omega_\downarrow^A$ from (7.4.2). The normalized vector $\varphi \in L^2(\mathbb{R}^3)$ occurring in the Hamiltonian H in (7.4.4) will be chosen as a rapidly decreasing $C^\infty(\mathbb{R}^3)$ function with compact support: $\varphi \in \mathcal{D}(\mathbb{R}^3) \subset \mathcal{H}_B \cap L^1(\mathbb{R}^3)$. To ensure a nontrivial interaction of the particle with the chain, let us assume that (cf. Lemma 7.4.4)

$$F_\varphi^0(\psi)(t) \equiv \langle \varphi | \exp(-it H_B) | \psi \rangle \neq 0, \quad t \in \mathbb{R}. \qquad (7.4.24)$$

We require, moreover, a condition on the upper bound of the interaction constant γ to be fulfilled:

$$0 < \|\gamma g\|_1 < 2, \qquad (7.4.25)$$

with $g \equiv F_\varphi^0(\varphi)$.

If these conditions are satisfied, then there exist, for all $a \in \mathfrak{A}$, $b \in \mathcal{L}(\mathcal{H}_B)$, the limits

$$\lim_{t \to \infty} \omega_t^{A\&B}(ab) = \left(w(\psi) \, \omega_\uparrow^A(a) + (1 - w(\psi)) \, \omega_\downarrow^A(a) \right) \omega_0^B(b), \qquad (7.4.26a)$$

with $\omega_t^{A\&B}(ab) := \langle \psi \otimes \Omega_0 | e^{itH} a \otimes b \, e^{-itH} | \Omega_0 \otimes \psi \rangle$, i.e.

$$w^*\text{-} \lim_{t \to \infty} \omega_0^{A\&B} \circ \tau_t \equiv w^*\text{-} \lim_{t \to \infty} \omega_t^{A\&B} = \left(w(\psi) \, \omega_\uparrow^A + (1 - w(\psi)) \, \omega_\downarrow^A \right) \otimes \omega_0^B.$$

$$(7.4.26b)$$

*The **probability of the detection** $w(\psi)$ is here positive: $w(\psi) > 0$, and, moreover, it depends on the initial state ψ of the particle as:*

$$\psi \mapsto \langle \psi | W | \psi \rangle \equiv \langle \psi | W_\gamma | \psi \rangle \equiv w(\psi), \tag{7.4.27}$$

where $W \equiv W_\gamma \in \mathcal{L}(\mathcal{H}_B)$ is a positive operator $0 < W_\gamma < I_{\mathcal{H}_B}$, independent of ψ. Moreover, for sufficiently small nonzero interaction constants $\gamma \in [-\gamma_0, \gamma_0] \subset \mathbb{R}$ it is $W_\gamma^2 \neq W_\gamma$, hence W_γ is not a projector.

Proof. Let us use the notation introduced in 7.4.3. We want to prove the existence of the limit (7.4.26) first. Let the state-vectors of the chain $|m\rangle$, $m = 0, 1, 2, \ldots \infty$ be defined as in (7.3.12) **with** $|0\rangle := \Omega_0$. The **Hilbert subspace** $\mathcal{K} \equiv (\mathcal{H}_A \oplus \mathcal{H}_{\{\emptyset\}}) \otimes \mathcal{H}_B$ **of the state-space of the (initial-state representation of the) compound system** "the spin half-chain & the particle" generated by vectors $|m\rangle \otimes |\psi\rangle$ ($m = 0, 1, \ldots$), $\psi \in \mathcal{H}_B$, is H-invariant, hence also invariant with respect to the time evolution $U_t \equiv \exp[-it(H_A + H_B + \gamma V_\varphi)]$. Let $P_\mathcal{K}$ be the orthogonal projector onto $\mathcal{K}$. Let us define the partial isometries P_{nm} in $\mathcal{K}$ by

$$P_{nm}|k\rangle \otimes |\psi\rangle = \delta_{mk}|n\rangle \otimes |\psi\rangle, \text{ for all } \psi \in \mathcal{H}_B, \; n, m, k = 0, 1, 2, \ldots \tag{7.4.28}$$

Let $P_n := P_{nn}$, $\forall n$. Denote also by P_ψ, $\psi \in \mathcal{H}_B$ ($\|\psi\|_2 = 1$), the one dimensional projector $|\psi\rangle\langle\psi|$ in $\mathcal{H}_B$. Clearly $P_0 \Omega_0^\psi = \Omega_0^\psi \equiv |0\rangle \otimes |\psi\rangle$, and for all $k, l, m, n \in \mathbb{Z}_+$ it is

$$P_{nm}^* = P_{mn}, \; P_{nk}P_{lm} = \delta_{kl}P_{nm}, \; \sum_{m=0}^{\infty} P_m = P_\mathcal{K}. \tag{7.4.29}$$

We shall write elements $x = a \otimes b \in \mathfrak{A} \otimes \mathfrak{B}$ as $x = ab$ (hence also $a \equiv a \otimes I_{\mathcal{H}_B}$, $b \equiv I_\mathfrak{A} \otimes b$), if a confusion would be improbable. So, we are looking for limits

$$\overline{\omega}(x) := \lim_{t \to \infty} \omega_t^{A\&B}(x), \; x \in \mathfrak{A} \otimes \mathfrak{B} = \mathfrak{C}. \tag{7.4.30}$$

We shall see that the limits (7.4.30) for $x \in \mathfrak{A} \subset \mathfrak{C}$ are expressible in terms of $\overline{\omega}(P_{mn})$. The very well known Dyson equation (7.4.31) expressing the unitary evolution group $U_t = \exp[-it(H_A + H_B + \gamma V_\varphi)]$ of a system with the interaction γV_φ in terms of this interaction and of the free system (without interaction) evolution group U_t^0 ($t \in \mathbb{R}$):

$$U_t = U_t^0 - i\gamma \int_0^t d\tau U_{t-\tau}^0 V_\varphi U_\tau, \tag{7.4.31}$$

with $U_t^0 := \exp[-it(H_A + H_B)]$, will be used repeatedly in our work here. We shall work in the Hilbert space $\mathcal{K}$ (for $t < \infty$). The restriction of the interaction Hamiltonian V_φ to the subspace $\mathcal{K}$ has the form

$$P_\mathcal{K} V_\varphi = (P_{01} + P_{10}) P_\varphi. \tag{7.4.32}$$

Due to the commutativity of U_t^0 with P_0, we obtain for $m \neq 0$ after the insertion from (7.4.32) into (7.4.31) :

$$P_m U_t P_0 = -i\gamma \int_0^t d\tau \, P_m U_{t-\tau}^0 P_{10} P_\varphi P_0 U_\tau P_0. \tag{7.4.33}$$

For $m = 0$, we obtain similarly:

$$P_0 U_t P_0 = P_0 U_t^0 - i\gamma \int_0^t d\tau \, U_{t-\tau}^0 P_{01} P_\varphi P_1 U_\tau P_0. \tag{7.4.34}$$

Substitution of (7.4.33) with $m = 1$ to this equation leads, after a linear change of integration variables, to an integral equation for $P_0 U_t P_0$:

$$P_0 U_t P_0 = P_0 U_t^0 - \gamma^2 \int_0^t dt' \int_0^{t-t'} d\tau \, U_{t-t'-\tau}^0 P_{01} P_\varphi U_\tau^0 P_\varphi P_{10} P_0 U_{t'} P_0. \tag{7.4.35}$$

Also the commutativity of P_φ with P_{mn} was used here. Since $U_t^A := e^{-it H_A}$ leaves the vector Ω_0 invariant, it is also $P_0 U_t^0 = P_0 \exp(-it H_B)$, and with (7.4.8) we have:

$$P_{01} P_\varphi U_\tau^0 P_\varphi P_{10} = \langle \varphi | e^{-i\tau H_B} | \varphi \rangle \langle 1 | U_\tau^A | 1 \rangle P_\varphi P_0 \equiv f(\tau) P_\varphi P_0. \tag{7.4.36}$$

The integral equation (7.4.35) can be rewritten now in the form:

$$P_0 U_t P_0 = P_0 e^{-it H_B} - \gamma^2 \int_0^t dt' \int_0^{t-t'} d\tau \, e^{-i(t-t'-\tau)H_B} P_\varphi f(\tau) P_0 U_{t'} P_0. \tag{7.4.37}$$

With the symbols from (7.4.6a) and (7.4.6b), by taking the matrix elements of both sides of this equation as in (7.4.6b), we can write the equation for $F(t)$, cf. Notation 7.4.3:

$$F(t) = F^0(t) - \gamma^2 \int_0^t dt' \int_0^{t-t'} d\tau \, g(t - t' - \tau) f(\tau) F(t'). \tag{7.4.38a}$$

If we take the restrictions of these functions to the values of the argument $t \geq 0$ according to (7.4.9), we can rewrite (7.4.38a) as a convolution equation, cf. also (7.4.10)[6]:

$$F_+ = F_+^0 - \gamma^2 g_+ * f_+ * F_+. \tag{7.4.38b}$$

We shall express now the quantities $\omega_t^{A\&B}(P_m)$ in terms of the Sect. 7.4.3, with a help of (7.4.33):

$$\omega_t^{A\&B}(P_m) = \gamma^2 \int_0^t dt' \int_0^t dt'' \, \bar{F}(t')F(t'')g(t'-t'')\bar{f}_m(t-t')f_m(t-t''), \quad m = 1, 2, \ldots. \tag{7.4.39}$$

To obtain a similar expression for $\omega_t^{A\&B}(P_0)$ we shall use completeness of the set of projections $\{P_m : m \in \mathbb{Z}_+\}$ in the subspace $\mathcal{K}$, cf. (7.4.29). We can sum over m in the argument of $\omega_t^{A\&B}(\cdot)$ in (7.4.39) because of normality of the state $\omega_t^{A\&B} \in \mathcal{S}(\mathfrak{C})$ for finite t. After the summation we can perform also $\lim_{t\to\infty}$. Summation over m in (7.4.39) can be performed under the integral signs due to Lebesgue dominated convergence theorem, cf. the definition of f_m. The completeness of the orthonormal basis $\{\,|m\rangle|\, m = 1, 2, \ldots\}$ in $\mathcal{H}_A$ gives also:

$$\sum_{m=1}^{\infty} \bar{f}_m(t-t')f_m(t-t'') = f_1(t'-t''). \tag{7.4.40}$$

We then obtain:

$$\omega_t^{A\&B}(P_0) = 1 - \gamma^2 \int_0^t dt' \int_0^t dt'' \, \bar{F}(t')F(t'')f(t'-t''). \tag{7.4.41}$$

To see the asymptotic properties of $\omega_t^{A\&B}(P_m)$ $(t \to +\infty)$, we shall need some properties of the solution $F(t)$ of (7.4.38). We shall obtain them by expressing the solution of the Volterra equation (7.4.38b) in the form of (Carl) Neumann series

$$F_+ = \sum_{n=0}^{\infty} (-\gamma^2 g_+ * f_+*)^n F_+^0, \tag{7.4.42}$$

converging uniformly on any bounded interval for any γ and any continuous f, g.[7] Since the free particle Hamiltonian $H_B := \hat{\mathbf{p}}^2$ has an absolute continuous spectrum,

---

[6]Note that, due to time-reflection symmetry of all the systems considered here, quite analogical equations and the corresponding results could be obtained also for the function $t \mapsto F(-t)$, $t \geq 0$.
[7]To see this, calculate $\sum_{n=0}^{\infty}(h_+*)^n(t)$ for $h \equiv const.$

the functions $F^0(t)$ and $g(t)$ from (7.4.6a) are continuous converging to 0 for $t \to \infty$. With our assumptions it is (see also [262, Sect. IX.4]) $|f(t)| \leq 1 \Rightarrow \|f\|_1 \leq \|g\|_1 = 2\|g_+\|_1$. This implies

$$\|\gamma^2 g_+ * f_+\|_1 < \gamma^2 \|g_+\|_1 \cdot \|f_+\|_1 < 1, \tag{7.4.43}$$

which is a sufficient condition for also the L^1-norm convergence of the series in (7.4.42). In this way we obtained (cf. also Footnote 6)

$$F \in L^2(\mathbb{R}) \cap C_0(\mathbb{R}). \tag{7.4.44}$$

We conclude from the preceding that

$$\lim_{t \to \infty} \omega_t^{A\&B}(P_m) = 0, \quad \text{for all } m \geq 1. \tag{7.4.45a}$$

The corresponding limit for $m = 0$ is obtained from (7.4.41). Written in the form of the scalar product $(\bullet, \circ) \in \mathbb{C}$ in $L^2(\mathbb{R})$, it has the form:

$$\lim_{t \to \infty} \omega_t^{A\&B}(P_0) = 1 - \gamma^2(F_+, F_+ * f). \tag{7.4.45b}$$

We can prove the assertion (7.4.26) of the Theorem now. Since the space $\mathcal{K}$ of the used representation of $\mathfrak{C}$ is time invariant with respect to our dynamics of the interacting systems, we shall restrict our work to investigation of the limits $\lim_{t \to \infty} \omega_t^{A\&B}(ab)$ for $a = P_{mn}$, $m, n \in \mathbb{Z}_+$, resp. $a = I_{\mathcal{H}_{vac}}$, and $b = |\psi_1\rangle\langle\psi_2|$, $\psi_j \in \mathcal{H}_B$, resp. $b = I_{\mathcal{H}_B}$; for possibly more details cf. [38].

Let $|\Omega_t^\psi\rangle := \exp(-itH)|0\rangle \otimes |\psi\rangle \equiv |\Omega_t(\psi)\rangle$, $|\Omega_0(\psi_j)\rangle := |0\rangle \otimes |\psi_j\rangle$. On the basis of the following elementary estimates:

$$|\omega_t^{A\&B}(P_{mn}b)| \equiv |\langle P_m \Omega_t^\psi | P_{mn} b P_n \Omega_t^\psi\rangle| \leq \|b\|\sqrt{\omega_t^{A\&B}(P_m)\omega_t^{A\&B}(P_n)} \tag{7.4.46}$$

we obtain from (7.4.45)

$$\lim_{t \to \infty} \omega_t^{A\&B}(P_{mn}b) = 0, \quad \text{for } m + n > 0. \tag{7.4.47}$$

Let us calculate now for arbitrary $\psi_{1,2} \in \mathcal{H}_B$

$$\omega_t^{A\&B}(P_0|\psi_1\rangle\langle\psi_2|) = \langle\Omega_t^\psi|\Omega_0(\psi_1)\rangle\langle\Omega_0(\psi_2)|\Omega_t^\psi\rangle. \tag{7.4.48}$$

We find, according to the notation from 7.4.3 (3.) (used now for arbitrary ψ', $\psi \in \mathcal{H}_B$), and according to the (7.4.38), that

$$\begin{aligned}
\langle\Omega_0(\psi')|\Omega_t^\psi\rangle_+ &\equiv \langle\Omega_0(\psi')|P_0 U_t P_0|\Omega_0(\psi)\rangle_+ \\
&= F_{\psi'}^0(\psi)_+(t) - \gamma^2 F_{\psi'}^0(\varphi)_+ * f_+ * F_+(t).
\end{aligned} \tag{7.4.49}$$

It follows from (7.4.44) that the right hand side of (7.4.49) converges with $t \to +\infty$ to zero, hence also the right hand side in (7.4.48) converges to zero (for all ψ_j). Hence

$$\lim_{t \to \infty} \omega_t^{A\&B}(P_0|\psi_1\rangle\langle\psi_2|) = 0, \quad \text{for all } \psi_{1,2} \in L^2(\mathbb{R}^3) = \mathcal{H}_B. \tag{7.4.50}$$

Let us note that a different situation appeared in the case $b := I_{\mathcal{H}_B}$ in which case the equation (7.4.45b) is valid.

It remained to find the limit of the expressions $\omega_t^{A\&B}(|\psi_1\rangle\langle\psi_2|) \equiv \omega_t^{A\&B}(I_{\mathcal{H}_{vac}} \otimes |\psi_1\rangle\langle\psi_2|)$. Because we are working in the time-invariant subspace $\mathcal{K} \subset \mathcal{H}_{vac} \otimes \mathcal{H}_B$, and the projection onto it is $P_{\mathcal{K}} = \sum_{m=0}^{\infty} P_m$, we shall write this sum instead of $I_{\mathcal{H}_{vac}}$ in $\omega_t^{A\&B}$. The summation over m in its argument should be done, however, before performing the limit $\lim_{t \to \infty} \omega_t^{A\&B}(P_m|\psi_1\rangle\langle\psi_2|)$.

With the help of (7.4.33), we can obtain

$$\omega_t^{A\&B}(P_m|\psi_1\rangle\langle\psi_2|) = \gamma^2 \int_0^t dt' \int_0^t dt'' \bar{F}(t')F(t'')F_\varphi^0(\psi_1)(t'-t)$$
$$\bar{F}_\varphi^0(\psi_2)(t''-t) \bar{f}_m(t-t')f_m(t-t''). \tag{7.4.51}$$

Let us introduce the functions $G(t', t'')$ **and** $g(\psi_1, \psi_2)(t', t'')$ **of** $\{t', t''\} \in \mathbb{R}^2$:

$$G(t', t'') := \bar{F}_+(t')F_+(t'');$$
$$g(\psi_1, \psi_2)(t', t'') := f_1(t''-t')F_\varphi^0(\psi_1)_-(-t')\bar{F}_\varphi^0(\psi_2)_-(-t'')$$

where e.g. $F_\varphi^0(\psi)_-(-t) := \theta(-t)F_\varphi^0(\psi)(-t)$.
A use of (7.4.40) leads us to:

$$\omega_t^{A\&B}((P_{\mathcal{K}} - P_0)|\psi_1\rangle\langle\psi_2|) = \gamma^2 \int_0^t dt' \int_0^t dt'' \bar{F}(t')F(t'')F_\varphi^0(\psi_1)(t'-t)$$
$$\bar{F}_\varphi^0(\psi_2)(t''-t) f_1(t'-t'')$$
$$= \gamma^2 G * g(\psi_1, \psi_2)(t, t), \tag{7.4.52}$$

where $*$ denotes the 2-dimensional convolution. From the given properties of the entering functions (cf. also our Lemma 7.4.6, and the L^p-estimates in [262, 298]), and with the use of (7.4.50), we obtain the desired result:

$$\lim_{t \to \infty} \omega_t^{A\&B}(|\psi_1\rangle\langle\psi_2|) = 0, \quad \psi_j \in \mathcal{H}_B. \tag{7.4.53}$$

The existence of a limit state $\bar{\omega} := w^*\text{-}\lim_{t \to +\infty} \omega_t^{A\&B}$ according to (7.4.26) is proved; its form as a product state (7.4.26) in $\mathcal{S}(\mathfrak{A} \otimes \mathfrak{B})$ can be seen by checking its values on elements of $\mathfrak{A} \otimes \mathfrak{B}$, cf. also [38] and [90, I.4.5. Proposition 2]. By comparing the definition in 7.4.3 of the no-particle state ω_0^B on $\mathfrak{B}$ with our results, and considering the results (7.4.45), (7.4.47), and (7.4.50) (together with (7.4.53)) we finally obtain:

$$\bar{w} := w^* \text{-} \lim_{t \to \infty} w_t^{A\&B} = (w \, w_\uparrow^A + (1 - w) \, w_\downarrow^A) \otimes w_0^B, \text{ with } w := \gamma^2(F_+, \ F_+ * f).$$

$$(7.4.54)$$

Let us show next that the probability w in (7.4.54) is positive and has the form

$$w = \langle \psi | W_\gamma | \psi \rangle, \text{ where } W_\gamma \in \mathcal{L}(\mathcal{H}_B), \quad 0 < W_\gamma \neq W_\gamma^2, \qquad (7.4.55)$$

where $\psi \in \mathcal{H}_B$ is the initial state-vector of the scattered particle.

Remember that the function f does not depend on the initial state ψ of the scattered particle, (7.4.8). The function $\psi \mapsto F(t) \equiv F_\varphi(\psi)(t)$, $\psi \in \mathcal{H}_B$ is, according to its definition (7.4.6b), a bounded linear functional of the initial state-vector ψ, and the same is valid for $F_+(t)$. Hence, the probability $w =: w(\psi)$ in (7.4.54) is a quadratic function of $\psi \in \mathcal{H}_B$. We can rewrite it, by applying to it the **polarization identity**, into a sesquilinear form dependent on two vectors ψ_1, $\psi_2 \in \mathcal{H}_B$ being "occasionally" chosen in the expression of $w(\psi)$ to be equal: $\psi_1 = \psi_2 \equiv \psi$. So, let us write $w(\psi) =: \mathcal{W}(\psi, \psi)$, and define:

$$\mathcal{W}(\psi_1, \psi_2) := \frac{1}{4} \sum_{\alpha = \pm i, \pm 1} \alpha \, w(\alpha \psi_1 + \psi_2) \qquad (7.4.56a)$$

which is the wanted bounded sesquilinear form on $\mathcal{H}_B$ depending on ψ_1 antilinearly; hence, it can be written as a matrix element of a bounded linear operator on $\mathcal{H}_B$. Let us denote this operator as W_γ:

$$\langle \psi_1 | W_\gamma | \psi_2 \rangle := \mathcal{W}(\psi_1, \psi_2) = \frac{1}{4} \sum_{\alpha = \pm i, \pm 1} \alpha \, \mathcal{W}(\alpha \psi_1 + \psi_2, \, \alpha \psi_1 + \psi_2), \, W_\gamma \in \mathcal{L}(\mathcal{H}_B),$$

$$(7.4.56b)$$

and we can write the probability w in the form of a diagonal element of $W \equiv W_\gamma$:

$$w \equiv w(\psi) := \gamma^2(F_\varphi(\psi)_+, \, F_\varphi(\psi)_+ * f) = \langle \psi | W_\gamma | \psi \rangle, \, \psi \in \mathcal{H}_B, \qquad (7.4.56c)$$

where the first bracket $(\cdot, \cdot)$ denotes the scalar product in $L^2(\mathbb{R})$, and the second one: $\langle \cdot | \bullet | \cdot \rangle$ is a matrix element in $\mathcal{H}_B = L^2(\mathbb{R}^3)$. If we notice that the function f from (7.4.8) entering (7.4.56c) is of positive type (because it is a diagonal matrix element of $\exp[-it(H_A + H_B)]$), cf. [262, Theorem IX.9], and if we reconsider the (commutative) convolution operation $f*$ in (7.4.56c) as a linear operator $f* \in \mathcal{L}(L^2(\mathbb{R}))$, we can immediately see that the operator W_γ is a positive operator on $\mathcal{H}_B$, $W_\gamma \geq 0$. It remains to check that the matrix element $\langle \psi | W_\gamma | \psi \rangle$ in (7.4.56c) is different from zero, if the assumptions of our Theorem are fulfilled.

To proceed further, let us rewrite the expression (7.4.56c) of w in terms of Fourier transforms.

Let us take Fourier transform of the (7.4.38b) for $F(t) = F_\varphi(\psi)(t)$. We shall use the notation[8]:

$$\hat{F}_+(u) \equiv \mathcal{F}(F_\varphi(\psi)_+(\bullet))(u) \equiv \mathcal{F}(\theta \cdot F_\varphi(\psi))(u), \qquad (7.4.57a)$$

and similarly for other functions $g_+ \mapsto \hat{g}_+$, $f_+ \mapsto \hat{f}_+$, or also

$$\mathcal{F}(F_+^0) \equiv (F_+^0)\hat{} \equiv \hat{F}_+^0 \equiv \mathcal{F}(F_\varphi^0(\psi)_+(\bullet)) \equiv \mathcal{F}(\theta \cdot F_\varphi^0(\psi)). \qquad (7.4.57b)$$

We obtain then from (7.4.38b) the transformed equation:

$$\hat{F}_+ = \hat{F}_+^0 - 2\pi\gamma^2\,\hat{g}_+\hat{f}_+\hat{F}_+, \qquad (7.4.58)$$

which can be solved immediately:

$$\hat{F}_+(u) = \frac{\hat{F}_+^0(u)}{1 + 2\pi\gamma^2\,\hat{g}_+(u)\hat{f}_+(u)}, \qquad u \in \mathbb{R}, \qquad (7.4.59a)$$

or in another form

$$\mathcal{F}(F_+)(u) = \frac{\mathcal{F}(F_+^0)(u)}{1 + 2\pi\gamma^2\,\mathcal{F}(g_+)(u)\mathcal{F}(f_+)(u)}. \qquad (7.4.59b)$$

This is the Fourier transform of the explicit expression (7.4.42) of the solution of (7.4.38) obtained with the help of Carl Neumann series.

Let us rewrite the expression (7.4.56c) for the probability $w(\psi) \equiv \langle\psi|W_\gamma|\psi\rangle$ with the help of (7.4.59) (remember the notation (7.4.6b)):

$$\langle\psi|W_\gamma|\psi\rangle = \gamma^2\,(\hat{F}_+, \hat{F}_+ \cdot \hat{f}) = \gamma^2\sqrt{2\pi}\int_\mathbb{R} du\,\hat{f}(u)\,\frac{|\hat{F}_+^0(u)|^2}{|1 + 2\pi\gamma^2\,\hat{g}_+(u)\hat{f}_+(u)|^2}. \qquad (7.4.60)$$

Let us investigate properties of the above integrand in some details. Let us express first the function $\hat{f}_+(u) = \mathcal{F}(f_1 \cdot g \cdot \theta)(u) = \sqrt{2\pi}\,\hat{f}_1 * \hat{g}_+(u) = \sqrt{2\pi}\mathcal{F}(f_1 \cdot \theta) * \hat{g}(u)$. The Fourier image $\hat{f}_1(u)$ of $f_1(t) \equiv \frac{1}{t}J_1(2t)$ can be obtained with a help of its integral representation taken from [129, 3.752-2]:

$$f_1(t) = \frac{1}{t}J_1(2t) = \frac{4}{\pi}\int_0^1 \cos(2tx)\sqrt{1 - x^2}\,dx. \qquad (7.4.61)$$

---

[8]This notation should not be confused with $\mathcal{F}(F)_+ := \theta \cdot \mathcal{F}(F) \equiv (\hat{F})_+$, differing by the place where the sign "+" occurs.

We can rewrite this expression to the forms

$$
\begin{aligned}
f_1(t) &= \frac{1}{2\pi} \int_{-2}^{2} e^{itu} \sqrt{4 - u^2}\, du \\
&= \frac{1}{2\pi} \int_{-\infty}^{+\infty} e^{itu}\, \theta(2 - |u|) \sqrt{4 - u^2}\, du \\
&= \left[\mathcal{F}^{-1} \left(\frac{1}{\sqrt{2\pi}} \theta(2 - |u|) \sqrt{4 - u^2} \right) \right](t),
\end{aligned}
\tag{7.4.62}
$$

hence, we obtain from (7.4.62) the wanted Fourier image immediately:

$$
\hat{f}_1(u) \equiv \mathcal{F}(f_1)(u) = \frac{1}{\sqrt{2\pi}} \theta(2 - |u|) \sqrt{4 - u^2}.
\tag{7.4.63}
$$

The expression (7.4.61) of f_1 leads, in agreement with its definition (7.4.7), to the estimates

$$
\begin{aligned}
|f_1(t)| &\leq \frac{4}{\pi} \int_0^1 dx\, |\cos(2tx)| \sqrt{1 - x^2} \leq \frac{4}{\pi} \int_0^1 dx\, \sqrt{1 - x^2} \\
&= \frac{4}{\pi} \int_0^{\frac{\pi}{2}} d\alpha\, \cos^2 \alpha = 1,
\end{aligned}
\tag{7.4.64}
$$

where we used the change of the integration variable $x := \sin \alpha$, the identity $\sin^2 \alpha + \cos^2 \alpha \equiv 1$, and the symmetry properties of the goniometric functions. Since both functions f_1, g are continuous, $g(t) = (\varphi, \exp(-it H_B)\varphi)$, $\varphi \in \mathcal{D}(\mathbb{R}^3) \Rightarrow$ the Fourier image $\hat{\varphi} \in \mathcal{S}(\mathbb{R}^3)$ is an entire analytic function of three complex variables [262, Theorem IX.12], the function $t \mapsto g(t) \neq 0$ (a.e. for $t \in \mathbb{R}$) according to Lemma 7.4.4, and the continuous function $f_1(t)$ is not constant, hence the function $|f_1(t)| < 1$ on certain intervals of $\mathbb{R}$, the estimate for L^1-norms gives:

$$
\|f\|_1 \equiv \|f_1 \cdot g\|_1 < \|g\|_1,
\tag{7.4.65}
$$

hence we have here obtained the sharp inequality. From the definition of the Fourier transformation it is seen that the following trivial inequality is valid for any function $h \in L^1(\mathbb{R})$:

$$
\|\hat{h}\|_\infty \leq \frac{1}{\sqrt{2\pi}} \|h\|_1.
\tag{7.4.66}
$$

These considerations give an estimate for the denominator in (7.4.60) by

$$
\begin{aligned}
\|2\pi\gamma^2 \hat{g}_+ \hat{f}_+\|_\infty &\leq 2\pi\gamma^2 \|\hat{g}_+\|_\infty \|\hat{f}_+\|_\infty \leq \gamma^2 \|g_+\|_1 \|f_+\|_1 \\
&= \frac{\gamma^2}{4} \|g\|_1 \|f\|_1 < \frac{\gamma^2}{4} \|g\|_1^2.
\end{aligned}
\tag{7.4.67}
$$

This proves, also due to the condition $\|\gamma g\|_1 < 2$ in (7.4.25), that the denominator of the integrand in (7.4.60) is everywhere different from zero and finite.

Another part of the integrand in (7.4.60) is the function $\hat{f} = \mathcal{F}(f_1 \cdot g) = \frac{1}{2\pi} \hat{f}_1 * \hat{g}$. The Fourier image of $g(t) \equiv \langle \varphi| \exp(-it H_B)\varphi \rangle$ is

$$
\begin{aligned}
\hat{g}(u) &= \frac{1}{\sqrt{2\pi}} \int_{-\infty}^{+\infty} e^{-itu} \langle \varphi | e^{-it H_B} \varphi \rangle \, dt \\
&= \frac{1}{\sqrt{2\pi}} \int_{-\infty}^{+\infty} dt \, e^{-itu} \int_{0}^{+\infty} d\lambda \, e^{-it\lambda} \langle \varphi | E_{H_B}(\lambda) | \varphi \rangle, \quad (7.4.68a)
\end{aligned}
$$

where $E_{H_B}(\lambda) \equiv E_{H_B}((-\infty, \lambda])$ is the projection-measure of the selfadjoint operator H_B. Because the spectrum of H_B is positive (and absolutely continuous with respect to Lebesgue measure on $\mathbb{R}$), and the function $g(t)$ is proportional to the Fourier image of $\lambda \mapsto \langle \varphi | E_{H_B}(\lambda) | \varphi \rangle$, one has

$$
\hat{g}(u) = \theta(-u)\mathcal{F}(g)(u) = \sqrt{2\pi} \langle \varphi | E_{H_B}(-u) | \varphi \rangle. \quad (7.4.68b)
$$

This can be rewritten in the "p-representation", which allows us to see better the dependence on the specific functions φ. We shall write the element of the solid angle ϕ in terms of the Euler angles θ, φ in $\mathbb{R}^3$ as $d\phi := \sin\theta \, d\theta \, d\varphi$, and the function $\hat{\varphi}(\mathbf{p}) \equiv \hat{\varphi}(p, \phi)$ ($p := |\mathbf{p}|$). It is

$$
\begin{aligned}
g(t) &= \langle \varphi | e^{-it H_B} \varphi \rangle = \int_{\mathbb{R}^3} d^3\mathbf{p} \, \overline{\hat{\varphi}(\mathbf{p})} e^{-itp^2} \hat{\varphi}(\mathbf{p}) \\
&= \int_{0}^{+\infty} dp \, p^2 e^{-itp^2} \int_{4\pi} d\phi \, |\hat{\varphi}(p, \phi)|^2, \quad (7.4.68c)
\end{aligned}
$$

which, after the change of variables $\lambda := p^2$, leads to

$$
g(t) = \frac{1}{2} \int_{0}^{+\infty} d\lambda \, \sqrt{\lambda} e^{-it\lambda} \int_{4\pi} d\phi \, |\hat{\varphi}(\sqrt{\lambda}, \phi)|^2; \quad (7.4.68d)
$$

this has the form of the Fourier image of

$$
\mathcal{F}^{-1}(g)(\lambda) := \theta(\lambda)\sqrt{\frac{\pi}{2}} \sqrt{\lambda} \int_{4\pi} d\phi \, |\hat{\varphi}(\sqrt{\lambda}, \phi)|^2, \quad (7.4.68e)
$$

and the Fourier image $\hat{g}$ has now the form

$$
\hat{g}(u) = \hat{g}(u)\,\theta(-u) = \mathcal{F}^{-1}(g)(-u) = \theta(-u)\sqrt{\frac{\pi}{2}} \sqrt{-u} \int_{4\pi} d\phi \, |\hat{\varphi}(\sqrt{-u}, \phi)|^2. \quad (7.4.68f)
$$

Similar considerations could be applied also to $f(t) \equiv \langle \Omega_1 \otimes \varphi | \exp(-it(H_A + H_B)) | \Omega_1 \otimes \varphi \rangle$; the spectrum of H_A from (7.4.1) acting on the Hilbert space $\mathcal{H}_{vac}$ of the used representation consists of a single eigenvalue $\{0\}$, and of absolutely continuous part consisting of the interval $[-2, +2] \subset \mathbb{R}$, which can be seen from the Sect. 7.3, and from [36]. So the function $f(t) = \langle \Omega_1 | \exp(-it H_A) | \Omega_1 \rangle \cdot \langle \varphi | \exp(-it H_B) | \varphi \rangle = f_1(t)g(t)$ has the Fourier image $\hat{f}(u) = (2\pi)^{-\frac{1}{2}} \hat{f}_1 * \hat{g}(u)$, which with the help of (7.4.63) and (7.4.68f) gives

$$\hat{f}(u) = \frac{1}{\sqrt{2\pi}} \hat{f}_1 * \hat{g}(u) = \frac{1}{\sqrt{2\pi}} \int d\tau \, \hat{f}_1(\tau) \hat{g}(u - \tau) \tag{7.4.69}$$

$$= \frac{1}{2\pi} \int_{-2}^{2} d\tau \sqrt{4 - \tau^2} \sqrt{\frac{\tau - u}{2}} \, \theta(\tau - u) \int_{4\pi} d\phi \, |\hat{\varphi}(\sqrt{\tau - u}, \phi)|^2.$$

Remember that $\varphi \in \mathcal{D}(\mathbb{R}^3)$, hence its Fourier image $\hat{\varphi} \in \mathcal{S}(\mathbb{R}^3)$ is an entire analytic function of three complex variables, so that the function $p \mapsto \int_{4\pi} d\phi$ $|\hat{\varphi}(p, \phi)|^2 > 0$, a.e. for $p > 0$. Then (7.4.69) implies that $\hat{f}(u) = 0$ for $u > 2$, and $\hat{f}(u) > 0$ for almost all $u < 2$.

For checking finally the conditions of the positivity of $w(\psi)$ from its expression (7.4.60), we have to check under which conditions it is $|\hat{F}_+^0(u)| \theta(2 - u) > 0$, $u \in S \subset \mathbb{R}$, for some S of positive Lebesgue measure.

Let us assume that $\hat{F}_+^0(u) \equiv 0$ in some nonzero interval: $u \in I \subset \mathbb{R}$. The function $\hat{F}_+^0(u) \equiv \frac{1}{\sqrt{2\pi}} \int_0^{+\infty} e^{-itu} \langle \varphi | \exp(-it H_B) | \psi \rangle \, dt$, cf. (7.4.6a), can be continued to a function analytic in the lower complex half plane $\text{Im } u < 0$ and continuous on the real axis $\mathbb{R}$. The identical vanishing of this function on an interval $I \subset \mathbb{R}$ would imply (with the help of the Schwarz Reflection Principle) its analyticity on I, and consequent vanishing everywhere in the analyticity domain, hence also on the whole real axis (i.e. vanishing also on the boundary of the analyticity domain). The identical vanishing $\hat{F}_+^0(u) \equiv 0$, $\forall u \in \mathbb{R}$, would imply, however, the identical vanishing $\langle \varphi | \exp(-it H_B) | \psi \rangle \equiv 0$, which contradicts (7.4.24). This proves that, for $\gamma^2 > 0$ satisfying (7.4.25), it is $\langle \psi | W_\gamma | \psi \rangle > 0$, iff ψ satisfies (7.4.24). Since the condition (7.4.24) does not depend on the parameter γ, the subspace of $\mathcal{H}_B$ consisting of those vectors ψ for which it is $\langle \psi | W_\gamma | \psi \rangle = 0$ does not depend on γ, hence also **its orthogonal complement $\mathcal{H}_W \subset \mathcal{H}_B$ is independent** of γ, cf. Lemma 7.4.4.

It remains to show that, at least for some values of $\gamma \in \mathbb{R}$, **it is $W_\gamma^2 \neq W_\gamma$**, i.e. that **the positive operator W_γ is not a projector**. For any nonzero orthogonal projector $P \in \mathcal{L}(\mathcal{H})$ there exists a subspace $P\mathcal{H} \equiv \mathcal{H}_P \subset \mathcal{H}$ such that for any normalized vector $\psi \in \mathcal{H}_P$ it is $\langle \psi | P | \psi \rangle = 1$, and for all vectors ψ from its orthogonal complement: $\psi \in \mathcal{H}_P^\perp := \mathcal{H} \ominus \mathcal{H}_P$, it is $\langle \psi | P | \psi \rangle = 0$. If an operator W_γ would be a nonzero projector, for all the normalized vectors $\psi \in \mathcal{H}_W$ it would be $\langle \psi | W_\gamma | \psi \rangle = 1$. Such a ψ would necessarily satisfy (7.4.24), and then $\langle \psi | W_\gamma | \psi \rangle > 0$ for any γ satisfying (7.4.25).

For any given normalized ψ satisfying (7.4.24), the numerical function $\gamma^2 \mapsto \langle \psi | W_\gamma | \psi \rangle$ expressed in (7.4.60) is continuous and monotonically increasing in a nonzero interval $\gamma^2 \in [0, \gamma_0^2] \subset \mathbb{R}$. For an arbitrary normalized $\psi \in \mathcal{H}_B$, it is $\langle \psi | W_{\gamma=0} | \psi \rangle = 0$, and it is $0 < \langle \psi | W_\gamma | \psi \rangle < 1$ for all sufficiently small $|\gamma| > 0$ and all normalized $\psi \in \mathcal{H}_B$. Hence, at least for sufficiently small nonzero $\gamma \in \mathbb{R}$, it is $\langle \psi | W_\gamma | \psi \rangle \neq 1$ for normalized $\psi \in \mathcal{H}_W$, so that $W_\gamma^2 \neq W_\gamma$, i.e. the positive operator W_γ is not a projector. The theorem is proved. $\square$

7.5 The X-Y Chain as a Measuring Device

7.5.1 The X-Y chain

Let us formulate first what we understand here under the "X-Y chain" (cf. [267], and also [35, 107, 271])—a special case of the Heisenberg spin chains:

It is again a model of one-dimensional spin chain with C^*-algebra of observables $\mathfrak{A}$ generated by spin creation-annihilation operators a_j^*, a_j ($j \in \mathbb{Z}$), as it was introduced in 7.3.2. The algebra $\mathfrak{A}$ is the C^*-inductive limit of the sequence of its local subalgebras $\mathfrak{A}_n$ ($n \in \mathbb{N}$), each generated by a_j^*, a_j ($|j| \leq n$). The dynamics in any subalgebra $\mathfrak{A}_n$ is given by the **local Hamiltonian** H_n (without interaction with external magnetic field):

$$H_n := \frac{\kappa}{2} \sum_{j=-n}^{n-1} (a_j^* a_{j+1} + a_{j+1}^* a_j), \tag{7.5.1}$$

where $\kappa \in \mathbb{R}$. These local Hamiltonians define the time-evolution $(t; x) \mapsto \tau_t^{(n)}(x)$ of local elements $x \in \mathfrak{A}_n$:

$$\tau_t^{(n)}(x) := e^{it H_n} x e^{-it H_n}, \quad x \in \mathfrak{A}_n, \; n \in \mathbb{N}, \; t \in \mathbb{R}. \tag{7.5.2}$$

The evolution in the whole algebra $\mathfrak{A}$ is obtained by taking first the limit $n \to \infty$ in norm of $\mathfrak{A}$ for any fixed $t \in \mathbb{R}$ and any local $x \in \mathfrak{A}$, and afterwards obtaining the result by the norm-continuity, extending it to all $x \in \mathfrak{A}$:

$$\tau_t(x) := \text{n-} \lim_{n \to \infty} \tau_t^{(n)}(x). \tag{7.5.3}$$

Note that the term "X-Y model" comes from the form of the hamiltonian if it is rewritten in the terms of **Pauli** σ-matrices : $\sigma_j^x := a_j^* + a_j$, $\sigma_j^y := ia_j - ia_j^*$, $\sigma_j^z := 2a_j^* a_j - 1$, i.e.

$$H = \frac{\kappa}{4} \sum_j (\sigma_j^x \sigma_{j+1}^x + \sigma_j^y \sigma_{j+1}^y). \tag{7.5.4}$$

We shall write often H instead of H_n, also without specifying the local characters of the entering algebraic elements x, or $A \in \mathfrak{A}$, ..., to simplify the notation and the corresponding comments; the reader could easily add the necessary specifications on his own.

We shall use the known formula to express the automorphism (7.5.3):

$$e^{itH} A e^{-itH} = \sum_{m=0}^{\infty} \frac{(it)^m}{m!} [H, A]^{(m)}, \tag{7.5.5}$$

where $[H, A]^{(0)} := A$, and higher elements are recurrently defined with a help of the commutator $[H, A]^{(1)} := [H, A] \equiv HA - AH$:

$$[H, A]^{(m+1)} := [H, [H, A]^{(m)}]. \tag{7.5.6}$$

The application of (7.5.5) to norm-bounded elements A (with also $H \hookrightarrow H_n$) makes no principal problems, but calculations of time evolved elements in (7.5.5) of e.g. $A \hookrightarrow a_j$ is technically complicated and it is much easier to work, instead with the spin operators a_j, with elements $b_j \in \mathfrak{A}$ satisfying the Fermi **canonical anticommutation relations** (*CAR*). This can be reached by the **Jordan-Wigner transformation** ([171], and also [106, Chaps. 3, 2]):

$$b_j := a_j \prod_{k=-n-1}^{j-1} (1 - 2a_k^* a_k), \ b_j^* := (b_j)^*, \tag{7.5.7}$$

for $|j| \leq n$. Although these elements become to be nonlocal with $n \to \infty$, their bilinear combinations remain local, and this is sufficient for our calculations. Note also that there is the inverse transformation expressing a_j in terms of b_j, which has the same form as (7.5.7) after the exchange $a_{j,k} \leftrightarrow b_{j,k}$.

The elements $b_j, b_k, \ j, k \in [-n, n]$ satisfy CAR:

$$[b_j, b_k]_+ \equiv 0, \ b_j b_k^* + b_k^* b_j =: [b_j, b_k^*]_+ = \delta_{jk}. \tag{7.5.8}$$

The local Hamiltonians H_n from (7.5.1) can be written now as

$$H_n = \frac{\kappa}{2} \sum_{j=-n}^{n-1} (b_j^* b_{j+1} + b_{j+1}^* b_j). \tag{7.5.9}$$

We can calculate now the time evolution of the elements $b_j \in \mathfrak{A}$. We shall need later the estimates for $\tau_t(a_j^* a_j)$, and due to equality $a_j^* a_j = b_j^* b_j$ the explicit expressions for $\tau_t(b_j)$ will be sufficient for us. We can use (7.5.5) to calculate $\tau_t(b_j)$. One easily checks that the multiple commutators have the form:

$$[H, b_j]^{(m)} = \sum_p c_j^{(m)}(p) b_p, \tag{7.5.10}$$

where the c-number coefficients $c_j^{(m)}(p)$ ($m \in \mathbb{Z}_+$, $j, p \in \mathbb{Z}$) satisfy following recurrent relations:

$$c^{(m+1)}(p) = -\frac{\kappa}{2}(c^{(m)}(p-1) + c^{(m)}(p+1)), \qquad (7.5.11)$$

$$\text{where} \quad c^{(0)}(p) = \delta_{0p}, \quad c^{(m)}(j-p) \equiv c_j^{(m)}(p).$$

It is seen that the coefficients $c_j^{(m)}(p)$ depend on $p - j$ only: they are expressible as linear combinations of the Kronecker deltas $\delta_{j,p+c}$. Notice also that $c^{(m)}(-p) = c^{(m)}(p)$, $\forall p \in \mathbb{Z}$. Note moreover that for each $m \geq 0$ only finite number of the coefficients $c^{(m)}(j-p)$ is nonzero.

From (7.5.5) and (7.5.10) we have:

$$\tau_t(b_j) = \sum_{k \in \mathbb{Z}} C_t(j-k)\, b_k, \qquad (7.5.12a)$$

where

$$C_t(r) := \sum_{m=0}^{\infty} \frac{(it)^m}{m!}\, c^{(m)}(r). \qquad (7.5.12b)$$

The Bessel functions of the first kind $J_r(t)$, $r \in \mathbb{Z}_+$, $t \in \mathbb{R}$, can be expressed by the power series:

$$J_r(t) = \sum_{k=0}^{\infty} (-1)^k \left(\frac{t}{2}\right)^{2k+r} \frac{1}{k!(r+k)!}. \qquad (7.5.13)$$

By calculation of coefficients $c^{(m)}(r)$ in (7.5.12b) with the help of (7.5.11) and by comparison of coefficients at equal powers t^m of the variable $t \in \mathbb{R}$ in the expressions (7.5.12b) for $C_t(r)$ and in (7.5.13) for $J_r(t)$, we can see that for $r \in \mathbb{Z}_+$ it is

$$C_t(r) \equiv (-i)^r J_r(\kappa t). \qquad (7.5.14)$$

After inserting this into (7.5.12a) (keep in mind that $C_t(-r) = C_t(r) = (-i)^{|r|} J_{|r|}(\kappa t)$) we obtain explicit expression for time evolution of elements $b_j \in \mathfrak{A}$, hence the time-automorphism group τ_t, $t \in \mathbb{R}$, of $\mathfrak{A}$ in terms of standard special functions J_r, $r \in \mathbb{Z}_+$.

7.5.2 Interaction with a small system. Let us use the just described X-Y spin chain to construction of an alternative "model of quantum measurement" now.

Let us represent the algebra $\mathfrak{A}$ in a subspace of the CTPS = $\otimes_{j \in \mathbb{Z}} \mathbb{C}_j^2$ (cf. 5.1.3) corresponding to the product-vector Ψ_0 defined as follows: Let the spins on our chain be well ordered and numbered by $j \in \mathbb{Z}$. Let $|\pm j\rangle$ be the states of the j-th spin being eigenvectors of the Pauli matrix σ_j^z corresponding to the up-, resp. down-orientations: $\sigma_j^z |\pm j\rangle = \pm |\pm j\rangle$. Let then

$$\Psi_0 := \bigotimes_{j \leq -1} | + j\rangle \otimes \bigotimes_{k \geq 0} | - k\rangle. \tag{7.5.15}$$

Let the Hamiltonian of this chain be

$$H_0 := \frac{\kappa}{2} \sum_{j \leq -2} (a_j^* a_{j+1} + a_{j+1}^* a_j) + \frac{\kappa}{2} \sum_{k \geq 0} (a_k^* a_{k+1} + a_{k+1}^* a_k), \tag{7.5.16}$$

which is the Hamiltonian of the X-Y model without the term $(a_{-1}^* a_0 + a_0^* a_{-1})$. This chain with the Hamiltonian H_0 will play for us the role of the "macroscopic (measuring) system". The state described by the vector Ψ_0 is stationary for this Hamiltonian:

$$H_0 \Psi_0 = 0. \tag{7.5.17}$$

The "measured microsystem" will be an additional $1/2$-spin (i.e. it does not belong to the chain) with the interaction Hamiltonian

$$V := P_+ \otimes \frac{\kappa}{2} (a_{-1}^* a_0 + a_0^* a_{-1}), \tag{7.5.18}$$

where P_+ is the projector in the state space $\mathbb{C}^2$ of the added spin-microsystem projecting onto the state $|+\rangle$ in which the spin "is pointing up": $\sigma^z |+\rangle = |+\rangle$.[9] If we write (in microsystem's state space $\mathbb{C}^2$) $P_- := I - P_+$, the total Hamiltonian $\tilde{H}$ of our compound system "micro & macro" reads:

$$\tilde{H} = H_0 + V = H P_+ + H_0 P_-, \tag{7.5.19}$$

where H is the total Hamiltonian of the X-Y model (7.5.4). Let the initial state of the compound system be

$$\Phi_0 := \varphi_0 \otimes \Psi_0, \quad \varphi_0 := c_+ |+\rangle + c_- |-\rangle, \tag{7.5.20}$$

where φ_0 is normalized: $|c_+|^2 + |c_-|^2 = 1$, and $|\pm\rangle$ are also normalized eigenvectors of $\sigma^z \in \mathcal{L}(\mathbb{C}^2)$:

$$P_\pm |\pm\rangle = |\pm\rangle, \quad P_+ P_- = 0. \tag{7.5.21}$$

Since, in accordance with (7.5.17),

$$\tilde{H}(|+\rangle \otimes \Psi_0) = |+\rangle \otimes H\Psi_0, \quad \tilde{H}(|-\rangle \otimes \Psi_0) = 0, \tag{7.5.22}$$

the time evolution looks like:

---

[9]We shall omit usually in the following the tensor-product symbol $\otimes$, according our preceding conventions.

$$\Phi_t := e^{-it\tilde{H}} \Phi_0 = c_+ |+\rangle \otimes e^{-itH} \Psi_0 + c_- |-\rangle \otimes \Psi_0. \qquad (7.5.23)$$

We shall show, similarly as in 7.3, that the pure state state vector Φ_t of the compound system converges in the limit $t \to \infty$ to the incoherent linear combination of two vectors, corresponding to two disjoint states of the compound system (as well as of the macrosystem-chain); hence this limit is a vector which describes a mixture of two macroscopically distinct states of the system. It is sufficient to check this assertion by calculation of the quantities

$$\tilde{\omega}_t(a_j^* a_j) := \langle \Phi_t | a_j^* a_j | \Phi_t \rangle \quad \text{for } j \in \mathbb{Z}, \qquad (7.5.24)$$

i.e. of

$$\tilde{\omega}_t(a_j^* a_j) = |c_+|^2 \, \langle \Psi_0 | \, \tau_t(a_j^* a_j) \, | \Psi_0 \rangle + |c_-|^2 \, \langle \Psi_0 | \, a_j^* a_j \, | \Psi_0 \rangle; \qquad (7.5.25)$$

here, the automorphisms τ_t are expressed in (7.5.12a).

It can be proved now that the limit $\bar{\omega}(A) := \lim_{t \to \infty} \langle \Psi_0 | \tau_t(A) | \Psi_0 \rangle$, $A \in \mathfrak{A}$, of a state from (7.5.25) exists, and the states $\bar{\omega}$, $\omega_0 \in \mathcal{S}(\mathfrak{A})$:

$$\bar{\omega}(A) := \lim_{t \to \infty} \omega_t(A) \equiv \lim_{t \to \infty} \langle \Psi_0 | \tau_t(A) | \Psi_0 \rangle, \quad \omega_0(A) := \langle \Psi_0 | A | \Psi_0 \rangle, \quad A \in \mathfrak{A}, \qquad (7.5.26)$$

are mutually disjoint and macroscopically distinct. We shall prove now existence of the limits (7.5.26) in (7.5.25) for $A = a_j^* a_j$. It is

$$\omega_0(a_j^* a_j) = \begin{cases} 1 & \text{for } j \leq -1, \\ 0 & \text{for } j \geq 0. \end{cases} \qquad (7.5.27)$$

Since according to (7.5.7) it is $a_j^* a_j = b_j^* b_j$, we can use (7.5.12a) to obtain:

$$\tau_t(a_j^* a_j) = \tau_t(b_j^*) \tau_t(b_j) = \sum_{r,s} \overline{C_t}(j - r) C_t(j - s) \, b_r^* b_s =$$

$$= \sum_{r,s} \overline{C_t}(j - r) C_t(j - s) \, a_r^* \left\{ \prod_{q=\min[r,s]}^{\max[r-1,s-1]} (1 - 2a_q^* a_q) \right\} a_s, \qquad (7.5.28)$$

where the products $\prod_n^m B_q := 1$ if $m < n$. Hence

$$\omega_0(\tau_t(a_j^* a_j)) = \sum_{r,s} \overline{C_t}(j - r) C_t(j - s) \, \omega_0 \left(a_r^* \left\{ \prod_{q=\min[r,s]}^{\max[r-1,s-1]} (1 - 2a_q^* a_q) \right\} a_s \right), \qquad (7.5.29)$$

and due to the properties (7.5.15) and (7.5.27) of w_0 and due to commutation properties of the a_j, a_k^* we see that the terms with $r \neq s$ are zeros. According to (7.5.14) we have:

$$w_t(a_j^* a_j) = \sum_{r=-\infty}^{+\infty} |C_t(j-r)|^2 w_0(a_r^* a_r) = \sum_{r=1}^{+\infty} |C_t(j+r)|^2$$

$$= \sum_{r=1}^{+\infty} J_{j+r}^2(\kappa t) \equiv \sum_{r=1}^{+\infty} J_{|j+r|}^2(\kappa t). \tag{7.5.30}$$

According to the known formula [180, (21.8-26)]:

$$1 = J_0^2(z) + 2 \sum_{k=1}^{+\infty} J_k^2(z), \tag{7.5.31}$$

and due to the asymptotic behaviour of Bessel functions

$$J_m(t) \asymp O(t^{-\frac{1}{2}}), \quad m \in \mathbb{Z}, \tag{7.5.32}$$

we have finally

$$\overline{w}(a_j^* a_j) := \lim_{t \to +\infty} w_t(a_j^* a_j) = \frac{1}{2}, \quad \text{for all} \quad j \in \mathbb{Z}. \tag{7.5.33}$$

Returning to the formulas (7.5.24) and (7.5.25) of our main interest, we have obtained:

$$\widetilde{w}_\infty(a_j^* a_j) := \lim_{t \to +\infty} \widetilde{w}_t(a_j^* a_j) = |c_+|^2 \overline{w}(a_j^* a_j) + |c_-|^2 w_0(a_j^* a_j). \tag{7.5.34}$$

The last formula describes an (incoherent) mixture of two mutually macroscopically distinct, hence disjoint states w_0, $\overline{w}$ on the C^*-algebra $\mathfrak{A}$ of the infinite spin chain. This can be checked in the explicit way by calculating values of a macroscopic observable in the states w_0, resp. $\overline{w}$, e.g. of the observable constructed from (7.3.29)

$$\gamma := w\text{-} \lim_{N \to \infty} \frac{1}{N} \sum_{n=1}^{N} a_n^* a_n \in \mathfrak{Z}(\mathfrak{A}^{**}) \subset \mathfrak{A}^{**}. \tag{7.5.35}$$

According to (7.5.33) and (7.5.27), it is:

$$\overline{w}(\gamma) = \frac{1}{2} \neq w_0(\gamma) = 0. \tag{7.5.36}$$

Hence, again here, a microscopic system interacting with the macroscopic X-Y chain changed the chain's initial state ω_0 into a new, macroscopically distinct state $\widetilde{\omega}_\infty = |c_+|^2\overline{\omega} + |c_-|^2\omega_0$. Here the probabilities $|c_\pm|^2$ of occurrence of the mutually *disjoint* states ω_0, $\overline{\omega}$ in the proper (resp. 'genuine', cf. 1.1.4) mixture $\widetilde{\omega}_\infty$ are exactly the probabilities of appearing of the states $|\pm\rangle$ of the microsystem in its initial state φ_0, cf. (7.5.20). This corresponds again to the *"ideal measurement"*, as it was discussed in 7.1.3, 7.3.6 and 7.4.2.

7.6 Radiating Finite Spin Chain

7.6.1 We shall present very briefly in this section, without proofs, the dynamics of a model of a large but finite system interacting with a Fermi field.[10] The system's initial state is stationary but unstable, as it was also the case of the models presented in the preceding sections. After an initial perturbation, the model evolves quickly into a new stationary state by simultaneous radiation of a Fermi particle, which escapes into infinity. The process is very quick in contrast to the time evolutions in the case of the models described in the previous Sects. 7.3, 7.4 and 7.5. The three preceding models might, however, serve as clear mathematical pictures of "quantum measurement" in the sense that the time evolution of a large system led with the time growing to infinity to the state "macroscopically different" from its initial state. The "macroscopic difference" between states of the system is mathematically expressed there as disjointness of the states on the C^*-algebra of observable quantities of the large system. The disjointness implies that if those states are represented as vectors in a Hilbert space, their mutual linear combinations do not lead to any interference (the C^*-algebra of observables representing all possible observations on the model system is fixed!) and such a linear combination is physically equivalent to a "proper mixture", or "genuine mixture" (cf. 1.1.4), i.e. to a classical statistical description of an ensemble in which the individual copies of the large system are distributed between the uniquely determined 'classical' states under consideration. This unique decomposability to pure states on the algebra of classical (macroscopic) observables is a consequence of the fact that the states of a classical system form a simplex. This differs from "mixed" quantum states described by density matrices of standard QM of finite-size systems having multiple convex decompositions to extremal (pure) states.

Since the model of a "large" system described in this section is finite (corresponding by physical intuition to that consisting of finite number of some "elementary" or "small" subsystems, each of them described by elementary QM in separable Hilbert space $\mathcal{H}_v$ with the algebra of its observables coinciding with the whole $\mathcal{L}(\mathcal{H}_v)$), there is no possibility of emergence of any disjoint states, hence there is no unambiguously

---

[10]The formulation and main features of the dynamics of this model were presented first time in [33]. The technical details are described in [39].

defined "macroscopic difference" between some of its states.[11] Of course, the infinite size of the previous models is a mathematical idealization, and there should be some empirical possibility of distinction between "microscopic" and "macroscopic", resp. between "quantum" and "classical", also in 'large but finite systems', as it is perceived in our everyday life.[12]

This distinction does not need to be, however, mathematically sharp. Such a possibility was sketched in [153]: In a verbal transcription it could be, perhaps, formulated so that it would be very improbable to construct such an observation device on states of large (however finite) system, which could "see" simultaneously sufficiently many atoms of the system to be able to detect some interference phenomenon. This could be considered as a rough 'definition' of the notion that some set of states of the (now finite) apparatus consists of elements being pairwise 'almost macroscopically different' (cf. also [153]).[13] To proceed in these considerations, one would need to build some (more) general theory of observational devices. E.g., as far as the present author knows, there were no published works paying attention to the fact that human observers come into contact with measuring apparatuses by electromagnetic interactions, and probably only by them. Shortly, according to the point of view proposed here: The formalized set of "observables" of any physical system should depend on the existing possibilities of the construction of measuring devices in accordance with physical laws and environmental conditions.

We have not stressed up to now, however, that the spin chain of our present model is also coupled to a Fermi particle (resp. to the Fermi field) representing a sort of **'environment'**. The particle occurs in the initial state of the system in its vacuum state, and afterwards it is radiated by the chain and subsequently escapes into infinity; the state of the Fermi field containing the radiated particle is in each finite time orthogonal to its vacuum state. This facilitates, in the intuitive sense of some sort of a 'decoherence program', cf. e.g. [124, 279, 343, 347], the possibility of interpretation of the effective absence of interference between the initial and final states of the spin chain in our model, as representing the two different 'macroscopically' distinguished 'pointer positions'.

---

[11] An exception consists in possible introduction 'by hand' by a theoretician some 'superselection rules' representing a model of 'macroscopic difference' and forbidding interference between vectors from specific subspaces of $\mathcal{H}_v$, cf. e.g. [167].

[12] Another possibility is some, up to now not clearly specified basic change of QM, as it was most urgently proposed by Penrose in several his publications, e.g. in [236–238]; the main motivation for these reformulations of QM was some inclusion of the usually postulated "reduction of wave packet" [226], called by Penrose the **"process R"**, into the dynamics of general QM systems.

[13] Let us illustrate briefly this idea on a long but finite spin-1/2 chain of the length N with the C^*-algebra $\mathfrak{A}$ of its observables generated by the spin creation-annihilation operators a_j, a_j^* ($j = 1, 2, \ldots N$) acting on the finite dimensional Hilbert space $\mathcal{H}_N := (C^2)^N$: If we are able to use apparatuses detecting the observables of this chain occurring in an arbitrary of the C^*-subalgebras $\mathfrak{B} \subset \mathfrak{A}$ generated by any of the fixed restricted set of operators $a_{j_m}, a_{j_m}^*$ ($m = 1, 2, \ldots K \ll N, 0 \leq j_m \leq N$) **only**, then the states $|\Psi\rangle, |\Phi\rangle$ from $\mathcal{H}_N$ for which it holds $\langle\Psi|B|\Phi\rangle \equiv 0 \,\forall B \in \mathfrak{B}$ could be considered as 'almost macroscopically different', resp. 'empirically disjoint'. This happens, e.g., if in the state $|\Psi\rangle$ all the spins are 'pointing up', and in the state $|\Phi\rangle$ all the spins are 'pointing down'.

We shall keep in mind such an idea to be able to believe that also our finite system described in this section can be considered as a model of "quantum measurement" process.

7.6.2 Let us look at the Quantum Domino from Sect. 7.3. We shall restrict here that model to finite number of degrees of freedom, hence the spin chain will be of finite length and its algebra of observables $\mathfrak{A}$ (with unity $I_{\mathfrak{A}}$) is generated by the spin-1/2 creation and annihilation operators a_j^*, a_j ($j = 0, 1, \ldots, N$) satisfying (7.3.1). This system will interact with the (nonrelativistic scalar) Fermi field, the **algebra** $\mathfrak{F}$ (with unity $I_{\mathfrak{F}}$) of which is generated by the **particle creation-annihilation operators** $b^*(\varphi)$, $b(\varphi)$ satisfying the relations

$$b(\varphi)^2 = 0, \ b(\varphi)b^*(\psi) + b^*(\psi)b(\varphi) = (\varphi, \psi)I_{\mathfrak{F}}, \ (\text{for all } \varphi, \psi \in L^2(\mathbb{R}^3, d^3x)), \tag{7.6.1}$$

with the linear dependence $\psi \mapsto b^*(\psi)$.

The dynamics is given by the Hamiltonian $H := H_0 + V$, where

$$H_0 := \left(\sum_{n=0}^{N-2} a_n^* a_n (a_{n+1}^* + a_{n+1}) a_{n+2} a_{n+2}^* - \varepsilon_0 a_N^* a_N \right) \otimes I_{\mathfrak{F}} + I_{\mathfrak{A}} \otimes d\Gamma(\boldsymbol{h}), \tag{7.6.2a}$$

$$V := v^2 \left(a_{N-1}^* a_{N-1} a_N^* \otimes b^*(\sigma) + a_{N-1}^* a_{N-1} a_N \otimes b(\sigma) \right). \tag{7.6.2b}$$

We can consider these algebras $\mathfrak{A}$ and $\mathfrak{F}$ as algebras of operators acting on the Hilbert space $\mathcal{H}_S := (\mathbb{C}^2)^{N+1}$, and on the Fermi Fock space $\mathcal{H}_F$ respectively, resp. on their tensor product $\mathcal{H} := \mathcal{H}_S \otimes \mathcal{H}_F$. In the above written formulas, the **symbol** $d\Gamma(\boldsymbol{h})$ **means the "second quantization"** (cf. [54, Sect. 5.2.1][14]) of the operator $\boldsymbol{h} \in \mathcal{L}(\mathfrak{h}) := \mathcal{L}(L^2(\mathbb{R}^3, d^3x))$ given by the function $\mathbf{p} \mapsto \varepsilon(\mathbf{p})$ of one-particle momentum $\mathbf{p}$, hence acting on the vectors of $\mathfrak{h} := L^2(\mathbb{R}^3, d^3x)$ "in the p-representation" as multiplication by $\varepsilon(\mathbf{p}) : (\boldsymbol{h}\psi)(\mathbf{p}) \equiv \varepsilon(\mathbf{p})\psi(\mathbf{p})$. The nonnegative function $\varepsilon(\mathbf{p})$, as well as the parameters $\varepsilon_0 > 0$, $v \in \mathbb{R}$, $\sigma \in L^2(\mathbb{R}^3, d^3x)$, will be specified later. In our expressions of action of elements of $\mathfrak{A}$, resp. $\mathfrak{F}$, on vectors of $\mathcal{H}_S \otimes \mathcal{H}_F$, the unity operators of the other algebra will be usually omitted, e.g. for $a \in \mathfrak{A}$, $|s\rangle \otimes |\varphi\rangle \in \mathcal{H}_S \otimes \mathcal{H}_F$, we shall write $a \otimes I_{\mathfrak{F}}(|s\rangle \otimes |\varphi\rangle) \equiv a(|s\rangle \otimes |\varphi\rangle) \equiv a|s\rangle \otimes |\varphi\rangle$.

Let Ω_0^F be the **Fermi vacuum** in $\mathcal{H}_F$, and $\Omega_0^S \in \mathcal{H}_S$ be the state of the spin chain "with all spins pointing down": $a_n \Omega_0^S = 0$, $\forall n$. Notice also that **here** $|n\rangle := a_0^* a_1^* \ldots a_n^* \Omega_0^S$, $n = 0, 1, \ldots N$. Let the **Hilbert subspace** $\mathcal{H}_{min} \subset \mathcal{H}$ be **generated by the vectors**

---

[14] The "second quantization" $d\Gamma(\boldsymbol{h})$ of the 'one-Fermi-particle-operator' $\boldsymbol{h}$ is the linear operator acting in the Fermi Fock space $\mathcal{H}_F := \oplus_{n=0}^{\infty} P_- \otimes_1^n \mathfrak{h}$, where P_- is the antisymmetrization operator, such that $d\Gamma(\boldsymbol{h})P_- \otimes_{k=1}^n \psi_k := P_- \sum_{j=1}^n \psi_1 \otimes \psi_2 \otimes \cdots \otimes \boldsymbol{h}\psi_j \otimes \cdots \otimes \psi_n$ for all $n \in \mathbb{Z}_+$.

$$\{\Omega_0 := \Omega_0^S \otimes \Omega_0^F \,, \ \beta_n := |n\rangle \otimes \Omega_0^F,$$
$$\beta_N(\psi) := |N\rangle \otimes b^*(\psi)\Omega_0^F; \ n = 0, 1, \dots N - 1, \ \psi \in L^2(\mathbb{R}^3, \mathrm{d}^3 x)\}.$$

Then it is valid:

7.6.3 Lemma. *The space $\mathcal{H}_{min}$ defined above is H-invariant: $H\mathcal{H}_{min} \subset \mathcal{H}_{min}$.*

A proof of this Lemma is presented in [39]. Hence the description of our process can be restricted to time evolution in the subspace $\mathcal{H}_{min} \subset \mathcal{H}$. We shall choose the parameters of the model, namely the operator $\boldsymbol{h}$ acting on $L^2(\mathbb{R}^3)$, and the quantities $\varepsilon_0 > 0$, $v \in \mathbb{R}$, $\sigma \in L^2(\mathbb{R}^3, \mathrm{d}^3 x)$, so that with our Hamiltonian given by (7.6.2a) the relation

$$\lim_{t \to \infty} \langle \beta_n | e^{itH} a_N^* a_N e^{-itH} | \beta_n \rangle = 1, \ n = 0, 1, \dots N - 1, \tag{7.6.3a}$$

or more specifically:

$$\langle \beta_n | e^{itH} a_N^* a_N e^{-itH} | \beta_n \rangle = 1 - o(t^{-m}), \ n = 0, 1, \dots N - 1, \ \text{for } t \to +\infty, \ \forall m \in \mathbb{N}, \tag{7.6.3b}$$

will be satisfied. The meaning of (7.6.3) is that the probability of emission of the Fermi particle and simultaneous transition of the spin chain to the stationary state β_N (i.e. all the spins in the chain "are pointing up" and the Fermi field is again in its vacuum state) approaches certainty 'almost exponentially quickly' if the time is growing to infinity.

The dynamics is investigated by a repeated use of **Fourier transform** $\mathcal{F}$, e.g. in [39, Lemma 2].

7.6.4 Lemma. *Let e^{-itH} be any (unitary) time evolution group. Then the Fourier transform of its (truncated) matrix elements for given $\phi, \psi \in \mathcal{H}$ is*

$$\mathcal{F}[\theta(t)\langle \phi, e^{itH}\psi \rangle](\xi) = \frac{i}{\sqrt{2\pi}} \langle \phi, R_H(\xi)\psi \rangle, \tag{7.6.4}$$

for $\xi \in \mathbb{C} : \text{Im } \xi < 0$.

The function θ is here the Heaviside function, and $R_H(\xi) \equiv (H - \xi I)^{-1}$ ($\xi \in \mathbb{C}$, $\xi \notin sp(H) \equiv$ *spectrum of* H) is the **resolvent of the** operator H.

Another useful result is that we obtain the resolvent $R_H(\lambda)$ as a solution of an operator equation, [39, Lemma 3].

7.6.5 Lemma. *Suppose $H = H_0 + V \in \mathcal{L}(\mathcal{H})$ and $\xi \notin sp(H) \cup sp(H_0)$. Then the resolvent $R_H(\xi)$ is the solution of the operator equation*

$$R_H(\xi) = R_{H_0}(\xi)(I - V R_H(\xi)). \tag{7.6.5}$$

Hence, the Fourier transform of the (truncated) matrix elements of the time evolution operator for $\text{Im } \xi < 0$ is given by:

$$\mathcal{F}[\theta(t)\langle\phi, e^{itH}\psi\rangle](\xi) = \frac{i}{\sqrt{2\pi}}\langle\phi, R_{H_0}(\xi)\psi\rangle - \frac{i}{\sqrt{2\pi}}\langle\phi, R_{H_0}(\xi)V R_H(\xi)\psi\rangle. \tag{7.6.6}$$

Important for the following analysis are the matrix elements

$$F_{mn} := \langle\beta_m, R_H(\xi), \beta_n\rangle, \tag{7.6.7}$$

since e.g.:

$$\mathcal{F}[\theta(t)\langle\beta_m, e^{itH}\beta_n\rangle](\xi) = \frac{i}{\sqrt{2\pi}}F_{mn}(\xi). \tag{7.6.8}$$

Now, the proper choice of the parameters of the model is, according to [39]:

$$\varepsilon(\mathbf{p}) := a|\mathbf{p}|^2, \ a > 0, \tag{7.6.9a}$$

$$\mathcal{F}(\sigma)(\mathbf{p}) = 0 \ (|\mathbf{p}| < b), \quad \mathcal{F}(\sigma)(\mathbf{p}) > 0 \text{ for all } |\mathbf{p}| > b > 0, \quad \sigma \in \mathcal{S}(\mathbb{R}^3), \tag{7.6.9b}$$

$$\varepsilon_0 > ab^2 + 2, \tag{7.6.9c}$$

where $\mathcal{S}(\mathbb{R}^3)$ is the set of all rapidly decreasing Schwartz complex valued functions on $\mathbb{R}^3$, the symbol $\mathcal{F}(\sigma)$ again means the Fourier transform (i.e. the transition to "p-representation"), and the constants a, b occurring in (7.6.9c) are the same as the ones occurring in (7.6.9a) and (7.6.9b).

After making this choice it is possible, after a series of considerations and calculations [39], to show that (cf. [39, (4.33)])

$$\mathcal{F}[\langle\beta_m, e^{itH}\beta_n\rangle](p) = -\sqrt{\frac{2}{\pi}}\lim_{\nu\to 0^+} \text{Im } F_{mn}(p - i\nu) \in \mathcal{S}(\mathbb{R}). \tag{7.6.10}$$

But the Schwartz set $\mathcal{S}(\mathbb{R})$ of rapidly decreasing smooth functions is invariant with respect to the Fourier transform, hence the function $t \mapsto \langle\beta_m, e^{itH}\beta_n\rangle$ also belongs to $\mathcal{S}(\mathbb{R})$, what proves the 'almost exponential decay' in time of this matrix element. This result is crucial for the proof of Theorem 7.6.6.

To formulate the main result as a theorem, let us introduce also the notation:

$$\mu((-\infty, \lambda]) := \int_{\varepsilon(\mathbf{p})<\lambda} |\mathcal{F}(\sigma)(\mathbf{p})|^2 \, d^3\mathbf{p}, \tag{7.6.11a}$$

$$\rho_\mu(\lambda) := \frac{d\mu((-\infty, \lambda])}{d\lambda}. \tag{7.6.11b}$$

7.6.6 Theorem. *In the above described model of finite spin chain QD interacting with nonrelativistic scalar Fermi field, with the parameters specified in (7.6.9), for either all such ε_0 with possibly one exception, or for all*

$$\varepsilon_0 > 2 + ab^2 + 2v^2 \int_{ab^2}^{\infty} \frac{\rho_\mu(\lambda)}{\lambda - ab^2} \, d\lambda,$$

the time evolution of the probability of all the $N + 1$ spins being turned up (realizing the wanted final state of the spin chain), if initially the Fermi field was in the vacuum state and the first n spins ($N - 1 \geq n \geq 0$) were turned up, approaches unity almost exponentially fast, i.e. the relation:

$$\langle \beta_n | e^{itH} a_N^* a_N e^{-itH} | \beta_n \rangle = 1 - o(t^{-m}), \text{ for all } 0 \leq n \leq N - 1, \text{ for any } m \in \mathbb{N},$$
$$(7.6.12)$$

is satisfied.

A detailed proof of this theorem can be found in [39].

7.6.7 Let us look at the result (7.6.12) from the point of view of the Sect. 7.7, to make it more intuitive as a relevant assertion with respect to the "measurement problem", cf. (7.7.2).

As the "measured system" in this model can be considered the single spin lying at the 'beginning' of the spin chain. Let its C^*-algebra of observables be generated by $\{a_0^*; a_0\}$ satisfying (7.3.1), and let $\varphi_\downarrow, \varphi_\uparrow$ be its normalized state vectors corresponding to the two opposite orientations of the spin. Let its initial normalized state vector be $\varphi_0 := c_\downarrow \varphi_\downarrow + c_\uparrow \varphi_\uparrow$, with $a_0^* \varphi_\downarrow = \varphi_\uparrow$, $a_0 \varphi_\uparrow = \varphi_\downarrow$.

The initial state of the whole composite system {*measured system* & *rest of the spin chain* & *Fermi field*} is then $\tilde{\Psi}_0 := c_\downarrow \Omega_0 + c_\uparrow \beta_0 = (c_\downarrow I_{\mathfrak{A}} + c_\uparrow a_0^*) \Omega_0^S \otimes \Omega_0^F$. The time evolved states $\tilde{\Psi}_t := \exp(-itH)\tilde{\Psi}_0$ can be written, due to the Lemma 7.6.3 as well as the stationarity of Ω_0, in the form

$$\tilde{\Psi}_t = c_\downarrow \Omega_0 + c_\uparrow e^{-itH} \beta_0.$$
$$(7.6.13)$$

The second term in (7.6.13) can be written, again due to the H-invariance of $\mathcal{H}_{min}$, cf. Lemma 7.6.3, in the form

$$e^{-itH} \beta_0 = \sum_{n=0}^{N-1} d_n(t) \beta_n + \beta_N(\psi(t)).$$
$$(7.6.14)$$

Since $a_N \beta_n = 0$, $(n = 0, 1, \ldots N - 1)$, and $a_N^* a_N \beta_N(\psi) = (1 - a_N a_N^*) \beta_N(\psi) = \beta_N(\psi)$, the expression from (7.6.12) with our $n = 0$ is

$$\langle \beta_0 | e^{itH} a_N^* a_N e^{-itH} | \beta_0 \rangle = \| a_N e^{-itH} \beta_0 \|^2 = \| \beta_N(\psi(t)) \|^2,$$
$$(7.6.15)$$

and this converges very quickly, according to (7.6.12), to unity. The vectors on the right hand side of (7.6.14) are mutually orthogonal and the whole right hand side has the constant norm equal to 1. Hence the norm of the sum on the right hand side of (7.6.14) quickly converges to zero. All the vectors $\beta_N(\psi)$ $(\psi \in L^2(\mathbb{R}^3, \mathrm{d}^3 x))$ describe the states of the *composite system*:

{ *the measured system* & *the rest of the spin chain* & *the Fermi field*}

in which all the $N + 1$ spins "are pointing up", which has to mimic the macroscopically different state from the initial state $(c_\downarrow I_{\mathfrak{A}} + c_\uparrow a_0^*)\Omega_0^S \otimes \Omega_0^F \equiv (c_\downarrow I_{\mathfrak{A}} + c_\uparrow a_0^*)\Omega_0$, as well as from Ω_0, of the compound system. For the wave function (7.6.13) of the compound system we obtain asymptotically for large times $t \to \infty$:

$$\widetilde{\Psi}_t = c_\downarrow \Omega_0 + c_\uparrow e^{-itH} \beta_0 \asymp c_\downarrow \Omega_0 + c_\uparrow \beta_N(\psi(t)), \qquad (7.6.16)$$

which has the form of the formula (7.7.2) for the (approximate expression of the) "measurement dynamics" in the conventional QM framework of considering of only finite systems (as measuring apparatuses). The probabilities of the two different "measurement results" corresponding to the states $\varphi_\downarrow$, resp. $\varphi_\uparrow$, occurring in the orthogonal decomposition of the initial state φ_0 of the measured system are, as it was expected, the numbers $|c_\downarrow|^2$, resp. $|c_\uparrow|^2$. By 'tracing out' the states of the *environment* $\equiv$ the Fermi field, we obtain the density matrix for the spin chain, and by tracing out the both {Fermi field & the spins 1,2,...N}, we obtain the *density matrix* $\varrho := |c_\downarrow|^2 P_{\varphi_\downarrow} + |c_\uparrow|^2 P_{\varphi_\uparrow}$, with $P_{\varphi_\uparrow} \equiv a_0^* a_0$ and $P_{\varphi_\downarrow} \equiv a_0 a_0^*$, in the state space of the measured system (i.e. of the spin placed in the point 0 of the chain), corresponding formally to the '*collapse of its wave packet*' $\varphi_0 := c_\downarrow \varphi_\downarrow + c_\uparrow \varphi_\uparrow$, i.e. of its initial state of the just described process. Neither of these density matrices can be, however, interpreted as describing a 'proper', or 'genuine' probability distribution of quantal states in the sense of classical statistics. To interpret them in that sense, and distinguish one decomposition of a density matrix as 'more relevant' (i.e. reflecting the classical-type statistics), some another additional assumption is needed. We have had in our interpretations of the infinite models in previous Sections the requirement of **disjointness** of mutually noninterfering states, and this was ensured by existence of a **macroscopic quantity** obtaining mutually different values in these states. For some alternative approaches, we could go back again to the attempts in the 'decoherence programs', [124, 279, 343, 347]. More detailed mathematical and interpretational considerations on decompositions of states of a C^*-algebra can be found in our 1.2.3, 1.3.3, 1.3.4, 1.4.3, and citations therein, e.g. [53, Chap. 4].

7.6.8 Notes on irreversibility. This model of a radiating multispin system can be also considered as a caricature reflecting one of the usual mechanisms of *irreversible behaviour* of large physical systems: Large systems usually (resp. 'almost always') are not isolated from their environment, and their interaction with (a 'relative stable', and a 'relative stationary') environment leads to their motion to more stable stationary, e.g. thermodynamic equilibrium, states. Some kind of radiation, as it was built in into our model, is a usual form of interactions of large systems with their environment.

   This approach reflects just one 'aspect' of irreversible behaviour of physical systems. Another often discussed 'aspect' of theoretical descriptions of irreversible behaviour of finite many-particle systems is their complicated mechanical motion even if they are isolated from any environment. Then we are dealing with such phenomena as various types of "chaos", and with "recurrences" in their (deterministic and time-reversible) mechanical motion. We shall not consider here such mechanical explanations of irreversibility, initiated by J. C. Maxwell and L. Boltzmann. As concerns some study on these topics in the case of classical systems, it might be interesting to look to nice conference or journal papers like, e.g. [332], but more elementary and also more complex information could be found in some books on the "theory of dynamical systems" listed in our Bibliography, e.g. [1, 7, 9, 10, 181, 248, 326]. However long are durations of the *Poincaré cycles* corresponding to the above mentioned recurrences in mechanical motions of isolated systems with several degrees of freedom (they are comparable with the lifetime of Universe [332]), an evolution during which the system approaches some stable stationary state cannot be reached in theoretical description of finite isolated mechanical systems. This does not exclude, however, effectiveness of the statistical physics, which does not deal with a unique phase-space trajectory of the considered system; here we have a certain physical reinterpretation of the mechanics of motions in the system's phase space. But full effectiveness of the statistical approach to description of behaviour of multiparticle systems, e.g. mathematically clear description of thermal equilibria and phase transitions, is again possible in the 'thermodynamical limit' of infinitely large systems only, e.g. [271].

It is seen that after making the finite quantum spin chain of our model to become an "open system" by adding to the Hamiltonian of the restricted QD the term corresponding to the radiation of a fermion, the speed of the motion to the limiting state was enormously increased in comparison with the infinite, but isolated, QD-chain, cf. (7.3.24) and (7.3.26b), i.e. with respect to (7.3.27). The finite-sized version of the isolated QD would behave, however, almost-periodically, cf. (7.3.23). The addition of interaction of the finite QD with Fermi field enabled us to obtain a system's state converging for $t \to \infty$ to a new stationary state. But a clear and unambiguous interpretation of some states of a finite system, e.g. the two states appearing in the sum on the right hand side of (7.6.16), as being approximately 'mutually macroscopically different' (hence their quantum interference being 'almost impossible'), is still open to discussion. We shall not further investigate here some other connections of these phenomena and questions.

7.7 On the "Measurement Problem" in QM

Let us add here several notes to the above mentioned "measurement problem", considered for a long time to be a fundamental problem of the conceptual structure of QM, cf. e.g. [63, 236, 237] and [238, Chap. 29]. These notes should be also supplemented by the notes in 7.6.1, esp. by the footnotes 12 and 13.

States of the physical systems are described in the mathematical theory of QM by mathematical objects like "wave functions", "density matrices", or "linear functionals ω on algebras of observable quantities" (which generalize the former two classes of objects). The "observable quantities" (represented by operators, resp. elements of an algebra) correspond to experimental, or observational, arrangements of empirical situations, in which the observer is able, after "installing" a specific state ω of the observed system, to perceive and appreciate by his human senses some well determined, in advance expected feelings (optical, auditory, acquired by touch or in another way) of some specific perceptions that are clearly distinguishable from others (e.g. when reading positions of a pointer, or hearing a characteristic sound from a counter,...), so that they can be formalized into a form suitable for further communication. A single observable A of a specific physical system appears in such an empirical situation through a specific instance of a set of such clearly distinguishable phenomena, each of which can be (and, as a rule, is) denoted by a number α_j $(\in \mathbb{R})$ called the **result of single measuring act in the state ω of a value of** A (not to be confused with "the value of A in ω "—different single measuring acts of the same observable on ω could lead to different results!). Many experiments on microscopic systems performed in the history of microphysics have shown that we are not able to prepare states of any microsystem in such a way that in a many times repeated measurement of an observable on the same (prepared each time anew) state ω one obtains the same measured value **for each observable** which can be chosen for these repeated measurements. To state it briefly: For any state of any microsystem there is some observable **which does not have any specific value** in that state. This is reflected mathematically in, e.g., Heisenberg uncertainty relations. On the other hand, to each value α_j of the given observable A there exists (for observables with discrete spectra) at least one state ω_j such that the repeated measurements of A on it give with certainty the same value α_j. The problem arises because there is (with certainty) some other observable B such that the repeated measurements of it on the same state ω_j give mutually different values $\beta_k \neq \beta_l \cdots \in \mathbb{R}$, i.e. the statistical dispersion of the measured values of B in that ω_j is nonzero. Sharp values (obtained consistently in the identical, many times repeated measurements) β_k of B can be obtained in other states ω'_k, for which, however, the measurements of some other observables $A, C, \ldots$ would have nonzero dispersions.

The existing very successful mathematical model of QM provides solution of this problem which consists in describing an arbitrarily chosen (but, by assumption,"pure") state ω_j as a **linear superposition** of some (again pure) states $\omega'_k, \omega'_l \ldots$, i.e., if we express all the states in the form of vectors in a Hilbert space $\mathcal{H}_v$, in writing the state in question as $\psi_j = \sum_k c_k \varphi_k$, where the correspondence with the values of the observables $A, B, \ldots$ described now as linear operators on $\mathcal{H}_v$, is such that the "state-vector" ψ_j, corresponding to the state ω_j, is an "eigenvector" of the operator A (a common practice is to use the same symbol for the operator as for the physical quantity represented by it): $A\psi_j = \alpha_j \psi_j$, and similarly the vectors φ_k corresponding to the states ω'_k are the eigenvectors of the operator $B :\ B\varphi_k = \beta_k \varphi_k$.

All this is, of course, very well known, and we have also briefly described it in our Sect. 1.2. We recall it here to stress the unusual intuition required when dealing with the phenomena described by the mathematical model of QM, in comparison with the intuition provided by the 'everyday life', whose formal reflection is contained in the mathematical models of classical physics.

One of the prominent results of the history of observations and measurements mentioned above is that QM is considered an *irreducibly statistical theory*; i.e., that the probabilistic results of the measurements with nonzero dispersions are not necessarily due to the presence of some *statistical ensembles* of systems in various states, as they are in the *classical statistical physics*, but that it is impossible to find any fully dispersionfree states even when considering individual (micro)systems. This is now (starting from 1920's) acceptable and included in a logically consistent manner into the description of our world. The resulting picture of the world is, however, not without problems, since its integral part is a class of counterintuitive phenomena encountered in QM. These are, pictorially expressed, the problems of the type of the well known "Schrödinger's cat paradox", which is just a popular representation of the "measurement problem" to be discussed further (the cat can be regarded here also as a measuring device).

We are measuring with some macroscopic apparatuses which belong to the same world as microsystems do, but seem to be correctly described by a theory that is very different from QM. Is QM a universal theory, or is there some borderline between the two differently behaving parts of the world? If so, it should be explained in the theory **where that borderline is located**. But the apparatuses are composite of many microsystems and (as far as the present author knows) no new aspect of microsystems was discovered which could effectively distinguish between them and macrosystems. Thus, let us regard the apparatuses as some quantum-mechanical systems. Then any measuring process should look as follows[15]:

If the initial state of the measured microsystem is described by the normalized vector φ_k corresponding to the value β_k of the observable B, and the initial state of the apparatus capable to measure the quantity B is described by the normalized vector Ψ_0 in its Hilbert space, installed independently of the measured state, then the unitary process $U(t)$ corresponding to the time evolution of the mutually interacting measured microsystem and apparatus will lead, after the 'time of the measurement' t_m, to the state

$$U(t_m)[\varphi_k \otimes \Psi_0] = \widetilde{\Psi}_k. \qquad (7.7.1)$$

Here, in the 'post-measurement state' $\widetilde{\Psi}_k$ of the compound system microsystem and apparatus, the "pointer position" of the apparatus corresponds to the value β_k of B. This is assumed to be valid for all β_k, hence for $\beta_k \neq \beta_j$ the pointer positions (i.e. certain macroscopic parameters) in the states $\widetilde{\Psi}_k$ and $\widetilde{\Psi}_j$ are different from each other. The same unitary evolution should lead, after the measurement by the same apparatus on the state $\psi := \sum_k c_k \varphi_k$, due to its linearity, to the state of the compound system

---

[15]We will work here with pure states (resp. vector states) only. In fact, it is not necessary to use density matrices in an analysis of the process of measurement in QM, as shown, e.g. by Wigner in [339].

$$U(t_m) \left[\left(\sum_{k \in J} c_k \, \varphi_k \right) \otimes \Psi_0 \right] \equiv \widetilde{\Psi} := \sum_k c_k \, \widetilde{\Psi}_k, \quad \sum_{k \in J} |c_k|^2 = 1. \qquad (7.7.2)$$

The 'macroscopic part of the world' appears here in the state $\widetilde{\Psi}$, expressed as a nontrivial linear superposition $\widetilde{\Psi}$ of the states $\widetilde{\Psi}_k$ corresponding to different values of some macroscopic parameter (different "pointer positions", distinguished here by the index k). Such superpositions in QM do not mean only a probability distribution with nonzero dispersion of the values of a macro-parameter corresponding to various β_k, but they should also allow (according to the principles of QM) a realization of measurements of some new observable having a sharp value in the state $\widetilde{\Psi}$ (on the statistical ensemble of equally prepared compound systems obtained in the process of the measurement of this new observable on the microsystem). The states $\widetilde{\Psi}$ are representing in such a way an **interference of different values of a macro-parameter** ('the cat is simultaneously dead and alive'). Thus, the apparent conceptual problem of QM does not consist in its probabilistic nature, it rather consists in the unanswered question of the existence of the very counterintuitive "**macroscopic interference**" we have just described, or/and in a dynamical explanation why they do not occur.

The widely accepted 'solution' of this "measurement paradox" (as termed by Penrose [238]) consists in accepting of so called "**reduction postulate**", consisting in the claim that there supposedly exists the phenomenon colloquially termed the "reduction (or also *collapse*) of the wave packet". This can be rephrased, in terms of our preceding considerations, in such a way that within some final phase of the process of measurement, either during or just after the measurement (e.g. such as is sketched in (7.7.2)) performed on the system, the system (i.e. either the measured system alone—this is the traditional point of view, or the apparatus, or—which seems to the present author as the most acceptable possibility—the compound system microsystem & apparatus) ends after each single run of the measurement in a specific state corresponding to the obtained value of the measured observable, and after many times repeated 'identical' measurements on such a state we arrive at a statistical mixture (in the sense of classical statistical physics, i.e. the "proper" or "genuine" mixture, cf. in 1.1.4) of the set of (systems occurring in the) states which, in the case of compound system, consists of

$$\{\widetilde{\Psi}_k : k \in J\} \text{ with probabilities } |(\widetilde{\Psi}, \widetilde{\Psi}_k)|^2 = |c_k|^2, \ k \in J. \qquad (7.7.3)$$

This transition from superpositions to classical mixtures of states with different "pointer positions" takes place, according to the reduction postulate, instantaneously, or in some "negligibly short time".

Many existing *theories of quantum measurements* which have appeared up to the present day analyze systematically possible results of various measurements (of corresponding observables) as well as their mutual relations like their mutual consistency or 'complementarity', see e.g. [63, 64, 84, 175]. These theories, called by their authors "operational", are purely phenomenological, built on the formal structure of quantum kinematics and usually manifested no interest in the description of specific dynamics of the considered processes. They are often mathematically highly elaborated, very elegant and probably also useful from the point of view of applications of QM. We were not concentrating ourselves here on these approaches and on the questions motivating them. The avoidance of the problems with the dynamics of the interaction of the measured microsystems with the measuring macroscopic apparatuses indicates that in these phenomenological works one assumes, at least implicitly, the existence of some unknown mechanism of the "*wave packet reduction*", or equivalently "wave packet collapse". This is acceptable from the 'practical point of view', because in the usual praxis of manipulations with microsystems (e.g. measurements on them) it is possible to deal with the results (e.g. the outcomes of the measurements) as if the "wave packet reduction" really happened. We are here, however, interested in the problem how this process can be included into a noncontradictory quantum theory. An extensive discussion of these problems by the leading physicists up to 1980's contains [331].

The last decades, on the other hand, have seen experiments whose results indicate that *the interference of macroscopically different states is possible in suitable conditions*, cf. e.g. [195, 196]. These 'suitable conditions' consist, first of all, in sufficient isolation of the considered quantum macro-system from any interactions with surrounding environment, then, of course, in the ability of experimenters to discover some suitable 'macrointerference detecting' observable quantity, and finally in the inventiveness of experimenters when constructing the desired measuring apparatus.

Our models described in the Sects. 7.3, 7.4 and 7.5 of this chapter, mainly inspired by the ideas published in [153], show that in the limit $t_m \to \infty$ the classical-like probability distributions of the measurement results (i.e. probability without mutual interferences of results) can be reached. In these models, apparatuses are treated as quantum collections of infinitely many "small" subsystems, and the time necessary for reaching the "reduction of the wave packet" is infinitely long; also, the convergence to the final states of the apparatuses of "proper mixtures"-type is in these simple models—contrary to the ideal requirements—very slow.

The last of our models described in Sect. 7.6 shows, however, that if we construct an "apparatus" as a large but finite collection of microsystems, interacting, moreover, with the environment by radiating a particle, the convergence proceeds fast enough— in the sense 'almost exponentially'. The problem here is nonvanishing possibility of interference of states with different pointer positions, although such a possibility would be for 'sufficiently large' apparatuses very improbable. Again, an opened question is the existence and location of a possible borderline for the validity of QM. A mathematically clear formulation of the dependence of possible interferences between macroscopic states of a "large system" on its size will be, probably, a subject of future investigations in theoretical physics. One cannot exclude, however, that there is no sharp borderline between QM and CM, and instead, there is a continuous transition from QM to CM dependent on more parameters than just the size of the measuring apparatus. Or, that there is no borderline at all, QM is a universal theory, but our understanding of its possible applications requires some completions.

Correction to Previously Published Version of Classical Systems in Quantum Mechanics

Correction to:
P. Bóna, *Classical Systems in Quantum Mechanics*,
https://doi.org/10.1007/978-3-030-45070-0

The original version of the book was inadvertently published with incorrect information in the book. The Correction to Previously Published Version of Classical Systems in Quantum Mechanics is given below.

Page	From top of page: line (or formula)	Chapter	Incorrect text [and/or explanation]	Correct text [and comment]
21	(2.1.2)	2	$:= \sqrt{2} \in \{...\}$	$:= \sqrt{2}\inf\{...\}$
87	7	5	$Ad^*(G)$ in G^*	$Ad^*(G)$ in $\mathfrak{g}^*$
123	2x in 8 and 4x in 9	6	["hats" omitted on the six symbols p_ω, p_ω^*]	$\hat{p}_\omega$, $\hat{p}_\omega^*$
130	4	6	[add "hat" on p_ω]	$\hat{p}_\omega$
139	18	6	$\omega \in p_g\mathcal{S}(\mathfrak{A})$	$\omega \in p_G\mathcal{S}(\mathfrak{A})$
161	9	6	$Ad^*(su(2))$	$Ad^*(SU(2))$
205	7	7	the operator h acting	the operator $\boldsymbol{h}$ acting

Additional Corrections

1. The following two-line equation on p. 83 [above (5.1.51)]:

$$P_G \in \mathfrak{Z}^{\#} := \text{the center of } \mathfrak{B}^{\#} := (\mathfrak{A}^\Pi)''$$
$$:= \text{the weak operator closure of } \mathfrak{A}^\Pi \text{ in } \mathcal{L}(\mathcal{H}_\Pi)$$

The updated version of the book can be found at
https://doi.org/10.1007/978-3-030-45070-0

© Springer Nature Switzerland AG 2020
P. Bóna, *Classical Systems in Quantum Mechanics*,
https://doi.org/10.1007/978-3-030-45070-0_8

has been corrected to read:

$$P_G \in \mathfrak{Z}^\# := \text{the center of } \mathfrak{B}^\#,$$
$$\mathfrak{B}^\# := (\mathfrak{A}^\Pi)'' := \text{the weak operator closure of } \mathfrak{A}^\Pi \text{ in } \mathcal{L}(\mathcal{H}_\Pi)$$

2. The symbols p_ω in Chap. 6 have been corrected to $\hat{p}_\omega$.
3. The title of Chap. 6 has been modified to the more appropriate "Dynamics of Quantum Mechanical Macroscopic Systems".

Bibliography

1. R. Abraham, J.E. Marsden, *Foundations of Mechanics*, 2nd edn. (Benjamin/Cummings, Reading, Mass., 1978)
2. S.T. Ali, E. Prugovečki, Physica **89A**, 501–521 (1977); S.T. Ali, G.G. Emch, J. Math. Phys. **15**, 176 (1974); S.T. Ali, J. Math. Phys. **21**, 818 (1980); F.E. Schroeck, Found. Phys. **12**, 825 (1982); S.T. Ali, E. Prugovečki, Acta Appl. Math. **6**, 1, 19, 47 (1986)
3. S.T. Ali, H.-D. Doebner, Phys. Rev. A **41**, 1199–1210 (1990)
4. S.T. Ali, J.-P. Antoine, J.-P. Gazeau, U.A. Mueller, Rev. Math. Phys. **7**, 1013–1104 (1995)
5. L.M. Alonzo, J. Math. Phys. **18**, 1577–1581 (1977); L.M. Alonzo, J. Math. Phys. **20**, 219–230 (1979)
6. H. Araki, J. Math. Phys. **4**, 1343 (1963); H. Araki, J. Math. Phys. **5**, 1 (1964)
7. V.I. Arnol'd, *Matematičeskije osnovy klassičeskoj mechaniki*, 3rd edn. (Nauka, Moscow, 1989); Eng. edition: V.I. Arnold, *Mathematical Methods of Classical Mechanics* (Springer, 2013)
8. V.I. Arnol'd, V.V. Kozlov, A.I. Neustadt, *Matematičeskije aspekty klassičeskoj i nebesnoj mechaniki, in Sovremennyje Problemy Matematiki-Fundamental'nyje napravlenija*, vol. 3 (VINITI, Moscow, 1985)
9. V.I. Arnol'd, *Dopol'nitel'nyje glavy teorii obyknovennych differencial'nych uravnenij* (Nauka, Moscow, 1978)
10. V.I. Arnol'd, A. Avez, *Ergodic Problems of Classical Mechanics* (W. A. Benjamin Inc., New York, Amsterdam, 1968)
11. A. Ashtekar, T.A. Schilling, *Geometrical Formulation of Quantum Mechanics* (1997). http://arxiv.org/abs/gr-qc/9706069
12. V. Bach, E.H. Lieb, M. Loss, J.P. Solovej, Phys. Rev. Lett. **72**, 2981–2983 (1994)
13. A.O. Barut, R. Raczka, *Theory of Group Representations and Applications* (PWN, Warsaw, 1977); Russ. edition: (Mir, Moscow, 1980); (World Scientific, 1986)
14. J.S. Bell, *Speakable and Unspeakable in Quantum Mechanics* (Cambridge University Press, Cambridge, 1988)
15. E.G. Beltrametti, S. Bugajski, J. Phys. A Math. Gen. **28**, 3329–3343 (1995)
16. I. Bengston, K. Życzkowski, *Geometry if Quantum States* (Cambridge University Press, Cambridge, 2008)
17. F.A. Berezin, Commun. Math. Phys. **63**, 131 (1978)
18. F.A. Berezin, Izvestija Akad. Nauk SSSR Ser. Matem. **36**, 1134 (1972)
19. F.A. Berezin, Izvestija Akad. Nauk SSSR Ser. Matem. **38**, 1116–1175 (1974)
20. J. Bialynicki-Birula, J. Mycielski, Ann. Phys. **100**, 62 (1976)
21. G. Birkhoff, S. Mac Lane, *Algebra* (Macmillan Co., New York, 1968)
22. M.S. Birman, M.Z. Solomyak, *Spektral'naya Teoriya Samosopryazhonnykh Operatorov v Gil'bertovom Prostranstve*, in Russ., *(Spectral Theory of Selfadjoint Operators in Hilbert Space)* (Leningrad University Press, Leningrad, Sankt Petersburg, 1980)

© Springer Nature Switzerland AG 2020, corrected publication 2020
P. Bóna, *Classical Systems in Quantum Mechanics*,
https://doi.org/10.1007/978-3-030-45070-0

23. N.N. Bogoliubov, Sov. Phys. JETP **34**, 58–73 (1958)
24. N. Bohr, Discussion with Einstein on epistemological problems in atomic physics, in *Albert Einstein: Philosopher-Scientist*, ed. by P.A. Schlipp. The Library of Living Philosophers (Evanston, 1949), pp. 200–241
25. N. Bohr, Phys. Rev. **48**, 696–702 (1935)
26. N. Bohr, *Izbrannyje naučnyje trudy v dvuch tomach* (NAUKA, Moskva, 1971); Russ. edition: *Selected Scientific Works in Two Volumes* (Nauka, Moscow, 1971)
27. N. Bohr, Kausalität und Komplementarität. Erkenntniss **6**, 293–303 (1937)
28. B.-Y. Chu, Trans. Am. Math. Soc. **197**, 145–159 (1974)
29. P. Bóna, *Quantum Mechanics with Mean-Field Backgrounds*, Preprint No. Ph10-91 (Comenius University, Faculty of Mathematics and Physics, Bratislava, October 1991). http://www.st.fmph.uniba.sk/~bona1/preprint%20Ph10-91.html
30. P. Bóna, On nonlinear quantum mechanics, in *Differential Geometry and Its Applications, Proceeding of the Conference on Opava (Czechoslovakia), August 24–28, 1992* (Silesian University, Opava, 1993), pp. 185–192. http://www.st.fmph.uniba.sk/~bona1/NLQM-Opava1992.pdf
31. P. Bóna, Czech. J. Phys. **B 33**, 837 (1983). http://www.st.fmph.uniba.sk/~bona1/QC.html. (This paper essentially coincides with the talk given by the author at VIIth International Congress on Mathematical Physics, Boulder, Colorado, USA, August 1–10, 1983)
32. P. Bóna, *Classical Projections and Macroscopic Limits of Quantum Mechanical Systems*. Unpublished monograph (Bratislava, 1984), revised version 1986. http://www.st.fmph.uniba.sk/~bona1/monograph.html
33. P. Bóna, Selfconsistency and objectification, in *Quantum Measurement, Irreversibility and the Physics of Information, Proceedings of the Symposium on the Foundations of Modern Physics*, ed. by P. Busch, P. Lahti, P. Mittelstaedt (World Scientific, 1993). http://www.st.fmph.uniba.sk/~bona1/selfconsist+object.pdf
34. P. Bóna, Acta Phys. Slov. **23**, 149 (1973). http://www.physics.sk/aps/pubs/1973/aps_1973_23_3_149.pdf; P. Bóna, *Interakcia makrosystému s mikroobjektom v kvantovej teórii (Interaction of Macrosystem with Microobject in Quantum Theory, in Slovak)*, dissertation, Comenius University, Bratislava, 1974. http://www.st.fmph.uniba.sk/~bona1/dissertation.html; P. Bóna, Acta Phys. Slov **27**, 101 (1977). http://www.physics.sk/aps/pubs/1977/aps_1977_27_2_101.pdf; P. Bóna, ACTA F.R.N. Univ. Comen.-PHYSICA **XX**, 65 (1980); P. Bóna, Selfconsistency and objectification, in *Quantum Measurement, Irreversibility and the Physics of Information, Proceedings of the Symposium on the Foundations of Modern Physics*, ed. by P. Busch, P. Lahti, P. Mittelstaedt (World Scientific, 1993). http://www.st.fmph.uniba.sk/~bona1/selfconsist+object.pdf
35. P. Bóna, *Interakcia makrosystému s mikroobjektom v kvantovej teórii. (Interaction of Macrosystem with Microobject in Quantum Theory, in Slovak)*, dissertation, Comenius University, Bratislava, 1974. http://www.st.fmph.uniba.sk/~bona1/dissertation.html
36. P. Bóna, Acta Phys. Slov. **27**, 101 (1977). http://www.st.fmph.uniba.sk/~bona1/aps/QuantumDomino(APS).pdf; resp. in the journal APS: http://www.physics.sk/aps/pubs/1977/aps_1977_27_2_101.pdf.
37. P. Bóna, Acta Phys. Slov. **50**, 1–198 (2000); A revised version containing also Index is in: P. Bóna, *Extended Quantum Mechanics*, http://arxiv.org/pdf/math-ph/9909022v7.pdf, or also in http://www.st.fmph.uniba.sk/~bona1/EQM/eqm8a.pdf; *Textbook*: P. Bóna: *Mathematical Physics: Quantities and Their Spectra in Quantum Theory*, http://davinci.fmph.uniba.sk/~bona1/MathForQM4.pdf.
38. P. Bóna, ACTA F. R. N. Univ. Comen. - PHYSICA **XX**, 65 (1980). http://davinci.fmph.uniba.sk/~bona1/Models-Q-Detect/AFRN-1980-QD.pdf.
39. P. Bóna, M. Širaň, *A Radiating Spin Chain as a Model of Irreversible Dynamics* (Bratislava, 2012). http://arxiv.org/abs/1211.6783
40. P. Bóna, J. Math. Phys. **29**, 2223 (1988). http://www.st.fmph.uniba.sk/~bona1/JMP-1-1988.pdf

41. P. Bóna, J. Math. Phys. **30**, 2994 (1989). http://www.st.fmph.uniba.sk/~bona1/JMP-2-1989. pdf

42. P. Bóna, Czech. J. Phys. **B37**, 482 (1987). http://www.st.fmph.uniba.sk/~bona1/CJP87/CJP-1987.pdf

43. P. Bóna, Geometric formulation of nonlinear quantum mechanics for density matrices, in *Trends in Quantum Mechanics, Proceedings of International Symposium, Goslar, Germany, 31 August–3 September 1998*, ed. by H.-D. Doebner, S. T. Ali, M. Keyl, R. F. Werner (World Scientific, Singapore, New Jersey, London, Hong Kong, 2000). arxiv:quant-ph/9910011

44. P. Bóna, *On Symmetries in Nonlinear Quantum Mechanics*. arxiv:quant-ph/9910012

45. P. Bóna, On Symmetries in mean-field theories, in *Selected Topics in Quantum Field Theory and Mathematical Physics*, ed. by J. Niederle, J. Fischer (World Scientific, Singapore, 1990). http://www.st.fmph.uniba.sk/~bona1/Sym-in-MFT-1989.pdf

46. P. Bóna, Phys. Rev. Lett. **90**(20), 208901 (2003) [Comment on "No-Signaling Condition and Quantum Dynamics".]

47. P. Bóna, J. Geom. Phys. **51**, 256–268 (2004). http://www.st.fmph.uniba.sk/~bona1/JGP-2004. pdf; arXiv:math-ph/0301007v4. Corrigendum to this paper is published in: P. Bóna, J. Geome. Phys. **130**, 349 (August 2018)

48. H.J. Borchers, R. Haag, B. Schroer, Nuovo Cimento **24**, 214 (1963); H.J. Borchers, Commun. Math. Phys. **1**, 281 (1965)

49. H.-J. Borchers, *Translation Group and Particle Representations in Quantum Field Theory* (Springer, Berlin, 1996)

50. N. Bourbaki, *Groupes et Algèbres de Lie* (Hermann, Paris, 1972)

51. N. Bourbaki, *Varietes differentielles et analytiques. Fascicule de resultats* (Hermann, Paris, 1967 and 1971); Russ. edition: (Mir, Moscow, 1975)

52. N. Bourbaki, *Espaces vectoriels topologiques*, Hermann, Paris; Russ. trnaslation: Izdatelstvo Inostrannoj Literatury, Moscow, 1959

53. O. Bratteli, D.W. Robinson, *Operator Algebras and Quantum Statistical Mechanics*, vol. I (Springer, New York, Heidelberg, Berlin, 1979)

54. O. Bratteli, D.W. Robinson, *Operator Algebras and Quantum Statistical Mechanics*, vol. II (Springer, New York, Heidelberg, Berlin, 1980)

55. T. Breuer, Phil. Sci. **62**, 197–214 (1995)

56. T. Breuer, Int. J. Theor. Phys. **37**, 235–240 (1998)

57. L. de Broglie, Phil. Mag. J. Sci. **XLVII**(CCLXXVIII), 446 (1924); (Cited according to [245].)

58. L. de Broglie, Ann. d. Phys. **III**(dixième série), 22–128 (1925); (Cited according to [245].)

59. L. de Broglie, *Non-Linear Wave Mechanics-A Causal Interpretation* (Elsevier, Amsterdam, 1950); D. Bohm, J. Bub, Rev. Mod. Phys. **38**, 453 (1966); P. Pearle, Phys. Rev. **D13**, 857 (1969)

60. J.-B. Bru, W. de Siqueira Pedra, The Papers Concerning Classical Dynamics, Long-Range Interactions and Self-Consistency Equations in Quantum Mechanics, e.g., in Google Scholar, or more specifically, e.g.: http://kleine.mat.uniroma3.it/c/19/19-37.pdf, http://kleine.mat.uniroma3.it/c/19/19-38.pdf, http://kleine.mat.uniroma3.it/c/19/19-39.pdf, http://kleine.mat.uniroma3.it/c/18/18-111.pdf

61. D. Buchholz, J. Yngvason, Phys. Rev. Lett. **73**, 613–616 (1994)

62. S. Bugajski, Phys. Lett. A **190**, 5–8 (1994)

63. P. Busch, P.J. Lahti, P. Mittelstaedt, *The Quantum Theory of Measurement* (Springer, Berlin, 1992 & 1996); P. Busch, M. Grabowski, P. Lahti, *Operational Quantum Physics* (Springer, 1995 & 1997); P. Busch, P. Lahti, J-P. Pellonpää, K. Ylinen, *Quantum Measurement* (Springer, 2016)

64. P. Busch, Int. J. Theor. Phys. **24**, 63–92 (1985)

65. P. Busch, P.J. Lahti, P. Mittelstaedt (eds.), Quantum measurement, irreversibility and the physics of information, in *Proceedings of the Symposium on the Foundations of Modern Physics* (World Scientific, 1993)

66. V. Bužek, R. Derka, G. Adam, P.L. Knight, Ann. Phys. **266**, 454–496 (1998)

67. V. Bužek, Phys. Rev. A **58**, 1723–1727 (1998)

68. V. Bužek, M. Hillery, R.F. Werner, quant-ph/99 01 053, to appear in Phys. Rev. Lett
69. V. Cantoni, Acad. Naz. dei Lincei **VIII, LXII/5**, 628 (1977); V. Cantoni, Commun. Math. Phys. **44**, 125 (1975)
70. V. Cantoni, Helvetica Phys. Acta **58**, 956–968 (1985)
71. H. Cartan, *Calcul différentiel. Formes différentielles* (Hermann, Paris, 1967)
72. P.R. Chernoff, J.E. Marsden, *Properties of Infinite Dimensional Hamiltonian Systems*. LNM 425 (Springer, New York, Heidelberg, Berlin, 1974)
73. G. Choquet, *Lectures on Analysis* (Benjamin, Reading, Mass., 1969)
74. Y. Choquet-Bruhat, C. DeWitt-Morette, M. Dillard-Bleick, *Analysis, Manifolds, and Physics*, Revised edition (North-Holland, New York, Amsterdam, Oxford, 1982)
75. R. Cirelli, P. Lanzavecchia, A. Maniá, J. Phys. A **16**, 3829 (1983); R. Cirelli, P. Lanzavecchia, Nuovo Cimento **B 79**, 271–283 (1984); M. C. Abbati, R. Cirelli, P. Lanzavecchia, A. Maniá, Nuovo Cimento **B 83**, 43 (1984); A. Heslot, Phys. Rev. D **31**, 1341 (1985)
76. R. Cirelli, A. Maniá, L. Pizzocchero, J. Math. Phys. **31**(2891–2897), 2898–2903 (1990)
77. R. Cirelli, A. Maniá, L. Pizzocchero, Int. J. Mod. Phys. A **6**, 2133–2146 (1991)
78. R. Cirelli, J. Math. Phys. **32**, 1235–1281 (1991)
79. R. Cirelli, A. Maniá, L. Pizzocchero, Rev. Math. Phys. **6**, 675–697 (1994)
80. R. Cirelli, M. Gatti, A. Maniá, J. Geom. Phys. **29**, 64–86 (1999)
81. A. Connes, *Noncommutative Geometry* (Academic Press, San Diego, 1994)
82. P. Curie, *Oeuvres* (Gauthier-Villars, Paris, 1908). Esp. J. de Phys. (III)**3**, 393 (1894), according to Chaps. IX–X of Russ. edition: (Nauka, Moscow, 1966)
83. M. Czachor, *"Entropic" framework for nonlinear quantum mechanics*, preprint, Warsaw, 1993; M. Czachor, Acta Phys. Slov. **48**, 1–6 (1998); M. Czachor, M. Marciniak, Phys. Lett. **A 239**, 353–358 (1998); M. Czachor, Phys. Rev. A **57**, 4122–4129 (1998); M. Czachor, M. Kuna, Phys. Rev. A **58**, 128–134 (1998); M. Czachor, Int. J. Theor. Phys. (1999); M. Czachor, Phys. Rev. A **57**, R2263–R2266 (1998)
84. E.B. Davies, *Quantum Theory of Open Systems* (Academic, New York, 1976)
85. E.B. Davies, Commun. Math. Phys. **55**, 231 (1977)
86. E.B. Davies: *Non-Linear Functionals in Quantum Mechanics*, Troisieme Cycle de la Physique, En Suisse Romande (Lausanne) Semestre d'été 1980
87. R. Derka, V. Bužek, A.K. Ekert, Phys. Rev. Lett. **80**, 1571–1575 (1998)
88. P.A.M. Dirac, *The Principles of Quantum Mechanics*, 4th edn. (Claredon, Oxford, 1958)
89. P.P. Divakaran, Phys. Rev. Lett. **79**, 2159 (1997)
90. J. Dixmier, *Les algèbres d'opérateurs dans l'espace Hilbertien*, 2nd edn., revised and completed (Gauthier-Villars, Paris, 1969)
91. J. Dixmier, *Les C*-algèbres et leurs représentations* (Gauthier-Villars, Paris, 1969); Russ. transl. Nauka, Moscow, 1974
92. H.-D. Doebner, G.A. Goldin, Phys. Rev. A **54**, 3764–3771 (1996)
93. H.D. Doebner, J. Tolar, J. Math. Phys. **16**, 975–984 (1975)
94. A. Dold, *Lectures on Algebraic Topology* (Springer, New York, Heidelberg, Berlin, 1980)
95. S. Doplicher, R. Haag, J. E. Roberts, Commun. Math. Phys. **13**, 1–23 (1969); S. Doplicher, R. Haag, J.E. Roberts, Commun. Math. Phys. **15**, 173–200 (1969)
96. S. Doplicher, R. Haag, J.E. Roberts, Commun. Math. Phys. **23**, 199–230 (1971); S. Doplicher, J.E. Roberts, Commun. Math. Phys. **28**, 331–348 (1972); S. Doplicher, R. Haag, J.E. Roberts, Commun. Math. Phys. **35**, 49–85 (1974)
97. S. Doplicher, Commun. Math. Phys. **85**, 73–86 (1982); S. Doplicher, R. Longo, Commun. Math. Phys. **88**, 399–409 (1983)
98. F. Dowker, A. Kent, Phys. Rev. Lett. **75**, 3038 (1995)
99. B.A. Dubrovin, S.P. Novikov, A.T. Fomenko, *Sovremennaja Geometrija-Metody i priloženia (Modern Geometry-Methods and Applications)* (Nauka, Moscow, 1979)
100. N.G. Duffield, R.F. Werner, Helv. Phys. Acta **65**, 1016–1054 (1992). http://davinci.fmph.uniba.sk/~bona1/Duffield-Werner.pdf; N.G. Duffield, R.F. Werner, Rev. Math. Phys. **4**, 383–424 (1992); G.A. Raggio, R.F. Werner, Helv. Phys. Acta **62**, 980–1003 (1989)

101. E. Duffner, A. Rieckers, Z. Naturforsch. **43a**, 321 (1988); W. Fleig, Acta Phys. Ausriaca **55**, 135 (1983); A. Rieckers, M. Ullrich, Acta Phys. Austriaca **56**, 131, 259 (1985); A. Rieckers, M. Ullrich, J. Math. Phys. **27**, 1082 (1986); E. Duffner, Z. Phys. **B 63**, 37 (1986)

102. F. J. Dyson, A. Lenard, J. Math. Phys. **8**, 423–434 (1967); F.J. Dyson, A. Lenard, J. Math. Phys. (9), 698 (1968)

103. A. Einstein, *The Meaning of Relativity* (Princeton UP, Princeton, NJ, 1955)

104. A. Einstein, Ann. d. Phys. **17**, 132 (1905); A. Einstein, Ann. d. Phys. **20**, 199 (1906)

105. A. Einstein, B. Podolsky, N. Rosen, Phys. Rev. **47**, 777 (1935)

106. G.G. Emch, *Algebraic Methods in Statistical Mechanics and Quantum Field Theory* (Wiley, New York, 1972)

107. G.G. Emch, C. Radin, J. Math. Phys. **12**, 2043–2046 (1971)

108. D.E. Evans, J.T. Lewis, *Dilations of Irreversible Evolutions in Algebraic Quantum Theory*, Communication of DIAS, Ser. A (Theoretical Physics), No. 24 (IAS-Dublin, 1977)

109. H. Everett III, Rev. Mod. Phys. **29**, 454–462; J.A. Wheeler, ibid, 463–465 (1957)

110. M. Fecko, Acta Phys. Slov. **44**, 445 (1994); M. Fecko, J. Math. Phys. **36**, 6709 (1995); ibid **36**, 1198–1207 (1995); ibid **38**, 4542–4560 (1997)

111. M. Fecko, *Differential Geometry and Lie Groups for Physicists* (Cambridge University Press, New York, 2006)

112. W. Feller, *An Introduction to Probability Theory and its Applications*, vols. I and II, Russ. edition (Mir, Moscow, 1984)

113. E. Fermi, Rev. Mod. Phys. **4**, 87 (1932)

114. M. Flato, D. Sternheimer, *Star Products, Quantum Groups, Cyclic Cohomology and Pseudodifferential calculus*. Lectures at the 10th Annual Joint Summer Research Conference (AMS–IMS–SIAM) on Conformal Field Theory, Topological Field Theory and Quantum Groups (Mount Holyoke College, South Hadley, MA; June 13–19, 1992), in *Contemporary Mathematics*, AMS-series (eds. M. Flato, J. Lepowsky, N. Reshetikin, P. Sally); D. Arnal, J.C. Cortet, M. Flato, D. Sternheimer, Star products and representations without operators, in *Field Theory, Quantization and Statistical Physics*, ed. E. Tirapegui (D. Reidel Pub. Co., Dordrecht-Holland, 1981)

115. G.W. Ford, J.T. Lewis, R.F. O'Connell, Phys. Rev. Lett. **55**, 2273 (1985); G.W. Ford, J.T. Lewis, R.F. O'Connell, Ann. Phys. (NY) **185**, 270–283 (1988); G.W. Ford, J.T. Lewis, R.F. O'Connell, Phys. Rev. A **37**, 4419 (1988), G.W. Ford, J.T. Lewis, R.F. O'Connell, J. Stat. Phys **53**, 439–455 (1988)

116. K. Fredenhagen, *Global Observables in Local Quantum Physics*, Preprint DESY 93-009 (Hamburg, 1993)

117. C. Fronsdal, Rep. Math. Phys. **15**, 111–145 (1978)

118. T.W. Gamelin, *Uniform Algebras* (Prentice-Hall, Englewood Cliffs, NJ, 1969)

119. M. Gatti, *Private Communications*, June 1992, and June 1993; M. Gatti, *From Quantum Phase Space to Classical Phase Space, a Draft of PhD Dissertation–A Concise Summary of the First 6 Chapters*, Università degli Studi di Milano, June 1993

120. I.M. Gel'fand, D.A. Rajkov, G.E. Šilov, *Kommutativnyje normirovannyje kol'ca (Commutative Normed Rings)* (GIFML, Moscow, 1960)

121. R. Gilmore, in *Symmetries in Science*, ed. by B. Gruber, R.S. Millman (Plenum, New York, 1980)

122. N. Gisin, Helv. Phys. Acta **62**, 363 (1989)

123. N. Gisin, Phys. Lett. A **143**, 1 (1990); J. Polchinski, Phys. Rev. Lett. **66**, 397 (1991); M. Czachor, Found. Phys. Lett. **4**, 351 (1991); H. Scherer, P. Busch, Phys. Rev. A **47**, 1647 (1993)

124. D. Giulini, E. Joos, C. Kiefer, J. Kupsch, I.-O. Stamatescu, H.D. Zeh, *Decoherence and the Appearance of a Classical World in Quantum Theory* (Springer, Berlin, Heidelberg, New York, 1996)

125. R. Glauber, Optical coherence and statistics of photons, in *Quantum Optics and Electronics* (Gordon and Breach, New York, 1965)

126. A.M. Gleason, Ann. Math. **56**, 193–212 (1952)

127. G.A. Goldin, Nonlin. Math. Phys. **4**, 6–11 (1997)
128. G.A. Goldin, R. Menikoff, D.H. Sharp, Phys. Rev. Lett. **51**, 2246 (1983)
129. I.S. Gradštein, I.M. Ryžik (engl. Gradshteyn, Ryzhik), *Tablicy integralov, summ, riadov, i proizvedenij*. Gosudarstvennoje izdatelstvo fiziko-matematièeskoj literatury (Moskva, 1963). [Table of Integrals, Series, and Products], there exist many new editions in various languages-cf. Internet
130. D.M. Greenberger, M.A. Horne, A. Shimony, A. Zeilinger, Am. J. Phys. **58**, 1131 (1990)
131. D.M. Greenberger, M.A. Horne, A. Zeilinger, Phys. Today, 22 (August 1993); D. Bouwmeester, J.-W. Pan, M. Daniell, H. Weinfurter, A. Zeilinger, Phys. Rev. Lett. **82**, 1345 (1999); A. Zeilinger, M.A. Horne, H. Weinfurter, M. Żukowski, Phys. Rev. Lett. **78**, 3031 (1997)
132. E. Grešák, *Communications in the Framework of PhD-Seminar on Mathematical Physics* (Comenius University, Bratislava, 1992)
133. H. Grosse, P. Prešnajder, Lett. Math. Phys. **28**, 239 (1993)
134. S. Gudder, Commun. Math. Phys. **29**, 249–264 (1973)
135. E.A. Guggenheim, *Thermodynamics, Classical and Statistical*, in Handbuch der Physik, Band III/2 (Springer, Berlin, Göttingen, Heidelberg, 1959)
136. V. Guillemin, S. Sternberg, *Geometric Asymptotics*, Mathematical Surveys 14, American Mathematical Society (Providence, Rhode Island, 1977)
137. M.C. Gutzwiller, *Chaos in Classical and Quantum Mechanics* (Springer, New York, 1991)
138. R. Haag, N.M. Hugenholtz, M. Winnink, Commun. Math. Phys. **5**, 215 (1967)
139. R. Haag, D. Kastler, J. Math. Phys. **5**, 848–861 (1964)
140. R. Haag, Nuovo Cimento **25**, 287–299 (1962)
141. R. Haag, *Local Quantum Physics* (Springer, New York, 1992)
142. R. Haag, Dan. Mat. Fys. Med. **29**(12) (1955)
143. R. Haag, B. Schroer, J. Math. Phys. **3**, 248 (1962)
144. G.A. Hagedorn, Commun. Math. Phys. **71**, 77–93 (1980)
145. P. Hájíček, J. Tolar, *Intrisic Properties of Quantum Systems*. Foundations of Physics **39**, 5 (2009) Springer
146. P.R. Halmos, *Introduction to the Theory of Hilbert Space and Spectral Multiplicity* (Chelsea Pub. Co., New York, 1957)
147. G. Hamel, Zs. Math. Phys. **50**, 1 (1904)
148. D.R. Hartree, *The Calculation of Atomic Srtuctures* (Wiley, New York, 1957)
149. T. Heinosaari, M. Ziman, Acta Phys. Slov. **58**, 487–674 (2008); T. Heinosaari, M. Ziman, *The Mathematical Language of Quantum Theory: From Uncertainty to Entanglement* (Cambridge University Press, 2012)
150. W. Heisenberg, *Physik und Philosophie* (Ullstein Bücher, West-Berlin, 1961); *Physics and Beyond: Encounters and Conversations* (Harper & Row Pub., New York, 1972)
151. G.C. Hegerfeldt, Phys. Rev. Lett. **72**, 596–599 (1994)
152. S. Helgason, *Differential Geometry and Symmetric Spaces* (Academic, New York, 1962)
153. K. Hepp, Helvetica Phys. Acta **45**, 237 (1972)
154. K. Hepp, Commun. Math. Phys. **35**, 265–277 (1974)
155. K. Hepp, E.H. Lieb, Helvetica Phys. Acta **46**, 573 (1973)
156. R. Hermann, J. Math. Phys. **6**, 1768 (1965)
157. E. Hewitt, K.A. Ross, *Abstract Harmonic Analysis I* (Springer, Berlin, Göttingen, Heidelberg, 1963)
158. T.L. Hill, *Statistical Mechanics-Principles and Selected Applications* (McGraw-Hill, New York, Toronto, London, 1956)
159. M.W. Hirsch, *Differential Topology* (Springer, New York, Heidelberg, Berlin, 1976)
160. A.S. Holevo, *Probabilistic and Statistical Aspects of Quantum Theory*, Russ. edition (Nauka, Moscow, 1980); A.S. Holevo, Rep. Math. Phys. **22**, 385–407 (1985); A.S. Holevo, *Statistical Structure of Quantum Mechanics and Hidden Variables*, in Russian, Matematika i Kibernetika, 6/1985 (Znanije, Moscow, 1985)
161. M. Horodecki, P. Horodecki, R. Horodecki, Phys. Rev. Lett. **80**, 5239 (1998)

162. R.M.F. Houtappel, H. Van Dam, E.P. Wigner, Rev. Mod. Phys. **37**, 595–632 (1965)
163. L.P. Hughston, Proc. Roy. Soc. Lond. A **452**, 953–979 (1996)
164. S.S. Horuzhy (Choruzij), *Vvedenie v algebraičeskuiyu kvantovuiyu teoriyu polia (Introduction to Algebraic Quantum Field Theory)* (Nauka, Moscow, 1986)
165. C.J. Isham, Topological and global aspects of quantum theory, in *The Proceedings of Les Houches XL (1983):Relativity, Groups and Topology II*, ed. by B.S. DeWitt, R. Stora (North-Holland, Amsterdam, New York, Tokyo, 1984), pp. 1059–1290
166. C. Itzykson, J.-B. Zuber, *Quantum Field Theory* (McGraw-Hill, New York, London, Toronto, 1980)
167. J.M. Jauch, *Foundations of Quantum Mechanics* (Reading, Mass., 1968)
168. F. Jelinek, Commun. Math. Phys. **9**, 169–175 (1968)
169. K.R.W. Jones, Ann. Phys. **233**, 295–316 (1994); K.R.W. Jones, Phys. Rev. A **48**, 822–825 (1993); K.R.W. Jones, Phys. Rev. A **50**, 1062–1070 (1994)
170. T.F. Jordan, Ann. Phys. (NY) **225**, 83 (1993)
171. P. Jordan, E. Wigner, Zs. f. Phys. **47**, 631 (1928)
172. K.G. Kay, Phys. Rev. A **42**, 3718–3725 (1990)
173. T.W.B. Kibble, Commun. Math. Phys. **65**, 189–201 (1979)
174. A.A. Kirillov, *Elementy teorii predstavleniyi (Elements of Representations Theory)*, 2nd edn. (Nauka, Moscow, 1978)
175. J. Kiukas, P. Lahti, J.P. Pellonpää, K. Ylinen, Found. Phys., **49** (2019). https://doi.org/10. 1007/s10701-019-00261-3
176. J.R. Klauder, J. Math. Phys. **4**, 1058 (1963)
177. F. Klein, Math. Ann. **43**, 63 (1893)
178. S. Kobayashi, K. Nomizu, *Foundations of Differential Geometry*, vols. 1 and 2, Interscience, New York, 1963 and 1969, Russ. edn. (Nauka, Moscow, 1981)
179. B.O. Koopman, Proc. Natl. Acad. Sci. **17**, 315–318 (1931)
180. G.A. Korn, T.M. Korn, *Mathematical Handbook for scientists and engineers-definitions, theorems and formulas for reference and review*, 2nd edn. (McGraw-Hill, 1968); Russ. transl. Nauka, Moscow, 1973
181. I.P. Kornfel'd, J.G. Sinaj, C.V. Fomin, *Ergodičeskaja Teorija (Ergodic Theory)* (Nauka, Moscow, 1980)
182. P. Kramer, M. Saraceno, *Geometry of the Time-Dependent Variational Principle in Quantum Mechanics*. LNP 140 (Springer, New York, 1981)
183. A. Kriegl, P.W. Michor, *Foundations of Global Analysis*. Manuscript of monograph (Institut für Mathematik der Universität Wien, Vienna, 1995)
184. T.S. Kuhn, *The Structure of Scientific Revolutions*, 2nd edn. (Chicago, 1970); Slovak translation: Pravda, Bratislava, 1982
185. L.D. Landau, E.M. Lifšic, *Mechanika (Mechanics)*. Theoretical Physics, vol. I (FM, Moscow, 1958)
186. L.D. Landau, E.M. Lifšic, *Statističeskaya Fizika (Statistical Physics)*. Theoretical Physics, vol. V, 2nd edn. (Nauka, Moscow, 1964)
187. L.D. Landau, E.M. Lifšic, *Kvantovaya Mechanika-nerelativistskaya teoriya (Nonrel. Quantum Mechanics)*. Theoretical Physics, vol. III, 2nd edn. (GIFML, Moscow, 1963)
188. L.D. Landau, E. M. Lifšic, *Teoriya Polya (Field Theory)*. Theoretical Physics, vol. II, 3rd edn. (GIFML, Moscow, 1960)
189. N.P. Landsman, *Mathematical Topics Between Classical and Quantum Mechanics*. Springer Monographs in Mathematics (Springer, New York, 1998)
190. N.P. Landsman, Rev. Math. Phys. **2**, 45 (1990); ibid **9**, 29–57 (1997); N.P. Landsman, Int. J. Mod. Phys. **A30**, 5349–5371 (1991); ibid **B10**, 1545–1554 (1996); N.P. Landsman, J. Geom. Phys. **15**, 285–319 (1995)
191. N.P. Landsman, *Spontaneous Symmetry Breaking in Quantum Systems: Emergence or Reduction?*. arxiv:1305.4473 [math-ph], 2 August 2013
192. K. Landsman, Foundations of Quantum Theory. *From Classical Concepts to Operator Algebras*. Fundamental Theories in Physics 188, Springer Open 2017

193. O.E. Lanford, D. Ruelle, Commun. Math. Phys. **13**, 194–215 (1969)
194. G. Lassner, Rep. Math. Phys. **3**, 279 (1972); G. Lassner, Algebras of unbounded operators and quantum dynamics, in *Proceedings of the VIIth International Congress on Mathematical Physics, Boulder, Colorado, USA, August 1–10, 1983*, ed. by W.E. Brittin, K.E. Gustafson, W. Wyss (North–Holland, Amsterdam, 1984), pp. 471–480
195. A.J. Leggett, J. Phys. Condens. Matter **14**, R415–R451 (2002); Topical Review: *Testing the limits of quantum mechanics: motivation, state of play, prospects*
196. A.J. Leggett, J. Phys. A Math. Theor. **40**, 3141–3149 (2007)
197. P. Leifer, *Nonlinear modification of quantum mechanics*, hep-th/9702160
198. M. Lewenstein, A. Sampera, Phys. Rev. Lett. **80**, 2261 (1998)
199. E.H. Lieb, Commun. Math. Phys. **31**, 327 (1973)
200. E.H. Lieb, R. Seiringer, *The Stability of Matter in Quantum Mechanics* (Cambridge University Press, 2010)
201. E. Lieb, T. Schultz, D. Mattis, Ann. Phys. **16**, 407 (1961); E. Lieb, D. Mattis, *Mathematical Physics in One Dimension* (Academic Press, New York and London, 1966)
202. G. Ludwig, Z. Phys. **135**, 483 (1953); G. Ludwig, Phys. Bl **11**, 489 (1955); G. Ludwig, Z. Phys. **152**, 98 (1958); G. Ludwig, Commun. Math. Phys. **4**, 331 (1967); G. Ludwig, Commun. Math. Phys. **9** (1968)
203. G. Ludwig, *An Axiomatic Basis of Quantum Mechanics*, vols. I and II (Springer, New York, 1985 & 1987)
204. W. Lücke, *Nonlinear schrödinger dynamics and nonlinear observables*, in *Nonlinear, Deformed, and Irreversible Quantum Systems*, ed. by H.-D. Doebner, V.K. Dobrev, P. Nattermann (World Scientific, Singapore, 1995); W. Lücke, *Gisin Nonlocality of the Doebner-Goldin 2-Particle Equation*, quant-ph/9710033 (1997); W. Lücke, R.F. Werner, *Inconsistency of Bialynicki-Birula and Mycielski's Nonlinear Quantum Mechanics*. A talk in the Doppler Institute (Czech Technical University, Prague, April 1997)
205. E. Mach, *Die Mechanik in ihrer Entwicklung* (Leipzig, 1933)
206. E. Mach, *Populär-Wissenschaftliche Vorlesungen* (Johann Ambrosius Barth, Leipzig, 1910)
207. G.W. Mackey, *Mathematical Foundations of Quantum Mechanics* (Benjamin, Reading, Mass., 1963)
208. G.W. Mackey, *Induced Representations* (Benjamin, Reading, Mass., 1968)
209. G.W. Mackey, *Unitary Representations in Physics, Probability, and Number Theory* (Benjamin, Reading, Mass., 1978)
210. N.H. March, W.H. Young, S. Sampanthar, *The Many-Body Problem in Quantum Mechanics* (Cambridge University Press, Cambridge, 1967)
211. L. Markus, K.R. Meyer, *Generic Hamiltonian Dynamical Systems are Neither Integrable nor Ergodic*. Memoirs of AMS, Number 144 (Providence, Rhode Island, 1974)
212. C.M. Marle, in *Bifurcation Theory, Mechanics and Physics*, ed. by C.P. Bruter, A. Aragnol, A. Lichnerowicz (D. Reidel, Dordrecht, Boston, Lancaster, 1983)
213. J.E. Marsden, *Geometric Methods in Mathematical Physics*. LNM 775 (Springer, New York, 1980)
214. J.E. Marsden, T.S. Ratiu, *Introduction to Mechanics and Symmetry* (Springer, New York, 1999)
215. J.E. Marsden, M. McCracken, *The Hopf Bifurcation and its Applications* (Springer, New York, 1976)
216. N.D. Mermin, Rev. Mod. Phys. **65**, 803–815 (1993)
217. A. Messiah: *Quantum Mechanics* (Russ. transl. from French) (Nauka, Moscow, 1978)
218. P.A. Meyer: *Probability and Potentials* (Blaisdell Pub. Co., Waltham, Mass., Toronto, London, 1966)
219. B. Mielnik, Commun. Math. Phys. **37**, 221–256 (1974)
220. D. Montgomery, L. Zippin, Ann. Math. **56**, 213–241 (1952)
221. G. Morchio, F. Strocchi, Commun. Math. Phys. **99**, 153 (1985)
222. G. Morchio, F. Strocchi: J. Math. Phys. **28**, (1987) 622; F. Strocchi , *Long-range dynamics and spontaneous symmetry breaking in many-body systems*, preprint ISAS-Trieste, 1987

223. M.A. Najmark, *Normirovannye kol'ca (Normed Rings)* (Nauka, Moscow, 1968)
224. J. Naudts, *C*-Multipliers, crossed product algebras, and canonical commutation relations*, preprint, Dept. Natuurkunde, Universiteit Antverpen UIA, July 1999
225. E. Nelson, *Dynamical Theories of Brownian Motion* (Princeton University Press, Princeton, NJ, 1967)
226. J. von Neumann, *Mathematische Grundlagen der Quantenmechanik* (Springer, Berlin, 1932)
227. J. von Neumann, Compos. Math. **6**, 1–77 (1938)
228. J. Niederle, J. Tolar, Czech. J. Phys. **B 29**, 1358–1368 (1979); P. Šťovíček, J. Tolar, Acta Polytechnica-Práce ČVUT, Prague, No. 6, 3–75(1984)
229. M.A. Nielsen, Phys. Rev. Lett. **79**, 2915 (1997)
230. E. Noether, Nachr. Königl. Ges. Wiss. Göttingen, Math. Phys. **K1**, 258 (1918)
231. A. Odzijewicz, T.S. Ratiu, Commun. Math. Phys. **243**, 1–54 (2003)
232. R. Omnès, Rev. Mod. Phys. **64**, 339–382 (1992); R.B. Griffiths, Phys. Rev. Lett. **70**, 2201 (1993); M. Gell-Mann, J.B. Hartle, Phys. Rev. D **47**, 3345–3382 (1993); L. Diósi, N. Gisin, J. Halliwell, I.C. Percival, Phys. Rev. Lett. **74**, 203 (1995); C.J. Isham, J. Math. Phys. **35**, 2157–2185 (1994); C.J. Isham, N. Linden, J. Math. Phys. **35**, 5452–5476 (1994); F. Dowker, A. Kent, Phys. Rev. Lett. **75**, 3038 (1995)
233. W. Pauli, *General Principles of Quantum Mechanics* (Springer, New York, 1980)
234. W. Pauli, *Teorija Otnositel'nosti (Theory of Relativity)* (translated from an English edition) (Nauka, Moscow, 1983)
235. G.K. Pedersen, *C*-algebras and their Automorphism Groups* (Academic Press, London, New York, San Francisco, 1979)
236. R. Penrose, *The Emperor's New Mind: Concerning Computers, Minds and the Laws of Physics* (Oxford University Press, Oxford, 1989)
237. R. Penrose, *Shadows of the Mind: An Approach to the Missing Science of Consciousness* (Oxford University Press, Oxford, 1994); R. Penrose, *The Large, the Small, and the Human Mind* (The Press Syndicate of the University of Cambridge, Cambridge, 1997) (Czech translation, Prague, 1999)
238. R. Penrose, *The Road to Reality: A Complete Guide to the Laws of the Universe* (Alfred A. Knopf, New York, 2005) (Orig.: London 2004)
239. A.M. Perelomov, Commun. Math. Phys. **26**, 222 (1972)
240. A.M. Perelomov, *Generalized Coherent States and Their Applications* (Springer, Berlin, 1986); J.R. Klauder, B.S. Skagerstam, *Coherent States. Applications in Physics and Mathematical Physics* (World Scientific, Singapore, 1984); S.T. Ali, G.G. Emch, J. Math. Phys. **27**, 2936–2943 (1986)
241. A. Peres, *Quantum Theory: Concepts and Methods* (Kluver Academic Publishers, Dordrecht, Boston, London, 1994)
242. A. Peres, Phys. Rev. Lett. **74**, 4571 (1995)
243. C. Piron, Helv. Phys. Acta **42**, 330 (1969); C. Piron, *Foundations of Quantum Physics* (Benjamin, Reading, Mass., 1976); A. Amann, U. Müller-Herold, Helv. Phys. Acta **59**, 1311 (1986); A. Amann, Helv. Phys. Acta **60**, 384 (1987); A. Amann: Fortschr. Phys. **34**, 167 (1986)
244. H. Poincaré, Compt. Rend. **140**, 1504 (1905)
245. L.S. Polak (ed.), *Variacionnyie Principy Mechaniki* (G.I. Fiz.-Mat. Lit, Moscow, 1959)
246. M. Polakovič, Int. J. Theor. Phys. **37**, 2923–2934 (1998)
247. L.S. Pontrjagin, *Nepreryvnyie Gruppy (Continuous Groups)* (Nauka, Moscow, 1973)
248. T. Poston, I. Stewart, *Catastrophe Theory and its Applications* (PITMAN, London, San Francisco, Melbourne, 1978)
249. S. Pulmannová, Commun. Math. Phys. **49**, 47–51 (1976)
250. K.R. Popper, *The Logic of Scientific Discovery* (Routledge, London, New York, 1994); Czech translation: *Logika vědeckého bádání* (OIKOIMENH, Prague, 1997). Completed from 10th German edition: *Logik der Forschung* (J. C. B. Mohr (P. Siebeck), Tübingen, 1994)
251. K.R. Popper, *Objective Knowledge (An Evolutionary Approach)*, Revised edition (Oxford UP, at the Claredon Press, Oxford, 1979)
252. A. Povzner, Trans. Am. Math. Soc. **51**, 189 (1966); (Cited from [1].)

253. H. Primas, *preprint ETH* (Zürich, 1972)
254. H. Primas, *Chemistry, Quantum Mechanics and Reductionism*. Lecture Notes in Chemistry, vol. 24 (Springer, Berlin, 1981)
255. J.P. Provost, G. Vallee, Commun. Math. Phys. **76**, 289 (1980)
256. E. Prugovečki, J. Math. Phys. **7**, 1680 (1966)
257. E. Prugovečki, Canad. J. Phys. **45**, 2173 (1967)
258. E. Prugovečki, Physica **91A**, 202 (1978)
259. E. Prugovečki, *Stochastic Quantum Mechanics and Quantum Spacetime* (Reidel, Dordrecht, 1984); J.A. Brooke, E. Prugovečki, Nuovo Cimento **89 A**, 126 (1985)
260. W.V.O. Quine, *From Stimulus to Science* (Harvard University Press, 1995); Czech transl. FILOSOFIA, Praha, 2002
261. M. Reck, A. Zeilinger, H.J. Bernstein, P. Bertani, Phys. Rev. Lett. **73**, 58 (1994)
262. M. Reed, B. Simon, *Methods of Modern Mathematical Physics*, vols. I and II (Academic Press, New York, London, 1972 and 1975)
263. C. Radin, J. Math. Phys. **11**, 2945–2955 (1970)
264. G. de Rham, *Variétés Différentiables* (Hermann, Paris, 1955)
265. A. Rieckers, J. Math. Phys. **25**, 2593–2601 (1984); R. Honegger, A. Rieckers, J. Math. Phys. **37**, 4292 (1996)
266. F. Riesz, B. Sz.- Nagy, *Vorlesungen über Functionalanalysis* (DVW, Berlin, 1956)
267. D.W. Robinson, Commun. Math. Phys. **7**, 337–348 (1968)
268. D.J. Rowe, Nucl. Phys. A **391**, 307 (1982)
269. D.J. Rowe, A. Ryman, G. Rosensteel, Phys. Rev. A **22**, 2362 (1980)
270. D. Ruelle, Commun. Math. Phys. **3**, 133 (1966)
271. D. Ruelle, *Statistical Mechanics-Rigorous Results* (W. A. Benjamin Inc., New York, Amsterdam, 1969)
272. D. Ruelle, *Thermodynamic Formalism: The Mathematical Structures of Classical Equilibrium Mechanics* (Addison-Wesley, Reading, Mass., London, 1978)
273. R.K. Sachs, H. Wu, *General Relativity for Mathematicians* (Springer, New York, Heidelberg, Berlin, 1977)
274. S. Sakai, *C*-algebras and W*-algebras* (Springer, New York, 1971)
275. S. Sakai, *Operator Algebras in Dynamical Systems* (Cambridge University Press, Cambridge, New York, Sydney, 1991)
276. H. Scherer, P. Busch, Phys. Rev. A **47**, 1647–1651 (1993)
277. L.I. Schiff, *Quantum Mechanics* (McGraw-Hill, New York, Toronto, London, 1955)
278. S. Schlieder, Commun. Math. Phys. **13**, 216–225 (1969)
279. M. Schlosshauer, Rev. Mod. Phys. **76**, 1267 (2005); M.A. Schlosshauer, *Decoherence: And the Quantum-To-Classical Transition* (Springer, Berlin, Heidelberg, Corrected Third Printing, 2008)
280. E. Schrödinger, Ann. d. Phys., vierte Volge **79**(6), 489–527 (1926); (Cited from [245].)
281. L.S. Schulman, Ann. Phys. (NY) **212**, 315–370 (1991)
282. J.T. Schwartz, *Nonlinear Functional Analysis* (Gordon and Breach, New York, 1969)
283. L. Schwartz, *Analyse mathèmatique* (Hermann, Paris, 1967)
284. S.S. Schweber, *An Introduction to Relativistic Quantum Field Theory* (Row-Peterson, New York, 1961)
285. I.E. Segal, Duke Math. J. **18**, 221–265 (1951)
286. G.L. Sewell, *Quantum Theory of Collective Phenomena* (Oxford Science Publ., Claredon Press, Oxford, 1989)
287. G.L. Sewell, J. Math. Phys. **26**, 2324–2334 (1985)
288. G.L. Sewell, *Macroscopic Quantum Electrodynamics of a Plasma Model: Derivation of the Vlasov Kinetics*, Texas Math.-Phys. Database: Preprint No. 96–179 (Queen Mary and Westfield College, London, 1996)
289. G.L. Sewell, *Quantum Mechanics and Its Emergent Macrophysics* (Princeton University Press, Princeton and Oxford, 2002)
290. R. Sikela, Thesis (in Slovak), Comenius University, Bratislava, 1991

291. B. Simon, Commun. Math. Phys. **71**, 247 (1980)

292. J.G. Sinaj, L.P. Šilnikov (eds.), *Strannye Attraktory (Strange Attractors)*. Mathematics in Science Abroad No. 22 (Mir., Moscow, 1981)

293. L. Sládeček, Thesis (in Slovak), Comenius University, Bratislava, 1991

294. J. Smital, *On Functions and Functional Equations* (Adam Hilger, Bristol and Philadelphia, 1988)

295. J. Śniatycki, *Geometric Quantization and Quantum Mechanics* (Springer, New York, 1980)

296. E.H. Spanier, *Algebraic Topology* (McGraw-Hill, New York, 1966)

297. H. Spohn, Rev. Mod. Phys. **53**, 569–615 (1980)

298. E.M. Stein, G. Weiss, *Introduction to Fourier Analysis on Euclidean Spaces* (Princeton UP, Princeton, NJ, 1971)

299. E. Störmer, J. Funct. Anal. **3**, 48 (1969)

300. R.F. Streater, Rep. Math. Phys. **33**, 203–219 (1993)

301. R.F. Streater, A.S. Wightman, *PCT, Spin, and Statistics* (Pergamon, London, 1964)

302. F. Strocchi, Rev. Mod. Phys. **38**, 36 (1966)

303. F. Strocchi (Lecture notes by F.S.), *An Introduction to the Mathematical Structure of Quantum Mechanics–A Short Course for Mathematicians*, 2nd edn. (World Scientific, Singapore, 2008)

304. W.T. Strunz, L. Diósi, N. Gisin, Phys. Rev. Lett. **82**, 1801 (1998)

305. M. Takesaki, *Tomita's Theory of Modular Hilbert Algebras and its Applications*. LNM 128 (Springer, Berlin, Heidelberg, New York, 1970)

306. M. Takesaki, *Theory of Operator Algebras I* (Springer, New York, 1979)

307. W. Thirring, *Lehrbuch der Mathematischen Physik, Band 1: Klassische Dynamische Systeme* (Springer, Wien, 1977)

308. W. Thirring, *Lehrbuch der Mathematischen Physik, Band 2: Klassische Feldtheorie* (Springer, Wien, 1978)

309. W. Thirring, *Lehrbuch der Mathematischen Physik, Band 3: Quantenmechanik von Atomen und Molekülen* (Springer, Wien, 1979)

310. W. Thirring, *Lehrbuch der Mathematischen Physik, Band 4: Quantenmechanik grosser Systeme* (Springer, Wien, 1980)

311. W. Thirring, Commun. Math. Phys. **7**, 181–189 (1968)

312. W. Thirring, A. Wehrl, Commun. Math. Phys. **4**, 303–314 (1967)

313. J. Tolar, Čs. Čas. Fyz. A **25**, 576–588 (1975) (in Czech)

314. J. Tolar, *Obecné Metody Kvantování (General Quantization Methods)*, DrSc Dissertation, FJFI ČVUT, Prague, 1985

315. J. Tolar, G. Chadzitaskos, J. Phys. A Math. Gen. **30**, 2509–2517 (1997)

316. G.E. Uhlenbeck, G.W. Ford, *Lectures in Statistical Mechanics* (AMS, Providence, Rhode Island, 1963)

317. T. Unnerstall, *Dynamische Beschreibungen und extensive physikalische Größen makroskopischer Quantensysteme mit Anwendung auf den Josephson-Kontakt*, Dissertation (Eberhard-Karls-Universität zu Tübingen), Tübingen, 1990

318. T. Unnerstall, J. Math. Phys. **31**, 680 (1990)

319. T. Unnerstall, Commun. Math. Phys. **130**, 237 (1990)

320. T. Unnerstall, Lett. Math. Phys. **20**, 183 (1990)

321. V.S. Varadarajan, *Geometry of Quantum Theory*, vols. I and II (Van Nostrand Reinhold, New York, 1970)

322. V. Vedral, M.B. Plenio, M.A. Rippin, P.L. Knight, Phys. Rev. Lett. **78**, 2275 (1997)

323. F. Verhulst, *Nonlinear Differential Equations and Dynamical Systems* (Springer, Berlin, Heidelberg, 1990)

324. V.S. Vladimirov, *Uravnenija Matematičeskoj Fiziki* (Nauka, Moscow, 1971); Engl. title: Equations of Mathematical Physics, cf. American Journal of Physics **39**, 1548 (1971)

325. V. Votruba, *Základy Speciální Teorie Relativity (Foundations of Special Relativity Theory)* (Academia, Prague, 1962)

326. P. Walters, *An Introduction to Ergodic Theory* (Springer, New York, 1982)

327. (a) R.L. Walsworth, I.F. Silvera, Phys. Rev. A **42**, 63 (1990); (b) M. Czachor, Found. Phys. Lett. **4**, 351 (1991)
328. (a) S. Weinberg, Phys. Rev. Lett. **62**, 485 (1989); (b) S. Weinberg, Ann. Phys. **194**, 336 (1989)
329. A. Weinstein, The local structure of poisson manifolds. J. Differ. Geom. **18**, 523–557 (1983)
330. H. Weyl, *The Theory of Groups and Quantum Mechanics* (Dover, New York, 1931)
331. J.A. Wheeler, W.H. Zurek, *Quantum Theory and Measurement* (Princeton University Press, 1983)
332. H. Wergeland, *Irreversibility in many-body systems*, in *The proceedings: Irreversibility in the Many-Body Problem*. Sitges International School of Physics, May 1972 (Plenum Press, New York, London, 1972), pp. 105-156
333. E.T. Whittaker, *A Treatise on the Analytical Dynamics of Particles and Rigid Bodies with an Introduction to the Problem of Three bodies* (Cambridge University Press, Cambridge, 1952)
334. B. Whitten-Wolfe, G.G. Emch, Helvetica Phys. Acta **49**, 45–55 (1976)
335. G.C. Wick, A.S. Wightman, E.P. Wigner, Phys. Rev. **88**, 101–105 (1952)
336. E.P. Wigner, *Gruppentheorie und ihre Anwendung auf die Quantenmechanik der Atomspektren* (Vieweg, Braunschweig, 1931)
337. E.P. Wigner, *Symmetries and Reflections* (Indiana University Press, Bloomington-London, 1970)
338. E.P. Wigner, in *Symmetries in Science*, ed. by B. Gruber, R.S. Millman (Plenum, New York, 1980)
339. E.P. Wigner, Am. J. Phys. **38**, 1005 (1968)
340. N. Woodhouse, *Geometric Quantization* (Claredon Press, Oxford, 1980)
341. S. Woronowicz, *Podstawy Aksiomatycznej Kwantowej Teorii Pola, I. Mechanika Kwantowa*, Preprint No. 82 (N. Copernicus University, Toruň, 1969)
342. L.G. Yaffe, Rev. Mod. Phys. **54**, 407–436 (1982)
343. H. Zeh, Found. Phys. **1**, 1 (1970)
344. H.D. Zeh, *The Problem of Conscious Observation in Quantum Mechanical Description*. arXiv:quant-ph/9908084 v3, 5 June 2000
345. M. Żukowski, A. Zeilinger, M.A. Horne, A.K. Ekert, Phys. Rev. Lett. **71**, 4287 (1993); A. Zeilinger, M.A. Horne, H. Weinfurter, M. Żukowski, Phys. Rev. Lett. **78**, 3031 (1997)
346. D.P. Želobenko, A.I. Stern, *Predstavlenija Grupp Li(Representations of Lie groups)* (Nauka, Moscow, 1983)
347. W. Żurek, Phys. Rev. D **24**, 1516 (1981); W. Żurek, Phys. Rev. D **26**, 1862 (1981); W. Żurek, Phys. Today **44**, 36 (1991); S. Habib, K. Shizume, W.H. Zurek, Phys. Rev. Lett. **80**, 4361 (1998)

Index

© Springer Nature Switzerland AG 2020, corrected publication 2020

P. Bóna, *Classical Systems in Quantum Mechanics*,

https://doi.org/10.1007/978-3-030-45070-0